MONOGRAPHS ON NUMERICAL ANALYSIS

MONOGRAPHS ON NUMERICAL ANALYSIS

An introduction to numerical linear algebra L. Fox

The algebraic eigenvalue problem J. H. Wilkinson

The numerical treatment of integral equations Christopher T. H. Baker

Numerical functional analysis Colin W. Cryer

Direct methods for sparse matrices I. S. Duff, A. M. Erisman, and J. K. Reid

Computer solution of linear programs J. L. Nazareth

Computer Solution of Linear Programs

J. L. NAZARETH

New York Oxford
OXFORD UNIVERSITY PRESS
1987

Oxford University Press

Oxford New York Toronto
Delhi Bombay Calcutta Madras Karachi
Petaling Jaya Singapore Hong Kong Tokyo
Nairobi Dar es Salaam Cape Town
Melbourne Auckland

and associated companies in
Beirut Berlin Ibadan Nicosia

Published by Oxford University Press, Inc.,
200 Madison Avenue, New York, New York 10016

Library of Congress Cataloging-in-Publication Data
Nazareth, J. L.
Computer solution of linear programs.
(Monographs on numerical analysis)
Bibliography: p. Includes index.
1. Linear programming—Data processing. I. Title. II Series.
T57.74.N38 1987 519.7′2 86-33108
ISBN 0-19-504278-6

9 8 7 6 5 4 3 2 1

Printed in the United States of America
on acid-free paper

Dedicated to my parents

Maria Monica and *John Maximian Nazareth*

and to

Abigail Stone Reeder

Preface

This book gives a systematic account of the main algorithms derived from the simplex method and the means by which they may be organized into effective procedures for solving practical linear programming (LP) problems on a computer. It is intended to be self-contained, given that the reader has a working knowledge of linear algebra and matrix theory. The book falls into three parts as follows:

Part I characterizes the problem and the method used to solve it.

Part II deals with the practicalities of the subject. Concerns of implementation, often relegated in introductory texts to a short concluding chapter, form an integral part of our presentation.

Part III discusses three fundamental principles of optimization —namely, duality, decomposition, and homotopy. In conjunction with the simplex method, they each lead to other key algorithms of linear programming.

Our subject lies at the confluence of mathematical programming, numerical analysis, and computer science, and this book, in the spirit of the mathematical sciences, seeks to bring together elements of these three areas. One point of departure from many linear programming texts lies in our detailed exploration of ideas and issues that center on the need to structure data suitably and to organize calculations in an efficient and numerically stable manner. This need arises, in turn, because practical linear programs are generally large and sparse, and because their solution involves long sequences of calculations in finite precision, floating-point arithmetic. We are concerned, however, with the fundamentals of sound implementation, *not* with the design of large-scale mathematical programming systems and associated topics like matrix generators and report writers. This is a highly specialized area that requires another giant step in elaboration and refinement of the basic implementation techniques discussed here, and owes much to structured (and skillful) programming, effective management of hierarchies of computer storage, careful tailoring of an implementation to the specific computing environment at hand, adaptive choice between different strategies depending on problem characteristics, and implementation of a wide range of user-oriented options. Such material is confined to a short concluding chapter of Part II.

A second point of departure from many linear programming texts lies in our overall perspective, which is grounded in nonlinear programming rather than combinatorics. Thus the reader should be able to make the transition to the subject of nonlinear programming (NLP) in a much more natural way. Indeed, recently, fresh life has been breathed into the subject of linear programming by Khachiyan [1979] and Karmarkar [1984] through NLP-based algorithmic approaches that take the special characteristics of the LP problem into consideration. They are likely, in the long run, to widen the repertoire of algorithms available for solving linear programs. However, after much debate, I chose not to include an introduction to these ideas in this *first* edition of the book because, though far reaching in their theoretical and practical implications, they are still very much in their nascency.

In addition to being familiar with the basic concepts of linear algebra and matrix theory, it would be helpful (but it is not essential) for the reader to have some exposure to programming a computer in a high-level language. He or she should know the meaning of floating-point computation and be aware of round-off error, but we do not expect any substantial background in numerical analysis or mathematical software.

For *self-study,* this book should be of interest to the following:

1. The casual user of linear programing techniques, whose primary concern is problem formulation and setup, but who wishes to know something about the off-the-shelf mathematical programming software that he or she uses to solve a linear programming problem. By not treating the software entirely as a "black box," such a user can arrive at a better understanding of the feasibility of particular LP formulations, of the potential numerical difficulties that can occur, and so on. The jungle of jargon that surrounds a mathematical programming system also becomes more penetrable; consider, for example, the terms multiple pricing, major and minor iterations, reinversion frequency.

2. The sophisticated user who wishes to devise a version of the simplex method that is tailored to a particular application. He or she may use the Dantzig–Wolfe decomposition principle, for example, to take advantage of special problem structure, or more generally, any advanced technique that involves the solution of coordinated sequences of linear programs. Such a user will need a detailed knowledge of the way the simplex method is implemented in practice. This class of users also includes researchers in optimization—in particular, large scale optimization—who are interested in implementing and testing new algorithmic ideas in a practical setting, not just on artificial examples.

As a *classroom text,* the book may be used as follows:

1. As a self-contained advanced course (possibly a second course) in linear programming.

2. As the second half of a semester course in linear programming, the first half of which covers LP formulation and modeling (discussed in many suitable texts) and the use of some available (black-box) LP solver.

Exercises are interspersed throughout and form an integral part of the text. Study of this book could also be accompanied by a classroom project in which the simplex method is implemented in a high-level language, using a suitable selection of the techniques discussed here, these being determined, perhaps, by individual student taste. It has been our experience that there is much to be gained when this project is carried out by students working in small groups rather than individually, because this distributes computer programming effort and because group members learn from one another.

Finally, there remains the pleasant task of thanking various individuals who have contributed, either directly or indirectly, to this effort. Warmest thanks to Beresford Parlett, who helped develop my interest in numerical computation at Berkeley and who has been, over the years, both a mentor and a friend. My thanks also to Stuart Dreyfus for his encouragement and always helpful advice. George Dantzig was a source of much inspiration, in particular, during the time I spent at his Systems Optimization Laboratory at Stanford. There I met Michael Saunders and John Tomlin, from both of whom I have learned a great deal. Gene Golub and Roger Wets intervened in my life at crucial junctures and helped open new opportunities for learning.

The book grew out of graduate courses I taught, during visiting appointments in the Operations Research Department at Stanford and the IEOR Department at Berkeley and two short series of lectures that I presented at the Tata Institute of Fundamental Research (TIFR), Bombay and the Instituto de Matematica Pura e Aplicada (IMPA), Rio de Janeiro. I am grateful to all the participants.

My thanks to Michael Saunders and Mukund Thapa for reading substantial portions of the manuscript and making many helpful suggestions. They are, of course, not implicated in any shortcomings and errors in the book, which are solely my responsibility. I also greatly appreciate the expert assistance of the editorial staff at OUP, in particular Donald Degenhardt and Joan Bossert.

Last, and therefore first, thanks to Abbey Reeder, for all her patience and encouragement during the writing of this book.

Berkeley, California
July 1987

J. L. N.

Contents

COMPUTER SOLUTION OF LINEAR PROGRAMS

Part I
Basic Theory and Method

We define a linear program (or linear programming problem) and characterize its solutions. We then describe the simplex method for solving linear programs and the organization of its calculations into the primal simplex algorithm. Our description is self-contained and sets the stage for subsequent chapters of this book. However, we assume a knowledge of linear algebra and matrix calculus, and the reader may find it helpful to supplement Part I with a presentation at a more elementary level. We suggest the book of Chvatal [1983].

A fascinating account of the origins of linear programming may be found in two articles by Dantzig [1982, 1985], and no serious student of linear programming should fail to have at least some acquaintance with Dantzig's classic [1963]; he is universally acknowledged to be the father of the subject.

1
Linear Programs and Their Solution

A linear program is an optimization problem defined over a finite-dimensional real vector space, whose solution is required to satisfy a given set of linear equations and linear inequalities (called *constraints*) and to minimize or maximize a given linear function (called the *objective*). In order to be more specific, let us consider the following simple illustration.

1.1. The Diet Problem

A budget-conscious consumer wishes to purchase, at minimum cost, suitable quantities of three basic foods, say, poultry, leafy spinach, and potatoes, so that his or her daily diet provides at least 65 grams of protein, 90 grams of carbohydrate (energy), 200 milligrams of calcium, 10 milligrams of iron, and 5000 international units of vitamin A. The nutritive food value contained in 100 grams of each basic food and its cost are summarized in Table 1.1.

Suppose that the consumer purchases $\bar{x}_1$ units (a unit is taken to be 100 grams) of poultry, $\bar{x}_2$ units of spinach, and $\bar{x}_3$ units of potatoes. Then his or her problem can be stated as follows:

$$\begin{aligned}
\text{minimize}\quad & z = 40\bar{x}_1 + 15\bar{x}_2 + 10\bar{x}_3 \\
\text{s.t.}\quad & 20\bar{x}_1 + 3\bar{x}_2 + 2\bar{x}_3 \geq 65 \\
& 3\bar{x}_2 + 18\bar{x}_3 \geq 90 \\
& 8\bar{x}_1 + 83\bar{x}_2 + 7\bar{x}_3 \geq 200 \\
& 1.4\bar{x}_1 + 2\bar{x}_2 + 0.6\bar{x}_3 \geq 10 \\
& 80\bar{x}_1 + 7300\bar{x}_2 \geq 5000 \\
& \bar{x}_1 \geq 0, \quad \bar{x}_2 \geq 0, \quad \bar{x}_3 \geq 0.
\end{aligned} \tag{1.1-1}$$

More generally, consider a diet problem involving $\bar{n}$ foods and $\bar{m}$ nutrients, with minimum daily requirements of, say, $\bar{b}_i$ units of nutrient i, $i = 1, \ldots, \bar{m}$. Let $\bar{a}_{ij}$ be the number of units of nutrient i in one unit of

Table 1.1 Nutritive Value of Foods

	Poultry	Spinach	Potatoes
Cost (cents)	40	15	10
Protein (g)	20	3	2
Carbohydrate (g)	0	3	18
Calcium (mg)	8	83	7
Iron (mg)	1.4	2	0.6
Vitamin A (I.U.)	80	7300	0

food j, where $i = 1, \ldots, \bar{m}$ and $j = 1, \ldots, \bar{n}$. (Each food will contain negligible amounts of certain nutrients, so many elements $\bar{a}_{ij}$ will have value zero.) Let $\bar{c}_j$ be the cost per unit of food j. Finally, suppose that the consumer wishes to purchase *at least* $\bar{l}_j$ units of food j and *at most* $\bar{u}_j$ units of food j. Then his problem can be formulated as the following linear program, whose variables $\bar{x}_j$ denote the quantity of food j that he or she purchases.

$$\begin{aligned} \text{minimize} \quad & \bar{c}_1\bar{x}_1 + \bar{c}_2\bar{x}_2 + \cdots + \bar{c}_{\bar{n}}\bar{x}_{\bar{n}} \\ \text{s.t.} \quad & \bar{a}_{11}\bar{x}_1 + \bar{a}_{12}\bar{x}_2 + \cdots + \bar{a}_{1\bar{n}}\bar{x}_{\bar{n}} \geq \bar{b}_1 \\ & \cdot \\ & \cdot \\ & \cdot \\ & \bar{a}_{\bar{m}1}\bar{x}_1 + \bar{a}_{\bar{m}2}\bar{x}_2 + \cdots + \bar{a}_{\bar{m}\bar{n}}\bar{x}_{\bar{n}} \geq \bar{b}_{\bar{m}} \\ & \bar{l}_j \leq \bar{x}_j \leq \bar{u}_j \qquad j = 1, \ldots, \bar{n}. \end{aligned} \tag{1.1-2}$$

1.2. The Linear Programming Problem

Linear programs (or linear programming problems) can be stated in general terms as follows:

$$\begin{aligned} \text{minimize (or maximize)} \quad & \bar{\mathbf{c}}^T\bar{\mathbf{x}} \\ \text{s.t.} \quad & \bar{\mathbf{A}}\bar{\mathbf{x}} \begin{bmatrix} \leq \\ = \\ \geq \end{bmatrix} \bar{\mathbf{b}} \\ & \bar{\mathbf{l}} \leq \bar{\mathbf{x}} \leq \bar{\mathbf{u}} \end{aligned} \tag{1.2-1}$$

where $\bar{\mathbf{A}}$ is an $\bar{m} \times \bar{n}$ matrix, $\bar{\mathbf{b}}$ is an $\bar{m}$-vector, and $\bar{\mathbf{c}}$, $\bar{\mathbf{x}}$, $\bar{\mathbf{l}}$, and $\bar{\mathbf{u}}$ are $\bar{n}$-vectors. The symbol $\begin{bmatrix} \leq \\ = \\ \geq \end{bmatrix}$ indicates that a particular row may be constrained in one of three possible ways, for example, $(\bar{\mathbf{a}}^i)^T\bar{\mathbf{x}} \leq \bar{b}_i$, where we use the notation $\bar{\mathbf{a}}^i$ to denote the vector corresponding to the ith *row* of $\bar{\mathbf{A}}$ and $\bar{\mathbf{a}}_j$ to denote the vector corresponding to the jth *column* of $\bar{\mathbf{A}}$.

$\bar{\mathbf{A}}\bar{\mathbf{x}}\begin{bmatrix}\leq\\=\\\geq\end{bmatrix}\bar{\mathbf{b}}$ are called the *general constraints* (usually abbreviated to *constraints*) and $\bar{\mathbf{l}}\leq\bar{\mathbf{x}}\leq\bar{\mathbf{u}}$ are called the *bound constraints* (usually abbreviated to *bounds*). We make the convention that $\bar{l}_j=-\infty$ when $\bar{x}_j$ is unbounded from below, i.e., can assume any negative value, and $\bar{u}_j=+\infty$ when $\bar{x}_j$ is unbounded from above. Note that bounds are a special case of the general constraints but are explicitly distinguished from them for practical reasons. The linear function $\bar{\mathbf{c}}^T\bar{\mathbf{x}}$ is termed the *objective function.*

A constraint of the form $(\bar{\mathbf{a}}^i)^T\bar{\mathbf{x}}\leq\bar{b}_i$ can be reexpressed as $(\bar{\mathbf{a}}^i)^T\bar{\mathbf{x}}+\bar{x}_{\bar{n}+i}=\bar{b}_i$, $\bar{x}_{\bar{n}+i}\geq 0$ and a constraint of the form $(\bar{\mathbf{a}}^i)^T\bar{\mathbf{x}}\geq\bar{b}_i$ can be reexpressed as $(\bar{\mathbf{a}}^i)^T\bar{\mathbf{x}}+\bar{x}_{\bar{n}+i}=\bar{b}_i$, $\bar{x}_{\bar{n}+i}\leq 0$—that is, we can convert inequalities to equalities by introducing a new *slack* variable $\bar{x}_{\bar{n}+i}$ and an appropriate bound on it. (For completeness, we may observe that $(\bar{\mathbf{a}}^i)^T\bar{\mathbf{x}}=\bar{b}_i$ may be rewritten in the form $(\bar{\mathbf{a}}^i)^T\bar{\mathbf{x}}+\bar{x}_{\bar{n}+i}=\bar{b}_i$, $0\leq\bar{x}_{\bar{n}+i}\leq 0$.) An objective function "maximize $\bar{\mathbf{c}}^T\bar{\mathbf{x}}$" can be reexpressed as "minimize $-\bar{\mathbf{c}}^T\bar{\mathbf{x}}$" and the sign reversed when this new objective is evaluated, in particular, at the optimum.

Thus, without loss of generality, we shall henceforth consider linear programs in the *standard* or *canonical form*

$$\begin{aligned}\text{minimize}\ & \mathbf{c}^T\mathbf{x}\\ \text{s.t.}\ & \mathbf{A}\mathbf{x}=\mathbf{b}\\ & \mathbf{l}\leq\mathbf{x}\leq\mathbf{u},\end{aligned}\tag{1.2-2}$$

where $\mathbf{x}$ is an n-vector that includes the slack variables, $\mathbf{A}$ is an $m\times n$ matrix with $m\leq n$, which includes the columns of the identity matrix corresponding to the slack variables and $\mathbf{c}$, $\mathbf{l}$, and $\mathbf{u}$ are n-vectors also suitably defined to incorporate the slacks. We shall assume, throughout this chapter, that $\mathbf{A}$ is of *full row rank,* i.e., that its rows are linearly independent.

Points $\mathbf{x}\in R^n$ that satisfy all constraints (including, of course, the bounds) are called *feasible* points. Other points are said to be *infeasible.* The subset of feasible points that minimize the objective function are called *optimal* points and they solve the above problem (1.2-2).

Figure 1.1 depicts the set of feasible points, P, for a problem involving a single general constraint $(\mathbf{a}^1)^T\mathbf{x}=b_1$, $\mathbf{x}\in R^3$ (which defines a *plane*), and a set of bound constraints (which define a three-dimensional box). The feasible region is a *bounded polygon* as shown in the figure. Figure 1.2 depicts the *unbounded* case. If a second linearly independent general constraint were present, say $(\mathbf{a}^2)^T\mathbf{x}=b_2$, and if the plane that it defines intersected the polygons shown, then the feasible region would be a line segment or, in the unbounded case, possibly a half-line that went to infinity. If the second plane did not intersect the polygons, then the

Figure 1.1 Bounded polyhedral set

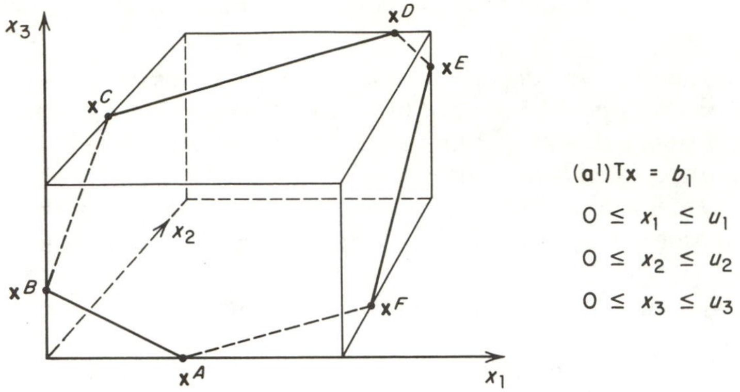
x_3
$\mathbf{x}^D$
$\mathbf{x}^E$
$\mathbf{x}^C$
x_2
$\mathbf{x}^B$
$\mathbf{x}^F$
$\mathbf{x}^A$
x_1
$(\mathbf{a}^1)^T\mathbf{x} = b_1$
$0 \leq x_1 \leq u_1$
$0 \leq x_2 \leq u_2$
$0 \leq x_3 \leq u_3$

problems would have no solutions that satisfy all constraints and bounds. Such problems are said to be *infeasible.*

Note that an equality constraint $(\mathbf{a}^i)^T\mathbf{x} = b_i$ can be converted into two inequality constraints, namely, $(\mathbf{a}^i)^T\mathbf{x} \leq b_i$ and $(\mathbf{a}^i)^T\mathbf{x} \geq b_i$, so that alternative forms to (1.2-2) are possible. However, not only does this transformation increase the number of general constraints, but it obscures the fact that feasible points are constrained to lie within the surface of a hyperplane (defined by the equality constraint). For computational purposes, as we shall presently see, the form (1.2-2) is generally the most convenient. Therefore we develop the main ideas using this form (1.2-2) and leave to exercises their elaboration within the setting of other forms.

Figure 1.2 Unbounded polyhedral set

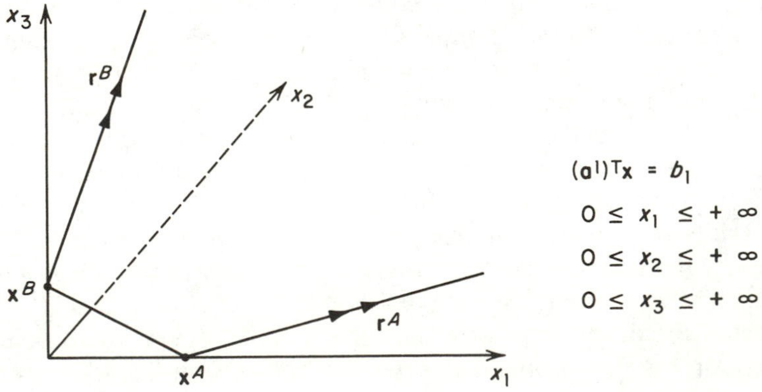
x_3
$\mathbf{r}^B$
x_2
$\mathbf{x}^B$
$\mathbf{r}^A$
$\mathbf{x}^A$
x_1
$(\mathbf{a}^1)^T\mathbf{x} = b_1$
$0 \leq x_1 \leq +\infty$
$0 \leq x_2 \leq +\infty$
$0 \leq x_3 \leq +\infty$

1.3. The Underlying Geometry

A picture is said to be worth a thousand words (or in computer jargon, a graphic image may be said to be worth a megabyte). Let us therefore, through examples, and without attempting any complete or rigorous treatment, introduce the geometric notions that underlie the algebraic linear programming model (1.2-2).

In n-dimensional space, each constraint of (1.2-2) of the form $(\mathbf{a}^i)^T\mathbf{x} = b_i$ restricts feasible points to lie within the surface of a *hyperplane.* (For example, in Figures 1.1 and 1.2, this corresponds to the usual notion of a plane, as depicted in R^3.) A set of constraints $\mathbf{Ax} = \mathbf{b}$ restricts the feasible points satisfying them to the intersection of a set of hyperplanes, formally called an *affine space,* say H; this is of dimension $n - m$, because each equality constraint removes one degree of freedom, so that collectively, they remove m degrees of freedom. The bound constraints define a "box" or *hyperrectangle* in R^n. Thus, in geometric terms, the feasible region consists of the points of H that lie within the hyperrectangle. This set of feasible points defines a *convex polyhedral set* and if it is bounded (i.e., if there exists a constant, say κ, such that $\|\mathbf{x}\|_2 \equiv (\mathbf{x}^T\mathbf{x})^{1/2} \leq \kappa$ for all $\mathbf{x}$ in the set), then it defines a *bounded convex polyhedral set.* In our earlier examples, see Figures 1.2 and 1.1, respectively, these are the polygons depicted. The term *convex* is used to characterize any set P with the following property: If $\mathbf{x}_1$ and $\mathbf{x}_2$ are both members of P, then all points on the line segment joining them are also members of P. The point

$$\mathbf{x} = (1 - s)\mathbf{x}_1 + s\mathbf{x}_2 = \mathbf{x}_1 + s(\mathbf{x}_2 - \mathbf{x}_1) \qquad s \in [0, 1] \qquad (1.3\text{-}1)$$

is called a *linear convex combination* of $\mathbf{x}_1$ and $\mathbf{x}_2$ and lies on the above line segment. More generally, if

$$\mathbf{x} = \sum_{i=1}^{k} s_i\mathbf{x}_i \qquad \sum_{i=1}^{k} s_i = 1 \qquad s_i \geq 0, \qquad (1.3\text{-}2)$$

then $\mathbf{x}$ is a linear convex combination of $\mathbf{x}_1, \ldots, \mathbf{x}_k$. A convex set contains all linear convex combinations of its members.

Exercise 1.3-1, Verify that the problem constraints of (1.2-2) define a convex set.

Points like $\mathbf{x}^A, \ldots, \mathbf{x}^F$ in Figure 1.1 are called *vertices.* Note that in this three-dimensional example, each vertex is given by the point where *three* problem constraints (general constraints or bounds) hold as equalities. For example, in Figure 1.1, at $\mathbf{x}^C$ we have $(\mathbf{a}^1)^T\mathbf{x} = b_1$, $x_1 = 0$, $x_3 = u_3$. For $\mathbf{x}^B$ in Figure 1.2 we have $(\mathbf{a}^1)^T\mathbf{x} = b_1$, $x_1 = 0$, $x_2 = 0$.

A constraint that holds as an equality at a given point is said to be *tight* at that point. In R^n, a *vertex* is the point of intersection of *n linearly independent tight* problem constraints that satisfies all the remaining problem constraints. A set of tight constraints that are *linearly independent* will be called *active*. All other constraints are called *inactive*. Since the m general constraints of (1.2-2) (assumed to be linearly independent) are always tight at any feasible point, we have $n - m$ linearly independent tight bounds at a vertex.

An alternative way of introducing a vertex is through the notion of an extreme point. An *extreme point* of a convex polyhedral set, say P, is one that *cannot* be expressed as a strict convex combination of two *distinct* points of P; i.e., it does *not* lie on any line segment joining two distinct points of P. There is a one-to-one correspondence between vertices and extreme points of P, as is easily demonstrated by the following theorem. (Throughout this book, proofs of theorems set in a smaller type may be skipped at a first reading.)

Theorem 1.3-1. Given a convex polyhedral set $P \equiv [\mathbf{x} \mid \mathbf{Ax} = \mathbf{b}, \mathbf{l} \le \mathbf{x} \le \mathbf{u}]$ defined by the problem constraints of (1.2-2), a necessary and sufficient condition that $\mathbf{x}^0$ be an extreme point of P is that $\mathbf{x}^0 \in P$ be the solution of a system of n linearly independent tight constraints, i.e., that $\mathbf{x}^0$ be a vertex.

Proof. (a) ($\Leftarrow$) Suppose $\mathbf{x}^0$ is a vertex. Then it is the solution of n linearly independent tight constraints $(\mathbf{a}^i)^T\mathbf{x}^0 = b_i$, $i = 1, \ldots, m$, $(\mathbf{e}^i)^T\mathbf{x}^0 = x_i^0 = l_i$, $i \in L$ and $(\mathbf{e}^i)^T\mathbf{x}^0 = u_i$, $i \in U$, where L and U denote index sets and $\mathbf{e}^i$ denotes the ith column of the $n \times n$ identity matrix. If $|L|$ and $|U|$ denote the number of indices in L and U, respectively, then $m + |L| + |U| = n$. Let us write this square system as $\boldsymbol{B}^0\mathbf{x}^0 = \boldsymbol{b}^0$ where $\boldsymbol{B}^0$ is an $n \times n$ nonsingular matrix. Suppose now that $\mathbf{x}^0$ is not an extreme point of P, so that there are two distinct points of P, say $\mathbf{x}_1^0$ and $\mathbf{x}_2^0$, for which

$$\mathbf{x}^0 = (1 - s)\mathbf{x}_1^0 + s\mathbf{x}_2^0 \qquad s \in (0, 1).$$

Obviously $(\mathbf{a}^i)^T\mathbf{x}_1^0 = b_i = (\mathbf{a}^i)^T\mathbf{x}_2^0$, $i = 1, \ldots, m$. For $i \in L$, $(\mathbf{e}^i)^T\mathbf{x}_1^0 \ge l_i$ and $(\mathbf{e}^i)^T\mathbf{x}_2^0 \ge l_i$. If one of these bound constraints holds as a *strict* inequality, say for $i = i'$, then

$$l_{i'} = (\mathbf{e}^{i'})^T\mathbf{x}^0 = (1 - s)(\mathbf{e}^{i'})^T\mathbf{x}_1^0 + s(\mathbf{e}^{i'})^T\mathbf{x}_2^0 > l_{i'}.$$

This is, of course, impossible; it follows that $(\mathbf{e}^i)^T\mathbf{x}_1^0 = l_i = (\mathbf{e}^i)^T\mathbf{x}_2^0$, $i \in L$. Similarly, $(\mathbf{e}^i)^T\mathbf{x}_1^0 = u_i = (\mathbf{e}^i)^T\mathbf{x}_2^0$, $i \in U$. Therefore $\boldsymbol{B}^0\mathbf{x}_1^0 = \boldsymbol{B}^0\mathbf{x}_2^0$. Since $\boldsymbol{B}^0$ is nonsingular, $\mathbf{x}_1^0 = \mathbf{x}_2^0$, which contradicts our assumption that $\mathbf{x}_1^0$ and $\mathbf{x}_2^0$ are distinct. Therefore $\mathbf{x}^0$ must be an extreme point.

(b) ($\Rightarrow$) Suppose that $\mathbf{x}^0$ is an extreme point of P. Suppose in addition to the m linearly independent general constraints $(\mathbf{a}^i)^T\mathbf{x}^0 = b_i$, $i = 1, \ldots, m$, that t bound constraints are tight, say $(\mathbf{e}^i)^T\mathbf{x}^0 = l_i$, $i \in \bar{L}$ and $(\mathbf{e}^i)^T\mathbf{x}^0 = u_i$, $i \in \bar{U}$. Let us denote this system of $(m + t)$ tight constraints by $\boldsymbol{B}\mathbf{x}^0 = \boldsymbol{b}$, where $\boldsymbol{B}$ is an $(m + t) \times n$ matrix. We now show that the columns of $\boldsymbol{B}$ are linearly independent.

Suppose that they are not linearly independent. Then there exists a vector $\mathbf{y} \ne \mathbf{0}$ such that $\boldsymbol{B}\mathbf{y} = \mathbf{0}$, i.e., $(\mathbf{b}^i)^T\mathbf{y} = 0$, $i = 1, \ldots, m + t$, where $\mathbf{b}^i$ denotes the ith row of $\boldsymbol{B}$. Consider a point $\mathbf{x} = \mathbf{x}^0 + s\mathbf{y}$. $\boldsymbol{B}(\mathbf{x}^0 + s\mathbf{y}) = \boldsymbol{B}\mathbf{x}^0 + s\boldsymbol{B}\mathbf{y} = \boldsymbol{b}$, so the tight constraints continue to be satisfied. All other bounds are assumed to be nontight at $\mathbf{x}^0$, i.e., to hold as strict inequalities. Therefore they will continue to be satisfied at $\mathbf{x}^0$ provided s is sufficiently small, i.e., there exists a $\delta > 0$ for which $\mathbf{x} \in P$ for all s such that $-\delta \leq s \leq \delta$. Since $\mathbf{x}^0 = \frac{1}{2}(\mathbf{x}^0 + \delta\mathbf{y}) + \frac{1}{2}(\mathbf{x}^0 - \delta\mathbf{y})$, this contradicts our assumption that $\mathbf{x}^0$ is an extreme point of P.

Therefore, the n columns of $\boldsymbol{B}$ must be linearly independent. It follows that there are $n - m$ linearly independent bound constraints that are tight at $\mathbf{x}^0$, i.e., $t \geq n - m$. Together with the m general constraints, these define an active set with n members. Therefore $\mathbf{x}^0$ is a vertex. This completes the proof. ■

Exercise 1.3-2. Construct an example to show there are points that are not contained in the polyhedral set P, and are therefore not vertices of P, which are also the solution of n linearly independent tight problem constraints. Also, construct an example of a polyhedral set with no vertices.

Henceforth we shall use the terms *vertex* and *extreme point* interchangeably. Since there are only a finite number of different ways of choosing n active problem constraints from among the $m + 2n$ problem constraints of (1.2-2), there are only a finite number of vertices of the associated polyhedral set. A bound on this number is $(m + 2n)C_n$, the number of combinations of $(m + 2n)$ objects taken n at a time. It is, in fact, overly pessimistic and much better bounds can be established; see Grunbaum [1967].

Sometimes additional problem constraints, in our particular case bound constraints, will be tight at a vertex, i.e., there may be $(n + t)$, $t > 0$, tight problem constraints at a vertex. Such constraints, in our case necessarily bounds that are linearly dependent on the active set, will be called *inactive but tight* or *inactive at bound* and the corresponding vertex is said to be *degenerate.* If there are no inactive but tight bounds at a vertex, it is said to be *nondegenerate.*

Two distinct (nondegenerate) vertices are adjacent when they have $n - 1$ active constraints in common. Feasible points on the line joining two adjacent vertices define an *edge* of P. When the polyhedral set is unbounded, as in Figure 1.2, there are directions like $\mathbf{r}^A$ and $\mathbf{r}^B$ that go off to infinity, instead of eventually meeting another vertex. These are examples of *extreme directions.* More generally, they can be defined in much the same way as extreme points, as follows: a *direction of a set,* say $\mathbf{d}$, is a nonzero vector such that for all $s \geq 0$, $\mathbf{x} + s\mathbf{d}$ is contained in the set, for each member $\mathbf{x}$ of the set. An *extreme direction* is one that *cannot* be expressed as a positive linear combination of two *distinct* directions of the set, i.e., as $t_1\mathbf{d}_1 + t_2\mathbf{d}_2, t_1 > 0, t_2 > 0,$ where $\mathbf{d}_1$ is not a positive multiple of $\mathbf{d}_2$.

In Figure 1.1, it should be intuitively evident that any point $\mathbf{x} \in P$ can be expressed as

$$\mathbf{x} = s_1\mathbf{x}^A + s_2\mathbf{x}^B + s_3\mathbf{x}^C + \cdots + s_6\mathbf{x}^F$$

where

$$s_1 + s_2 + s_3 + \cdots + s_6 = 1 \qquad s_i \geq 0 \qquad i = 1, \ldots, 6$$

i.e., $\mathbf{x}$ can be expressed as a linear convex combination of the extreme points. Similarly, it should be evident in Figure 1.2 that any point $\mathbf{x} \in P$ can be expressed as

$$\mathbf{x} = s_1\mathbf{x}^A + s_2\mathbf{x}^B + t_1\mathbf{r}^A + t_2\mathbf{r}^B$$

where

$$s_1 + s_2 = 1 \qquad s_1 \geq 0 \qquad s_2 \geq 0 \qquad t_1 \geq 0 \qquad t_2 \geq 0$$

i.e., $\mathbf{x}$ can be expressed as a linear convex combination of $\mathbf{x}^A$ and $\mathbf{x}^B$ plus a positive linear combination of $\mathbf{r}^A$ and $\mathbf{r}^B$.

The previous examples are particular instances of a general representation theorem, whose validity should be intuitively self-evident. The detailed proof, when one starts from scratch, is a little tedious and we shall therefore content ourselves here with simply stating the result. Later, in Chapter 12, we shall have available more powerful tools that will enable us to give a very simple proof.

Theorem 1.3-2. Representation Theorem. Let $P = [\mathbf{x} \mid \mathbf{Ax} = \mathbf{b}, \mathbf{l} \leq \mathbf{x} \leq \mathbf{u}]$ be a nonempty convex polyhedral set. Then the set of vertices is finite, say $\mathbf{x}^1, \ldots, \mathbf{x}^{n_p}$. The set of extreme directions is nonempty if and only if P is unbounded. If P is unbounded, the set of extreme directions is finite, say $\mathbf{r}^1, \ldots, \mathbf{r}^{n_r}$.

$\mathbf{x} \in P$ if and only if it can be represented as

$$\mathbf{x} = \sum_{j=1}^{n_p} s_j\mathbf{x}^j + \sum_{j=1}^{n_r} t_j\mathbf{r}^j \qquad \sum_{j=1}^{n_p} s_j = 1 \qquad s_j \geq 0,\ t_j \geq 0 \text{ for all } j. \tag{1.3-3}$$

Proof. Will be given in Chapter 12, Sec. 12.2. ■

Let us next consider $\mathbf{c}^T\mathbf{x}$, the objective function in (1.2-2), from a geometric standpoint. Suppose that the objective function assumes some value, say, z^0. Then the equation $\mathbf{c}^T\mathbf{x} = z^0$ defines a hyperplane whose *normal* is the vector $\mathbf{c}$. $\mathbf{c}$ points in the direction of *increasing* values of the objective function and all points that lie in the hyperplane have objective value z^0. These points must, of course, include at least one feasible point of P, if they are to be of interest. If the objective hyperplane is now translated parallel to itself and in the opposite direction to which $\mathbf{c}$ points, we obtain new points $\mathbf{x}$ that lie within it and have a *decreased* objective value. Suppose that the objective hyperplane is moved over the feasible region in this fashion. (To be specific, consider the examples of Figures 1.1 and 1.2.) Then it will either be able to go to infinity while continuously decreasing the value of the objective function without violating feasibility, or at some point of translation, no further move is possible without P and the objective hyperplane having an empty intersection. Intuitively, therefore, the optimal value of (1.2-2) will either be unbounded from below, i.e., will go to $-\infty$, or will occur at a vertex (or set of vertices and all linear convex combinations of them). The representation theorem quoted above enables us to easily prove this intuitive assertion of optimality.

Theorem 1.3-3. Fundamental Theorem of Linear Programming. Suppose that the polyhedral set P defined by the problem constraints of the linear program (1.2-2) has a feasible solution. Then one of the following two conditions must hold:

(a) The optimal solution is unbounded from below.
(b) There exists a vertex (extreme point) that is optimal for the linear program.

Proof. Using the expression for $\mathbf{x} \in P$ given by the representation theorem, namely (1.3-3), we can express the linear program (1.2-2) in the following form:

$$\begin{aligned} \text{minimize} \quad & \sum_{j=1}^{n_p} (\mathbf{c}^T\mathbf{x}^j)s_j + \sum_{j=1}^{n_r} (\mathbf{c}^T\mathbf{r}^j)t_j \\ \text{s.t.} \quad & \sum_{j=1}^{n_p} s_j = 1 \\ & s_j \geq 0 \qquad j = 1, \ldots, n_p \qquad t_j \geq 0 \qquad j = 1, \ldots, n_r. \end{aligned} \tag{1.3-4}$$

Suppose $(\mathbf{c}^T\mathbf{r}^{j'}) < 0$, for some index j'. Then the objective function can be made as small as desired by making $t_{j'}$ sufficiently large. Thus the optimal value is unbounded from below. Suppose, therefore, that $(\mathbf{c}^T\mathbf{r}^j) \geq 0$, $j = 1, \ldots, n_r$. Then by setting $t_j = 0$, $j = 1, \ldots, n_r$, we obtain the least value of the second term in the objective function of (1.3-4). To minimize the first term in this objective function, subject to the convexity constraint, namely,

$$\sum_{j=1}^{n_p} s_j = 1 \qquad s_j \geq 0,$$

find any index j'' for which

$$(\mathbf{c}^T\mathbf{x}^{j''}) = \min_{j=1,\ldots,n_p} (\mathbf{c}^T\mathbf{x}^j)$$

and set $s_{j''} = 1$ and $s_j = 0$, $j \neq j''$. The minimum is then attained at the extreme point (vertex) $\mathbf{x}^{j''}$. We have shown, therefore, that either (a) or (b) above must hold, and the theorem is proved. ■

The foregoing proof is unfortunately incomplete, because it relies so heavily on the representation theorem, which we have merely stated. In the next chapter, a direct and constructive proof of the fundamental theorem of linear programming will be a consequence of the method of solution introduced there.

Exercise 1.3-3. Study the underlying geometry within the setting of the alternative form "Find $\bar{\mathbf{x}} \in R^n$ subject to $\bar{\mathbf{A}}\bar{\mathbf{x}} \leq \bar{\mathbf{b}}$ and $\bar{\mathbf{c}}^T\bar{\mathbf{x}}$ is maximized." Show that all the geometric concepts and results just developed apply equally well within this setting.

This completes our brief introduction to the geometry underlying the linear program (1.2-2). This geometric development can be pursued more rigorously and in greater depth through the references given in the notes at the end of this chapter. We return to our more rigorous algebraic characterization, in particular of optimal points.

1.4. Algebraic Characterization of Optimality

We now come to a central question. Given a point $\mathbf{x}^*$ that is feasible for the linear program (1.2-2) (i.e., $\mathbf{x}^* \in P$), how do we recognize whether it is optimal? (To emphasize the fact that we are concerned in this section with characterization of an optimal point, we shall attach the symbol $*$ to $\mathbf{x}$.) Let us make the following definition. A *feasible direction* at $\mathbf{x}^*$ is a direction $\mathbf{d}$ such that $(\mathbf{x}^* + s\mathbf{d}) \in P$ for all sufficiently small $s > 0$. Thus $\mathbf{d}$ is a feasible direction if there exists an $\bar{s} > 0$ such that $(\mathbf{x}^* + s\mathbf{d}) \in P$ for all s satisfying $\bar{s} \geq s \geq 0$.

Let $D(\mathbf{x}^*)$ be the set of all feasible directions at $\mathbf{x}^*$, namely,

$$D(\mathbf{x}^*) = [\mathbf{d} \mid \text{there exists } \bar{s} > 0 \text{ such that } (\mathbf{x}^* + s\mathbf{d}) \in P \text{ whenever } \bar{s} \geq s \geq 0]. \quad (1.4\text{-}1)$$

Note in the foregoing definition that $\bar{s}$ will, in general, depend on $\mathbf{d}$.

Lemma 1.4-1. Suppose $\mathbf{x}^*$ is optimal for (1.2-2). Then

$$\mathbf{c}^T\mathbf{d} \geq 0 \quad \text{for all} \quad \mathbf{d} \in D(\mathbf{x}^*). \quad (1.4\text{-}2)$$

Proof. Suppose that there exists a particular feasible direction, say $\mathbf{d}^* \in D(\mathbf{x}^*)$ such that $\mathbf{c}^T\mathbf{d}^* < 0$. Then there exists $s > 0$ for which $(\mathbf{x}^* + s\mathbf{d}^*) \in P$ and

$$\mathbf{c}^T(\mathbf{x}^* + s\mathbf{d}^*) = \mathbf{c}^T\mathbf{x}^* + s(\mathbf{c}^T\mathbf{d}^*) < \mathbf{c}^T\mathbf{x}^*.$$

Therefore $\mathbf{x}^*$ cannot be an optimal point.

It follows, when $\mathbf{x}^*$ is optimal, that $\mathbf{c}^T\mathbf{d} \geq 0$ for all $\mathbf{d} \in D(\mathbf{x}^*)$. This completes the proof. ■

Given a feasible point $\mathbf{x}^*$, how do we algebraically characterize $D(\mathbf{x}^*)$ in terms of the constraints of (1.2-2)? In seeking to do so, we need only consider the problem constraints that are *tight* at $\mathbf{x}^*$; for any direction $\mathbf{d}$, problem constraints that are nontight at $\mathbf{x}^*$ continue to be nontight at $(\mathbf{x}^* + s\mathbf{d})$, provided s is sufficiently small.

The general constraints are always tight when $\mathbf{x}^*$ is feasible. Let L and

U be (ordered) *index sets* of tight lower and upper bounds, respectively, at $\mathbf{x}^*$. Let $|L|$ and $|U|$ denote the number of elements in L and U, respectively.

$$\begin{aligned} L &= \{j \mid x_j^* = l_j \qquad j = 1, \ldots, n\} \\ U &= \{j \mid x_j^* = u_j \qquad j = 1, \ldots, n\}. \end{aligned} \tag{1.4-3}$$

Without loss of generality, we may assume that $l_j < u_j, j = 1, \ldots, n$. (Whenever $l_{j'} = u_{j'}$, for some j', then $x_{j'}^*$ is fixed and may be eliminated from the problem.) Thus L and U may be assumed to be disjoint. Let $\mathbf{e}^j$ denote the jth column of the $n \times n$ identity matrix. Then the tight constraints at $\mathbf{x}^*$ are

$$\begin{aligned} (\mathbf{a}^i)^T\mathbf{x}^* &= b_i \qquad i = 1, \ldots, m, \\ (\mathbf{e}^j)^T\mathbf{x}^* &= l_j \qquad j \in L, \\ (\mathbf{e}^j)^T\mathbf{x}^* &= u_j \qquad j \in U. \end{aligned} \tag{1.4-4}$$

Let us define the following set of directions at $\mathbf{x}^*$.

$$\begin{aligned} \boldsymbol{D}(\mathbf{x}^*) = [\mathbf{d} \mid (\mathbf{a}^i)^T\mathbf{d} &= 0 \qquad i = 1, \ldots, m \\ (\mathbf{e}^j)^T\mathbf{d} &\geq 0 \qquad j \in L \\ (\mathbf{e}^j)^T\mathbf{d} &\leq 0 \qquad j \in U]. \end{aligned} \tag{1.4-5}$$

The next lemma gives the more convenient algebraic characterization of the set of feasible directions.

Lemma 1.4-2.

$$D(\mathbf{x}^*) = \boldsymbol{D}(\mathbf{x}^*).$$

Proof. (a) ($\Rightarrow$). Suppose $\mathbf{d} \in D(\mathbf{x}^*)$. Then for $s > 0$ and sufficiently small (so as not to violate the nontight constraints), $(\mathbf{x}^* + s\mathbf{d})$ is feasible.

$(\mathbf{a}^i)^T(\mathbf{x}^* + s\mathbf{d}) = b_i, i = 1, \ldots, m.$ Since $(\mathbf{a}^i)^T\mathbf{x}^* = b_i$ and $s > 0$, it follows that $(\mathbf{a}^i)^T\mathbf{d} = 0, i = 1, \ldots, m$.

$(\mathbf{e}^j)^T(\mathbf{x}^* + s\mathbf{d}) \geq l_j, j \in L$. Since $(\mathbf{e}^j)^T\mathbf{x}^* = l_j$ and $s > 0$, it follows that $(\mathbf{e}^j)^T\mathbf{d} \geq 0, j \in L$. Similarly, $(\mathbf{e}^j)^T\mathbf{d} \leq 0, j \in U$.

Therefore, $\mathbf{d} \in \boldsymbol{D}(\mathbf{x}^*)$.

(b) ($\Leftarrow$) The proof of the converse is very similar. It is left as the following exercise and completes the proof. ∎

Exercise 1.4-1. Show that $\mathbf{d} \in \boldsymbol{D}(\mathbf{x}^*)$ implies that $\mathbf{d} \in D(\mathbf{x}^*)$.

Lemma 1.4-3. If $\mathbf{x}^*$ is feasible for (1.2-2) then

$$\mathbf{c}^T\mathbf{d} \geq 0 \quad \text{for all} \quad \mathbf{d} \in \boldsymbol{D}(\mathbf{x}^*). \tag{1.4-6}$$

Proof. This is a direct consequence of Lemmas 1.4-1 and 1.4-2. ∎

We now reformulate (1.4-6) by appealing to the following classical result. Its proof is somewhat technical and is included for completeness;

the reader may prefer to convince himself or herself of its validity by considering specific examples and skip the proof at a first reading. (Throughout this book, proofs of theorems set in smaller type may be skipped at a first reading.)

Lemma 1.4-4. Minkowski–Farkas. Consider a given set of vectors $\mathbf{v}_i \in R^n$, $i = 1, \ldots, k$, and a vector $\mathbf{c} \in R^n$ for which the following statement holds true.

"For any $\mathbf{d} \in R^n$, $\mathbf{v}_i^T\mathbf{d} \geq 0, i = 1, \ldots, k$ implies that $\mathbf{c}^T\mathbf{d} \geq 0$."

Then

$$\mathbf{c} = \sum_{i=1}^{k} \alpha_i \mathbf{v}_i \quad \text{with} \quad \alpha_i \geq 0 \qquad i = 1, \ldots, k. \tag{1.4-7}$$

Proof. The above statement in quotes can be equivalently formulated as follows.

$$\mathbf{v}_i^T\mathbf{d} \geq 0 \qquad i = 1, \ldots, k \quad \text{and} \quad \mathbf{c}^T\mathbf{d} < 0 \text{ has } no \text{ solution } \mathbf{d} \in R^n. \tag{1.4-8}$$

Let us call this our *assumption*.

We shall show that if we cannot find $(\alpha_1, \alpha_2, \ldots, \alpha_k) \geq \mathbf{0}$ for which

$$\mathbf{c} = \sum_{i=1}^{k} \alpha_i \mathbf{v}_i,$$

then we contradict our assumption.

(a) Suppose that

$$\mathbf{c} = \sum_{i=1}^{k} \alpha_i \mathbf{v}_i$$

has no solution, i.e., that $\mathbf{c}$ does not lie in the space spanned by $\mathbf{v}_1, \ldots, \mathbf{v}_k$. Let us denote this space by V. Let $\mathbf{c} = \tilde{\mathbf{c}} + \bar{\mathbf{c}}$ where $\tilde{\mathbf{c}} \in V$ and $\bar{\mathbf{c}} \in V^+$, the orthogonal complement of V. Thus $\tilde{\mathbf{c}}^T\bar{\mathbf{c}} = 0$.

Let $\mathbf{d} \equiv -\bar{\mathbf{c}}$. Then $\mathbf{v}_i^T\mathbf{d} = 0, i = 1, \ldots, k$ and $\mathbf{c}^T\mathbf{d} = (\tilde{\mathbf{c}} + \bar{\mathbf{c}})^T(-\bar{\mathbf{c}}) = -\bar{\mathbf{c}}^T\bar{\mathbf{c}} < 0$. Thus our assumption is contradicted. Our supposition that

$$\mathbf{c} = \sum_{i=1}^{k} \alpha_i \mathbf{v}_i$$

has no solution α_i, $i = 1, \ldots, k$ is therefore false.

(b) Suppose, therefore, that

$$\mathbf{c} = \sum_{i=1}^{k} \alpha_i \mathbf{v}_i$$

has a solution *but no nonnegative solution* $(\alpha_1, \ldots, \alpha_k)$. We now prove the lemma by induction on the number k of elements in $\mathbf{v}_1, \ldots, \mathbf{v}_k$.

(i) Let $k = 1$. Suppose $\mathbf{c} = \alpha_1\mathbf{v}_1$ has a solution $\alpha_1 < 0$. Let $\mathbf{d} = -\mathbf{c}$. Then

$$\mathbf{v}_1^T\mathbf{d} = -\mathbf{v}_1^T\mathbf{c} = (-\alpha_1)\mathbf{v}_1^T\mathbf{v}_1 \geq 0 \quad \text{and} \quad \mathbf{c}^T\mathbf{d} = -\mathbf{c}^T\mathbf{c} < 0.$$

This contradicts our assumption (1.4-8). Therefore our supposition that $\alpha_1 < 0$ is false, and the lemma is proved for $k = 1$.

(ii) Let us make the *induction hypothesis* that the lemma holds for $k = 1, 2, \ldots, (k' - 1)$, i.e., given that (1.4-8) holds for $k = 1, 2, \ldots, (k' - 1)$ then

(1.4-7) holds. We now show that the induction hypothesis holds for $k = 1, 2, \ldots, k'$.

Given our assumption (1.4-8) for $k = 1, 2, \ldots, k'$ suppose that (1.4-7) is false so that

$$\mathbf{c} = \sum_{i=1}^{k'} \alpha_i \mathbf{v}_i \quad \text{and} \quad (\alpha_1, \ldots, \alpha_{k'}) \geq \mathbf{0} \tag{1.4-9}$$

has no solution. Henceforth we shall call this our *premise*.

There exists no nonnegative solution to

$$\mathbf{c} = \sum_{i=1}^{k'-1} \alpha_i \mathbf{v}_i, \tag{1.4-10a}$$

because, if this were not true, we could use $\alpha_1, \ldots, \alpha_{k'-1}$ from (1.4-10a) and $\alpha_{k'} = 0$, to give a nonnegative solution to (1.4-9), leading to a contradiction. Thus by the induction hypothesis, there is a $\bar{\mathbf{d}}$ such that

$$\mathbf{v}_i^T \bar{\mathbf{d}} \geq 0 \qquad i = 1, \ldots, (k'-1) \quad \text{and} \quad \mathbf{c}^T \bar{\mathbf{d}} < 0. \tag{1.4-10b}$$

(ii.1) Suppose that $(\mathbf{v}_{k'})^T \bar{\mathbf{d}} \geq 0$. Then $\bar{\mathbf{d}}$ is a solution to $\mathbf{v}_i^T \bar{\mathbf{d}} \geq 0, i = 1, \ldots, k'$, and $\mathbf{c}^T \bar{\mathbf{d}} < 0$ and we have contradicted our assumption (1.4-8). The premise (1.4-9) is therefore false and the lemma follows by induction. (We have already shown it to be true for $k = 1$.)

(ii.2) Suppose that $(\mathbf{v}_{k'})^T \bar{\mathbf{d}} < 0$. Define vectors $\mathbf{v}_i'$ and $\mathbf{c}'$ that are orthogonal to $\bar{\mathbf{d}}$ as follows.

$$\mathbf{v}_i' = \mathbf{v}_i + \lambda_i \mathbf{v}_{k'} \quad \text{where} \quad \lambda_i = -(\mathbf{v}_i^T \bar{\mathbf{d}})/(\mathbf{v}_{k'}^T \bar{\mathbf{d}}) \geq 0 \qquad i = 1, \ldots, (k'-1), \tag{1.4-11a}$$

$$\mathbf{c}' = \mathbf{c} + \lambda_0 \mathbf{v}_{k'} \quad \text{where} \quad \lambda_0 = (-\mathbf{c}^T \bar{\mathbf{d}})/(\mathbf{v}_{k'}^T \bar{\mathbf{d}}) < 0. \tag{1.4-11b}$$

The inequalities in (1.4-11a–b) follow directly from (1.4-10b).

The system

$$\sum_{i=1}^{k'-1} \alpha_i' \mathbf{v}_i' = \mathbf{c}'$$

can be written, using (1.4-11a–b), as follows.

$$\sum_{i=1}^{k'-1} \alpha_i' \mathbf{v}_i + \left(\sum_{i=1}^{k'-1} \lambda_i \alpha_i' - \lambda_0 \right) \mathbf{v}_{k'} = \mathbf{c}.$$

If this system had a nonnegative solution in α_i', then we would contradict the premise (1.4-9). Therefore by the induction hypothesis, there must exist a vector $\bar{\mathbf{d}}'$ such that

$$(\mathbf{v}_i')^T \bar{\mathbf{d}}' \geq 0 \qquad i = 1, \ldots, (k'-1) \quad \text{and} \quad (\mathbf{c}')^T \bar{\mathbf{d}}' < 0.$$

Define the vector $\mathbf{d}$, orthogonal to $\mathbf{v}_{k'}$ as follows.

$$\mathbf{d} = \bar{\mathbf{d}}' - (\mathbf{v}_{k'}^T \bar{\mathbf{d}}' / \mathbf{v}_{k'}^T \bar{\mathbf{d}}) \bar{\mathbf{d}}. \tag{1.4-11c}$$

The definitions (1.4-11a–c) required $(\mathbf{v}_i')^T \bar{\mathbf{d}} = 0, i = 1, \ldots, (k'-1)$, $(\mathbf{c}')^T \bar{\mathbf{d}} = 0$ and $\mathbf{v}_{k'}^T \mathbf{d} = 0$. These, in turn, then imply the following.

$$\mathbf{v}_i^T \mathbf{d} = (\mathbf{v}_i')^T \mathbf{d} = (\mathbf{v}_i')^T \bar{\mathbf{d}}' \geq 0 \qquad i = 1, \ldots, (k'-1),$$

and

$$\mathbf{c}^T\mathbf{d} = (\mathbf{c}')^T\mathbf{d} = (\mathbf{c}')^T\bar{\mathbf{d}}' < 0.$$

Therefore we have found a vector $\mathbf{d} \in R^n$ for which

$$\mathbf{v}_i^T\mathbf{d} \geq 0 \qquad i = 1, \ldots, k' \quad \text{and} \quad \mathbf{c}^T\mathbf{d} < 0.$$

But this contradicts our assumption (1.4-8). The premise (1.4-9) must therefore be false and the induction hypothesis must hold for $k = 1, 2, \ldots, k'$. Since we have shown that it holds for $k = 1$, the proof by induction is complete. ■

Exercise 1.4-2. Show that the converse statement in Lemma 1.4-4 is also true, namely, when (1.4-7) holds then any $\mathbf{d} \in R^n$ satisfying $\mathbf{v}_i^T\mathbf{d} \geq 0, i = 1, \ldots, k$ implies that $\mathbf{c}^T\mathbf{d} \geq 0$.

Exercise 1.4-3. Give a geometric interpretation of Lemma 1.4-4.

Lemma 1.4-5. Given vectors $\mathbf{v}_i \in R^n, i = 1, \ldots, k$ and $\mathbf{c} \in R^n$, one and only one of the following two statements holds.

(a) There exists a vector $\mathbf{d} \in R^n$ satisfying $\mathbf{v}_i^T\mathbf{d} \geq 0, i = 1, \ldots, k$ and $\mathbf{c}^T\mathbf{d} < 0$.
(b) $\mathbf{c} = \sum_{i=1}^k \alpha_i \mathbf{v}_i$, $\alpha_i \geq 0$, $i = 1, \ldots, k$.

Proof. Follows immediately from Lemma 1.4-4 and Exercise 1.4-2. ■

Armed with this important result, let us now combine it with Lemma 1.4-3. To do so, we rewrite $\boldsymbol{D}(\mathbf{x}^*)$ in (1.4-5) as follows.

$$\boldsymbol{D}(\mathbf{x}^*) = [\mathbf{d} \mid (\mathbf{a}^i)^T\mathbf{d} \geq 0 \quad i = 1, \ldots, m, \quad -(\mathbf{a}^i)^T\mathbf{d} \geq 0 \quad i = 1, \ldots, m,$$
$$(\mathbf{e}^j)^T\mathbf{d} \geq 0 \quad j \in L, \quad -(\mathbf{e}^j)^T\mathbf{d} \geq 0 \quad j \in U]. \qquad (1.4\text{-}12)$$

If we identify $\mathbf{v}_i$, $i = 1, \ldots, k$ in Lemma 1.4-4, with the set of vectors

$$[\mathbf{a}^i \quad i = 1, \ldots, m, \qquad -\mathbf{a}^i \quad i = 1, \ldots, m, \qquad \mathbf{e}^j \quad j \in L, \qquad -\mathbf{e}^j \quad j \in U];$$

and if we define

$$\pi_i^+, \quad \pi_i^- \quad i = 1, \ldots, m, \qquad \lambda_j^* \quad j \in L \quad \text{and} \quad \mu_j^* \quad j \in U,$$

to play the role of the quantities α_i, $i = 1, \ldots, k$, then we can apply Lemma 1.4-4 directly to (1.4-6) and (1.4-12), to obtain

$$\mathbf{c} = \sum_{i=1}^{m} \pi_i^+\mathbf{a}^i + \sum_{i=1}^{m} \pi_i^-(-\mathbf{a}^i) + \sum_{j\in L} \lambda_j^*\mathbf{e}^j + \sum_{j\in U} \mu_j^*(-\mathbf{e}^j) \qquad (1.4\text{-}13)$$

and

$$\pi_i^+ \geq 0, \quad \pi_i^- \geq 0 \quad i = 1, \ldots, m, \quad \lambda_j^* \geq 0 \quad j \in L, \quad \mu_j^* \geq 0 \quad j \in U.$$

Let us define $\pi_i^* = \pi_i^+ - \pi_i^-$, $i = 1, \ldots, m$. Then π_i^* has *arbitrary* sign and we can express (1.4-13) as

$$\mathbf{c} = \sum_{i=1}^{m} \pi_i^*\mathbf{a}^i + \sum_{j\in L} \lambda_j^*\mathbf{e}^j - \sum_{j\in U} \mu_j^*\mathbf{e}^j \qquad (1.4\text{-}14)$$

and

$$\lambda_j^* \geq 0 \quad j \in L, \qquad \mu_j^* \geq 0 \quad j \in U.$$

We have therefore shown when $\mathbf{x}^*$ is optimal that condition (1.4-14) must hold, i.e., (1.4-14) is a *necessary* condition for optimality of $\mathbf{x}^*$.

We now show that (1.4-14) is also a sufficient condition for optimality of $\mathbf{x}^*$. Suppose that $\mathbf{x}^*$ is feasible ($\mathbf{x}^* \in P$) but *not* optimal and that the tight constraints at $\mathbf{x}^*$ are given by (1.4-4). Then there exists some other point, say $\mathbf{x}^{**} \in P$ with a lower objective value. Along the line joining $\mathbf{x}^*$ and $\mathbf{x}^{**}$, all points are feasible and the objective function decreases continuously, as can be easily seen as follows.

Let $\mathbf{d} \equiv \mathbf{x}^{**} - \mathbf{x}^*$. For $t \in (0, 1]$,

$$\mathbf{c}^T(\mathbf{x}^* + t\mathbf{d}) = \mathbf{c}^T((1-t)\mathbf{x}^* + t\mathbf{x}^{**}) < (1-t)\mathbf{c}^T\mathbf{x}^* + t\mathbf{c}^T\mathbf{x}^* = \mathbf{c}^T\mathbf{x}^*.$$

Thus $t\mathbf{c}^T\mathbf{d} < 0$. Since $t > 0$ it follows that $\mathbf{c}^T\mathbf{d} < 0$. Since $\mathbf{d}$ is a feasible direction, we also have $\mathbf{d} \in \boldsymbol{D}(\mathbf{x}^*)$. It then follows from Lemma 1.4-5 that (1.4-14) has *no* solution with $\lambda_j^* \geq 0, j \in L$, and $\mu_j^* \geq 0, j \in U$.

We can summarize these results by the following theorem.

Theorem 1.4-1. Optimality Criterion. A necessary and sufficient condition that a feasible point $\mathbf{x}^*$ be optimal for the linear program (1.2-2) is that there exist constants $\pi_i^*, i = 1, \ldots, m$, $\lambda_j^* \geq 0, j \in L$, and $\mu_j^* \geq 0, j \in U$ (called Lagrange multipliers) such that

$$\mathbf{c} = \sum_{i=1}^{m} \pi_i^* \mathbf{a}^i + \sum_{j \in L} \lambda_j^* \mathbf{e}^j - \sum_{j \in U} \mu_j^* \mathbf{e}^j, \tag{1.4-15}$$

where L and U denote the index sets of tight lower and upper bounds, respectively, at $\mathbf{x}^*$.

Corollary 1.4-1. Consider a linear program in the form

$$\begin{aligned} \text{minimize } & \mathbf{c}^T\mathbf{x} \\ \text{s.t. } & \mathbf{A}\mathbf{x} \geq \mathbf{b} \\ & \mathbf{x} \geq \mathbf{0}. \end{aligned}$$

Let $\mathbf{x}^*$ be a feasible point. Let I be the index set of tight constraints $(\mathbf{a}^i)^T\mathbf{x}^* = b_i, i \in I$ and L be the index set of tight bounds $x_j^* = 0, j \in L$. A necessary and sufficient condition that $\mathbf{x}^*$ be optimal is that there exist constants $\pi_i^* \geq 0, i \in I$, and $\lambda_j^* \geq 0, j \in L$ (Lagrange multipliers) such that

$$\mathbf{c} = \sum_{i \in I} \pi_i^* \mathbf{a}^i + \sum_{j \in L} \lambda_j^* \mathbf{e}^j.$$

Exercise 1.4-4. Prove Corollary 1.4-1 and formulate and prove the optimality criterion for a linear program in the form

$$\begin{aligned} \text{maximize } & \bar{\mathbf{c}}^T\bar{\mathbf{x}} \\ \text{s.t. } & \bar{\mathbf{A}}\bar{\mathbf{x}} \leq \bar{\mathbf{b}}. \end{aligned}$$

(See also the Notes on Sec. 1.2. at the end of this chapter, regarding the choice of notation for this exercise.)

Exercise 1.4-5. Give a geometric interpretation of the optimality criterion (1.4-15).

1.5. Associating Multipliers with a Vertex and Their Interpretation

In the previous section, we did not restrict $\mathbf{x}^*$ to be a vertex. The optimality criterion given there is an existence result, i.e., for a given *point* $\mathbf{x}^*$ to be optimal, there must exist multipliers satisfying (1.4-14) with $\lambda_j^* \geq 0, j \in L$ and $\mu_j^* \geq 0, j \in U$.

Consider now *any* given *vertex*, say $\mathbf{x}^0 \in P$, which is defined in the usual way by n active problem constraints $(\mathbf{a}^i)^T\mathbf{x}^0 = b_i$, $i = 1, \ldots, m$; $(\mathbf{e}^j)^T\mathbf{x}^0 = l_j$, $j \in L$ and $(\mathbf{e}^j)^T\mathbf{x}^0 = u_j$, $j \in U$, where $m + |L| + |U| = n$. Let us define a square matrix $\boldsymbol{B}^0$ and a vector $\boldsymbol{b}^0$ as follows.

$$\boldsymbol{B}^0 \equiv \begin{bmatrix} (\mathbf{a}^i)^T, i = 1, \ldots, m \\ (\mathbf{e}^j)^T, j \in L \\ (\mathbf{e}^j)^T, j \in U \end{bmatrix}$$

$$\boldsymbol{b}^0 \equiv \begin{bmatrix} b_i, i = 1, \ldots, m \\ l_j, j \in L \\ u_j, j \in U \end{bmatrix}.$$

By definition, the vertex $\mathbf{x}^0$ is the solution of $\boldsymbol{B}^0\mathbf{x}^0 = \boldsymbol{b}^0$ and $\boldsymbol{B}^0$ is nonsingular.

Let us *define* multipliers, π_i^0, $i = 1, \ldots, m$, $\lambda_j^0, j \in L$, and $\mu_j^0, j \in U$, associated with the vertex $\mathbf{x}^0$, as the solution of the $n \times n$ system corresponding to (1.4-14), namely,

$$\mathbf{c} = \sum_{i=1}^{m} \pi_i^0 \mathbf{a}^i + \sum_{j \in L} \lambda_j^0 \mathbf{e}^j - \sum_{j \in U} \mu_j^0 \mathbf{e}^j. \tag{1.5-1}$$

This is a nonsingular system of n equations in n unknowns and has a unique solution, but the quantities computed from it need not satisfy the conditions

$$\lambda_j^0 \geq 0 \quad j \in L \quad \text{and} \quad \mu_j^0 \geq 0 \quad j \in U. \tag{1.5-2}$$

If these conditions are satisfied, then $\mathbf{x}^0$ would, of course, be an optimal point. Henceforth we consistently use the notation $\mathbf{x}^*$, π_i^*, λ_j^*, and μ_j^* for an optimal point and its associated multipliers and we drop the symbol * attached to these quantities when we refer to an arbitrary point, in particular, an arbitrary vertex.

What do the quantities λ_j^0 and μ_j^0 obtained from (1.5-1) tell us about the linear program? To answer this question, consider a feasible direction

$\mathbf{d}$ such that $\mathbf{d}^T\mathbf{e}^{j'} > 0$ for some index $j' \in L$. Also assume $\mathbf{d}^T\mathbf{e}^j = 0, j \in L, j \neq j'$; $\mathbf{d}^T\mathbf{e}^j = 0, j \in U$ and $\mathbf{d}^T\mathbf{a}^i = 0, i = 1, \ldots, m$. We see that $\mathbf{d}$ is a step off the lower bound $(\mathbf{e}^{j'})^T\mathbf{x}^0 = l_{j'}$ and that all other active constraints at $\mathbf{x}^0$ remain active. Since $n-1$ problem constraints thus remain active, we are left with *one* degree of freedom. This defines $\mathbf{d}$ uniquely and represents an edge or extreme direction of the feasible set P. Let $\mathbf{x} = \mathbf{x}^0 + t\mathbf{d}, t > 0$. Then using (1.5-1) and the restrictions on $\mathbf{d}$,

$$\Delta z \equiv \mathbf{c}^T\mathbf{x} - \mathbf{c}^T\mathbf{x}^0 = t(\mathbf{c}^T\mathbf{d}) = t(\mathbf{d}^T\mathbf{e}^{j'})\lambda_{j'}^0. \qquad (1.5\text{-}3)$$

Since $t(\mathbf{d}^T\mathbf{e}^{j'}) > 0$, we see that $\lambda_{j'}^0$ is a measure of the change in the value of the objective function, when a step is taken off the lower bound for the component $x_{j'}$. In particular, $\lambda_{j'}^0 < 0$ implies that $\Delta z < 0$, so that, under these circumstances, it would be advantageous to relax the active bound corresponding to j'.

Exercise 1.5-1. Show by a similar argument, when $\mathbf{d}^T\mathbf{e}^{j'} < 0$ for some index $j' \in U$, and all other active constraints continue to remain active, that

$$\Delta z = -t(\mathbf{d}^T\mathbf{e}^{j'})\mu_{j'}^0, \qquad (1.5\text{-}4)$$

and $\mu_{j'}^0 < 0, j' \in U$ implies that $\Delta z < 0$.

Exercise 1.5-2. Form the expressions corresponding to (1.5-1), this time for the linear program in the form: Maximize $\bar{\mathbf{c}}^T\bar{\mathbf{x}}$ s.t. $\bar{\mathbf{A}}\bar{\mathbf{x}} \leq \bar{\mathbf{b}}$ (see also Exercise 1.4-4).

We have now assembled all the material that is needed to develop a method for solving the linear program (1.2-2), which is the subject of the next chapter.

Notes

Sec. 1.2. Expression (1.2-2) is the canonical form most suited to computation. Rather than first develop results for the simpler bounds $\mathbf{x} \geq \mathbf{0}$ and then generalize to $\mathbf{l} \leq \mathbf{x} \leq \mathbf{u}$, we prefer to characterize the solutions for the form (1.2-2), right from the outset. A second (canonical) form of the linear programming problem will be consistently distinguished throughout the book as maximize $\bar{\mathbf{c}}^T\bar{\mathbf{x}}$ s.t. $\bar{\mathbf{A}}\bar{\mathbf{x}} \leq \bar{\mathbf{b}}$. In particular, this will be of relevance to Chapter 11, which deals with duality theory.

Sec. 1.3. For a further discussion of the geometric characterization see, for example, Simonnard [1966], Grunbaum [1967], and Bazaraa and Jarvis [1977].

Sec. 1.4. The proof of Lemma 1.4-4 follows Simonnard [1966]. The overall development in this section is the counterpart of Zangwill [1969] and permits a natural transition to the nonlinear programming problem, where the optimality conditions of Theorem 1.4-1 are a special case of the so-called Karush–Kuhn–Tucker conditions.

2
The Simplex Method

A vertex $\mathbf{x}^0, \mathbf{x}^0 \in R^n$ of a linear program is a point where all problem constraints are satisfied and where n problem constraints are active (tight and linearly independent). This vertex is the solution of an associated square system of linear equations. Also associated with the set of n active constraints are a set of Lagrange multipliers, which are derived from the *transpose* of the foregoing square system of linear equations. The signs of these multipliers then tell us whether $\mathbf{x}^0$ is an optimal point (see Sec. 1.5). When $\mathbf{x}^0$ is not optimal, the multipliers can be used to identify an active problem constraint that would lead to a reduction in the value of the objective function when it is relaxed. Relaxing this constraint, but imposing the requirement that the other $n-1$ currently active constraints continue to remain tight, defines a unique direction. A feasible step along it leads to an improving vertex (unless the problem has an unbounded optimal solution). The procedure can then be repeated. This is the essence of the *simplex method* for solving a linear program.

In order to gain a thorough understanding of the simplex method, we look at the foregoing iterative process in some detail and from three different perspectives (in Secs. 2.1 through 2.3). These are all closely interrelated and, indeed, are merely different aspects of the same *method.* The salient point is that there is just one simplex method, which can be arrived at in different ways. Its particular expression and detailed computational prescription *within a specific context* then leads to a particular *simplex algorithm,* i.e., a method is a high-level description, which can be formulated in a variety of ways, to give rise to one or more algorithms or computational procedures. Of these, the most important is the *primal simplex algorithm.* (It is common practice to omit the word *primal* and to refer to it as *the simplex algorithm.*) This is formulated in Sec. 2.4 and its convergence and complexity discussed in Sec. 2.5. (In Chapters 11 through 13, we shall see how the simplex method, in other contexts, gives rise to the *dual simplex algorithm,* the *decomposition algorithm,* and the *self-dual simplex algorithm.*)

2.1. The Simplex Method on the Canonical Form

Consider the linear program (1.2-2), namely,

$$\begin{aligned} \text{minimize } & \mathbf{c}^T\mathbf{x} \\ \text{s.t. } & \mathbf{A}\mathbf{x} = \mathbf{b} \\ & \mathbf{l} \le \mathbf{x} \le \mathbf{u} \end{aligned}$$

where **A**, **b**, **c**, **l**, and **u** have been defined earlier. Throughout this chapter, for convenience, **A** *is assumed to be of full row rank, i.e., to have linearly independent rows, and each variable* $x_j, j = 1, \ldots, n$ *is assumed to have at least one finite bound.* (Relaxing these requirements is a technicality that we can ignore, at present.) Let $\mathbf{x}^0 \in P$ be an initial nondegenerate vertex of the polyhedral set of feasible solutions. $\mathbf{x}^0$ is defined by n active problem constraints, namely, the m general constraints $(\mathbf{a}^i)^T\mathbf{x}^0 = b_i, i = 1, \ldots, m$ and the $n - m$ bound constraints $(\mathbf{e}^j)^T\mathbf{x}^0 = l_j, j \in L,$ $(\mathbf{e}^j)^T\mathbf{x}^0 = u_j, j \in U$. The $n - m$ variables with indices $j \in L \cup U$ are called the *nonbasic* variables. The remaining m variables are called the *basic* variables. Let $N \equiv L \cup U$ denote the index set of the nonbasic variables and B denote the index set of the basic variables. Also, without loss of generality, we can assume that L and U are disjoint. Fixed variables (with $l_j = u_j$) can be eliminated from the problem by incorporating them into the right-hand side **b**. Let us also assume, for convenience, that the index sets B and N are ordered, and that the first m variables are the basic variables, and the last $n - m$ variables are the nonbasic variables, i.e., $B = \{1, \ldots, m\}$ and $N = \{m + 1, \ldots, n\}$. L is a subset of N, and U is the subset consisting of the remaining indices of N.

Under these assumptions, **A** can be partitioned as follows.

$$\mathbf{A} = [\mathbf{B}^0 \mid \mathbf{N}^0] \tag{2.1-1}$$

where $\mathbf{B}^0$ is an $m \times m$ matrix corresponding to the basic variables (called the *basic columns* of **A**) and $\mathbf{N}^0$ is an $m \times (n - m)$ matrix corresponding to the *nonbasic columns* of **A**. Let $\mathbf{x}_B^0$ denote the components of $\mathbf{x}^0$ that correspond to the basic variables and $\mathbf{x}_N^0$ denote the components that correspond to the nonbasic variables. Thus the vertex $\mathbf{x}^0$ is given by the solution of the following system of linear equations.

$$\begin{bmatrix} \mathbf{B}^0 & \mathbf{N}^0 \\ \mathbf{0} & \mathbf{I}_{n-m} \end{bmatrix} \begin{bmatrix} \mathbf{x}_B^0 \\ \mathbf{x}_N^0 \end{bmatrix} = \begin{bmatrix} \mathbf{b} \\ \mathbf{b}_N \end{bmatrix}. \tag{2.1-2}$$

$\mathbf{b}_N$ is an $n - m$ vector defined by the corresponding active lower and upper bounds $l_j, j \in L,$ and $u_j, j \in U$ taken in the order defined by N. Since the problem constraints defining the vertex $\mathbf{x}^0$ are linearly independent, $\mathbf{B}^0$ must be nonsingular. $\mathbf{B}^0$ is called the *basis matrix.* (Note also that $\mathbf{B}^0$ corresponds to the variables *whose bounds are inactive.*) Thus $\mathbf{x}^0$

can be found by solving (2.1-2) as follows.

$$\begin{aligned} \mathbf{x}_N^0 &= \mathbf{b}_N \\ \mathbf{B}^0\mathbf{x}_B^0 &= \mathbf{b} - \mathbf{N}^0\mathbf{b}_N. \end{aligned} \tag{2.1-3}$$

Since we have assumed that $\mathbf{x}^0 \in P$, the solution $\mathbf{x}^0$ must necessarily be feasible, i.e., must satisfy

$$(\mathbf{l}_B)_i \le (\mathbf{x}_B^0)_i \le (\mathbf{u}_B)_i \qquad i = 1, \ldots, m, \tag{2.1-4}$$

where $\mathbf{l}_B$ and $\mathbf{u}_B$ have components in $\mathbf{l}$ and $\mathbf{u}$ corresponding to indices in B, i.e., for our assumed ordering, $\mathbf{l}_B = (l_1, \ldots, l_m)$, $\mathbf{u}_B = (u_1, \ldots, u_m)$. $\mathbf{x}^0$ is also termed a *basic feasible solution* of the linear program. If $\mathbf{x}^0$ is the solution of a system of n linearly independent tight problem constraints, but the corresponding $\mathbf{x}_B^0$ does *not* satisfy (2.1-4) (see also Exercise 1.3-2), then $\mathbf{x}^0$ is a *basic infeasible solution.* The set of basic feasible and basic infeasible solutions are simply referred to as *basic* solutions. When the inequalities of (2.1-4) hold strictly, then the corresponding basic feasible solution is *nondegenerate,* which we have assumed to be the case.

Exercise 2.1-1. Verify the equivalence of nondegeneracy as defined in Chapter 1, Sec. 1.3, and as defined above.

Let us now compute the Lagrange multipliers associated with the n active constraints defining the vertex (basic feasible solution) $\mathbf{x}^0$, as in Sec. 1.5, in particular (1.5-1). It is convenient to define

$$\sigma_j^0 \equiv \lambda_j^0 \quad j \in L \qquad \sigma_j^0 \equiv -\mu_j^0 \quad j \in U.$$

Let $\boldsymbol{\sigma}_N^0$ be the vector with components $\sigma_j^0, j \in N$, i.e., $(\boldsymbol{\sigma}_N^0)_k = \sigma_{m+k}^0$, $k = 1, \ldots, n-m$. (We shall consistently use the index k when we refer to the sequence $1, \ldots, n-m$, and the index j when we refer to the sequence $m+1, \ldots, n$.) From these definitions and our assumed ordering of the index sets defining basics and nonbasics, the multipliers are given by

$$\begin{bmatrix} (\mathbf{B}^0)^T & \mathbf{0} \\ (\mathbf{N}^0)^T & \mathbf{I}_{n-m} \end{bmatrix} \begin{bmatrix} \boldsymbol{\pi}^0 \\ \boldsymbol{\sigma}_N^0 \end{bmatrix} = \begin{bmatrix} \mathbf{c}_B \\ \mathbf{c}_N \end{bmatrix} \tag{2.1-5}$$

where $\mathbf{c} = \begin{bmatrix} \mathbf{c}_B \\ \mathbf{c}_N \end{bmatrix}$ conforms to the partition of the variables. Thus $\boldsymbol{\pi}^0$ and $\boldsymbol{\sigma}_N^0$ can be obtained from the following system of equations.

$$\begin{aligned} (\mathbf{B}^0)^T\boldsymbol{\pi}^0 &= \mathbf{c}_B \\ \boldsymbol{\sigma}_N^0 &= \mathbf{c}_N - (\mathbf{N}^0)^T\boldsymbol{\pi}^0. \end{aligned} \tag{2.1-6}$$

If $\sigma_j^0 \ge 0, j \in L$ and $\sigma_j^0 \le 0, j \in U,$ then the solution $\mathbf{x}^0$ of (2.1-3) is optimal. If not, there is a multiplier $\sigma_s^0 < 0$ and associated active lower bound $x_s \ge l_s$ (or $\sigma_s^0 > 0$ and associated active upper bound $x_s \le u_s$) that violates the conditions of the optimality theorem 1.4-1. Note with the ordering of variables assumed throughout this section that $s \geq m+1$. When this lower (or upper) bound is dropped from the active set, with all other bounds in the current active set remaining active, the resulting direction $\mathbf{d}$ from $\mathbf{x}^0$ must satisfy

$$(\mathbf{a}^i)^T\mathbf{d} = 0 \quad i = 1, \ldots, m \qquad (\mathbf{e}^j)^T\mathbf{d} = 0 \quad j \in N \quad j \neq s.$$

(See also Sec. 1.5.) Finally, since the length of $\mathbf{d}$ is immaterial and its direction is determined by the sign of σ_s^0, we may impose the condition

$$(\mathbf{e}^s)^T\mathbf{d} = -\text{sign}(\sigma_s^0).$$

For example, if σ_s^0 came from a lower bound, it would be negative and we would thus ensure a feasible move (increase from lower bound) of the variable x_s. With the usual partition of $\mathbf{d}$ into basic and nonbasic components, these conditions are equivalent to

$$\begin{bmatrix} \mathbf{B}^0 & \mathbf{N}^0 \\ \mathbf{0} & \mathbf{I}_{n-m} \end{bmatrix} \begin{bmatrix} \mathbf{d}_B \\ \mathbf{d}_N \end{bmatrix} = -\text{sign}(\sigma_s^0)\mathbf{e}^s. \tag{2.1-7}$$

Therefore

$$\begin{aligned} d_s &= -\text{sign}(\sigma_s^0) \\ d_j &= 0 \qquad j \in N \quad \text{and} \quad j \neq s \\ \mathbf{B}^0\mathbf{d}_B &= \text{sign}(\sigma_s^0)\mathbf{a}_s. \end{aligned} \tag{2.1-8}$$

$\mathbf{a}_s$ is the column of $\mathbf{A}$ corresponding to variable x_s. Let us define $\tilde{\mathbf{a}}_s$ as the solution of $\mathbf{B}^0\tilde{\mathbf{a}}_s = \mathbf{a}_s$, and let $t > 0$ be a step along $\mathbf{d}$ from $\mathbf{x}^0$ to a new point, say $\mathbf{x}$, so that $\mathbf{x} = \mathbf{x}^0 + t\mathbf{d}$. Then

$$(\mathbf{x}_B)_i = (\mathbf{x}_B^0)_i + t(\mathbf{d}_B)_i = (\mathbf{x}_B^0)_i + t(\text{sign}(\sigma_s^0)(\tilde{\mathbf{a}}_s)_i) \qquad i = 1, \ldots, m \tag{2.1-9a}$$

$$x_s = x_s^0 - t(\text{sign}(\sigma_s^0)) \tag{2.1-9b}$$

$$x_j = x_j^0 \qquad j \in N \quad j \neq s. \tag{2.1-9c}$$

Finally, we may observe from (1.5-3) and (1.5-4), with $j' \equiv s$ and $\lambda_{j'}^0$ and $\mu_{j'}^0$ replaced by σ_s^0, that the change in objective value Δz is given by

$$\Delta z = t(\mathbf{d}^T\mathbf{e}^s)\sigma_s^0 = t(-\text{sign}(\sigma_s^0))\sigma_s^0 = -t\,|\sigma_s^0|. \tag{2.1-10}$$

Since the decrease in objective value is proportional to t, we wish, of course, to make t as large as possible. By how much can t be increased before some currently feasible basic variable violates a bound? It will be useful to consider a simple example.

Example 2.1-1. Consider a linear program with just four basic variables and with the incoming variable x_s at its lower bound ($\sigma_s^0 < 0$) with all bounds being finite, as depicted in Figure 2.1. Assume that the basic variables are strictly between their bounds and that the elements of $\tilde{\mathbf{a}}_s$ satisfy

$$(\tilde{\mathbf{a}}_s)_1 > 0 \qquad (\tilde{\mathbf{a}}_s)_2 < 0 \qquad (\tilde{\mathbf{a}}_s)_3 > 0 \qquad (\tilde{\mathbf{a}}_s)_4 = 0.$$

Figure 2.1 Choosing the exiting variable for feasible basis

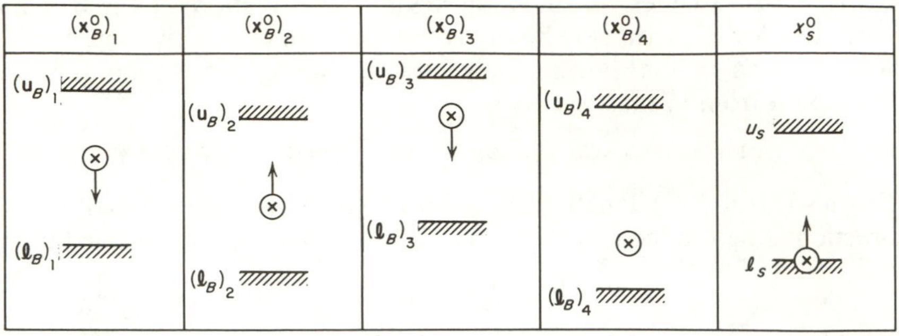

From (2.1-9a)

$$(\mathbf{x}_B)_i = (\mathbf{x}_B^0)_i - t(\tilde{\mathbf{a}}_s)_i \qquad i = 1, \ldots, 4.$$

This implies the following:

$(\tilde{\mathbf{a}}_s)_1 > 0$ (or $(\tilde{\mathbf{a}}_s)_3 > 0$) implies that $(\mathbf{x}_B)_1$ (or $(\mathbf{x}_B)_3$) decreases as t increases from zero.
$(\tilde{\mathbf{a}}_s)_2 < 0$ implies that $(\mathbf{x}_B)_2$ increases as t increases from zero.
$(\tilde{\mathbf{a}}_s)_4 = 0$ implies that $(\mathbf{x}_B)_4$ is fixed at $(\mathbf{x}_B^0)_4$.
$(\mathbf{x}_B)_1$ will meet its lower bound when $t = t_1$ given by

$$t_1 = ((\mathbf{x}_B^0)_1 - (\mathbf{l}_B)_1)/(\tilde{\mathbf{a}}_s)_1 > 0.$$

Similarly for $(\mathbf{x}_B)_3$,

$$t_3 = ((\mathbf{x}_B^0)_3 - (\mathbf{l}_B)_3)/(\tilde{\mathbf{a}}_s)_3 > 0.$$

$(\mathbf{x}_B)_2$ will meet its upper bound when $t = t_2$ given by

$$t_2 = ((\mathbf{x}_B^0)_2 - (\mathbf{u}_B)_2)/(\tilde{\mathbf{a}}_s)_2 > 0.$$

Since either t_1, t_2, or t_3 may exceed $(u_s - l_s)$, the *minimum* of these four quantities will determine which variable first meets the bound toward which it moves, as t increases from zero. Let θ_s denote this minimum value.

Consider also the case when $(\mathbf{l}_B)_1 = (\mathbf{l}_B)_3 = -\infty$ and $(\mathbf{u}_B)_2 = u_s = +\infty$. Then there would be *no* blocking variable and t could be increased indefinitely.

This completes the example.

The general situation is clear. We must determine which basic variables move toward a bound (lower or upper) and find the first one, if any, that meets a bound. This is called the *blocking variable.* If no variable is blocking, then the linear program has an unbounded optimal solution, because t can be increased indefinitely; from (2.1-10), Δz then goes to $-\infty$.

There are three cases to consider in (2.1-9a):

1. If x_s^0 is at its lower bound (i.e., $\sigma_s^0 < 0$), then as t increases from zero:
   $(\tilde{\mathbf{a}}_s)_i > 0$ implies that $(\mathbf{x}_B)_i$ moves toward its lower bound $(\mathbf{l}_B)_i$.
   $(\tilde{\mathbf{a}}_s)_i < 0$ implies that $(\mathbf{x}_B)_i$ moves toward its upper bound $(\mathbf{u}_B)_i$.
2. If x_s^0 is its upper bound (i.e., $\sigma_s^0 > 0$), then as t increases from zero:
   $(\tilde{\mathbf{a}}_s)_i > 0$ implies that $(\mathbf{x}_B)_i$ moves toward its upper bound $(\mathbf{u}_B)_i$.
   $(\tilde{\mathbf{a}}_s)_i < 0$ implies that $(\mathbf{x}_B)_i$ moves toward its lower bound $(\mathbf{l}_B)_i$.
3. When $(\tilde{\mathbf{a}}_s)_i = 0$, then $(\mathbf{x}_B)_i$ remains at value $(\mathbf{x}_B^0)_i$.

With the usual convention that variables unbounded from below (above) have bounds $l_i = -\infty$ $(u_i = +\infty)$, the rule for choosing the blocking variable, often known as the *min-ratio rule,* is to find index p and ratio $\theta_s > 0$ as follows.

$$p = \underset{i=1,\dots,m}{\operatorname{argmin}} \begin{cases} \dfrac{((\mathbf{x}_B^0)_i - (\mathbf{l}_B)_i)}{|(\tilde{\mathbf{a}}_s)_i|} & \text{when } (\sigma_s^0 < 0 \ \& \ (\tilde{\mathbf{a}}_s)_i > 0) \\ & \text{or } (\sigma_s^0 > 0 \ \& \ (\tilde{\mathbf{a}}_s)_i < 0) \\ \dfrac{((\mathbf{u}_B)_i - (\mathbf{x}_B^0)_i)}{|(\tilde{\mathbf{a}}_s)_i|} & \text{when } (\sigma_s^0 < 0 \ \& \ (\tilde{\mathbf{a}}_s)_i < 0) \\ & \text{or } (\sigma_s^0 > 0 \ \& \ (\tilde{\mathbf{a}}_s)_i > 0) \\ (u_s - l_s). & \end{cases} \tag{2.1-11}$$

θ_s is defined by the above expression (2.1-11) when "argmin" is replaced by "min."

If there is no blocking variable, i.e., if $\theta_s = \infty$ (this can only occur when some basics are unbounded from above or below), then the optimal value is unbounded. If $p = s$, $\theta_s = u_s - l_s$, then the variable x_s meets its opposite bound before any basic variable does so. In this case, the set of basic variables is unchanged, but the solution is revised by (2.1-9) with $t \equiv \theta_s = (u_s - l_s)$. Otherwise, the first basic variable to meet a bound, say x_p (where ties in the choice of candidate in the min-ratio rule are arbitrarily resolved) implies that one of the bound constraints $x_p = l_p$ or $x_p = u_p$ becomes tight. In this case the roles of x_p and x_s are interchanged, so that x_p becomes nonbasic and x_s becomes basic. B is revised to the indices

$$\{1, 2, \dots, p-1, s, p+1, \dots, m\}.$$

N is revised to the indices of the remaining variables and the solution is revised by (2.1-9) with $t \equiv \theta_s$. We can now reorder the variables so basics precede nonbasics, and begin afresh. This constitutes a cycle of the simplex method.

Exercise 2.1-2. Show that the revised set of tight constraints, defined by N, constitute an active set.

2.1-1. Convergence of the Simplex Method

Given a starting nondegenerate basic feasible solution (vertex) and assuming that all subsequent basic feasible solutions (vertices) generated are also nondegenerate, it is easy to see that the simplex method converges to an optimal solution (or the problem is found to have an unbounded optimal value) in a *finite* number of steps. This follows from the following three observations:

1. When $(\mathbf{l}_B)_i < (\mathbf{x}_B^0)_i < (\mathbf{u}_B)_i$, $i = 1, \ldots, m$ then $\theta_s > 0$ in (2.1-11).
2. From (2.1-10), $\theta_s > 0$ implies that $\Delta z < 0$, so there is a *strict reduction* in the objective value from one vertex to the next one in the sequence generated by the simplex method.
3. There are only a finite number of vertices.

Since no vertex in the sequence can be repeated, the simplex method must arrive at an optimal vertex (or discover that the problem has an unbounded optimum) in a finite number of steps. We shall discuss convergence when the nondegeneracy assumption is removed, in Sec. 2.5.

2.2. The Simplex Method: Column-Oriented Description

We can develop the simplex method from first principles and with practically no prior assumptions. This approach is often taken in introductory texts and is, of course, *implicitly,* very closely related to the development in Sec. 2.1. We shall point out correspondences as we go along.

Consider the m general constraints $\mathbf{Ax} = \mathbf{b}$ (where $\mathbf{A}$ is assumed to be of full row rank) and partition the variables into basics and nonbasics, i.e.,

$$\mathbf{x} = \begin{bmatrix} \mathbf{x}_B \\ \mathbf{x}_N \end{bmatrix} \quad \text{and} \quad \mathbf{A} = [\mathbf{B}^0 \mid \mathbf{N}^0].$$

Again, without loss of generality, we may assume that $B = \{1, \ldots, m\}$ and $N = \{m+1, \ldots, n\}$. $\mathbf{B}^0$, the $m \times m$ basis matrix, is assumed to be nonsingular. Similarly, partition $\mathbf{c}$, $\mathbf{l}$, and $\mathbf{u}$ to conform to the partition of $\mathbf{x}$, i.e.,

$$\mathbf{c} = \begin{bmatrix} \mathbf{c}_B \\ \mathbf{c}_N \end{bmatrix} \qquad \mathbf{l} = \begin{bmatrix} \mathbf{l}_B \\ \mathbf{l}_N \end{bmatrix} \quad \text{and} \quad \mathbf{u} = \begin{bmatrix} \mathbf{u}_B \\ \mathbf{u}_N \end{bmatrix}.$$

The foregoing partition immediately implies that

$$\mathbf{B}^0\mathbf{x}_B + \mathbf{N}^0\mathbf{x}_N = \mathbf{b} \tag{2.2-1}$$

$$\mathbf{x}_B = (\mathbf{B}^0)^{-1}\mathbf{b} - ((\mathbf{B}^0)^{-1}\mathbf{N}^0)\mathbf{x}_N. \tag{2.2-2}$$

Suppose that the nonbasic variables are fixed at one of their finite bounds given by the vector $\mathbf{x}_N^0$, whose components are chosen appropriately from $\mathbf{l}_N$ and $\mathbf{u}_N$ (see remarks at the beginning of Sec. 2.1) and let us assume that $\mathbf{x}_B^0$, as defined by (2.2-2), is *feasible* when $\mathbf{x}_N$ is set to $\mathbf{x}_N^0$. This, of course, implies that $\mathbf{x}^0$ is a vertex, but we do not have to explicitly identify it as such, if we are pursuing a development from first principles.

Using (2.2-2), let us write the linear program (1.2-2) as follows.

$$\begin{aligned} \text{minimize } & \mathbf{c}_B^T\mathbf{x}_B + \mathbf{c}_N^T\mathbf{x}_N \\ \text{s.t. } & \mathbf{x}_B = (\mathbf{B}^0)^{-1}\mathbf{b} - ((\mathbf{B}^0)^{-1}\mathbf{N}^0)\mathbf{x}_N \\ & \mathbf{l}_B \le \mathbf{x}_B \le \mathbf{u}_B \\ & \mathbf{l}_N \le \mathbf{x}_N \le \mathbf{u}_N. \end{aligned} \tag{2.2-3}$$

When we eliminate $\mathbf{x}_B$ from the objective function and bounds and therefore, in effect, from the problem, we obtain

$$\begin{aligned} \text{minimize } & (\mathbf{c}_N^T - \mathbf{c}_B^T(\mathbf{B}^0)^{-1}\mathbf{N}^0)\mathbf{x}_N + \mathbf{c}_B^T(\mathbf{B}^0)^{-1}\mathbf{b} \\ \text{s.t. } & \mathbf{l}_B \le (\mathbf{B}^0)^{-1}\mathbf{b} - ((\mathbf{B}^0)^{-1}\mathbf{N}^0)\mathbf{x}_N \le \mathbf{u}_B \\ & \mathbf{l}_N \le \mathbf{x}_N \le \mathbf{u}_N. \end{aligned} \tag{2.2-4}$$

Let us now consider a change $\Delta\mathbf{x}_N$ in the nonbasic variables $\mathbf{x}_N$ *relative* to $\mathbf{x}_N^0$, their current setting, i.e., $\mathbf{x}_N = \mathbf{x}_N^0 + \Delta\mathbf{x}_N$. Then (2.2-4) can be written as follows.

$$\begin{aligned} \text{minimize } & (\mathbf{c}_N^T - \mathbf{c}_B^T(\mathbf{B}^0)^{-1}\mathbf{N}^0)\Delta\mathbf{x}_N + (\mathbf{c}_B^T\mathbf{x}_B^0 + \mathbf{c}_N^T\mathbf{x}_N^0) \\ \text{s.t. } & \mathbf{l}_B \le \mathbf{x}_B^0 - ((\mathbf{B}^0)^{-1}\mathbf{N}^0)\Delta\mathbf{x}_N \le \mathbf{u}_B \\ & \mathbf{l}_N \le \mathbf{x}_N^0 + \Delta\mathbf{x}_N \le \mathbf{u}_N. \end{aligned} \tag{2.2-5}$$

It will prove to be useful to make the following reinterpretation of (2.2-5). Observe that the (bound) constraints of (2.2-5) can be expressed as

$$\begin{bmatrix} \mathbf{l}_B \\ \mathbf{l}_N \end{bmatrix} \le \begin{bmatrix} \mathbf{x}_B^0 \\ \mathbf{x}_N^0 \end{bmatrix} + \begin{bmatrix} -(\mathbf{B}^0)^{-1}\mathbf{N}^0 \\ \mathbf{I}_{n-m} \end{bmatrix} \Delta\mathbf{x}_N \le \begin{bmatrix} \mathbf{u}_B \\ \mathbf{u}_N \end{bmatrix}. \tag{2.2-6}$$

Define

$$\mathbf{Z} = \begin{bmatrix} -(\mathbf{B}^0)^{-1}\mathbf{N}^0 \\ \mathbf{I}_{n-m} \end{bmatrix} \tag{2.2-7}$$

and

$$(\boldsymbol{\sigma}_N^0)^T \equiv (\mathbf{c}_N^T - \mathbf{c}_B^T(\mathbf{B}^0)^{-1}\mathbf{N}^0) = \mathbf{c}^T\mathbf{Z}. \tag{2.2-8}$$

Also define

$$z^0 \equiv \mathbf{c}_B^T\mathbf{x}_B^0 + \mathbf{c}_N^T\mathbf{x}_N^0 = \mathbf{c}^T\mathbf{x}^0.$$

Then (2.2-5) can be expressed as the *equivalent reduced problem*

$$\begin{aligned} \text{minimize } & (\boldsymbol{\sigma}_N^0)^T\Delta\mathbf{x}_N + z^0 \\ \text{s.t. } & \mathbf{l} \le \mathbf{x}^0 + \mathbf{Z}\Delta\mathbf{x}_N \le \mathbf{u}. \end{aligned} \tag{2.2-9}$$

To see the significance of the matrix $\mathbf{Z}$, observe that the general constraints $(\mathbf{a}^i)^T\mathbf{x} = b_i$, $i = 1, \ldots, m$, remove m degrees of freedom from R^n. Given some point $\mathbf{x}^0$ that satisfies these constraints, a move from $\mathbf{x}^0$ to another feasible point $\mathbf{x}$ must satisfy $(\mathbf{a}^i)^T(\mathbf{x} - \mathbf{x}^0) = 0, i = 1, \ldots, m$. Thus $\Delta\mathbf{x} \equiv \mathbf{x} - \mathbf{x}^0$ must be orthogonal to $\mathbf{a}^i$, $i = 1, \ldots, m$. Now from the definition of $\mathbf{Z}$, we see that

$$\mathbf{AZ} = [\mathbf{B}^0 \mid \mathbf{N}^0]\begin{bmatrix} -(\mathbf{B}^0)^{-1}\mathbf{N}^0 \\ \mathbf{I}_{n-m} \end{bmatrix} = \mathbf{0}. \tag{2.2-10}$$

Thus the $n - m$ columns of $\mathbf{Z}$ are orthogonal to $\mathbf{a}^i$, $i = 1, \ldots, m$ and they are clearly linearly independent. They therefore span the orthogonal complement of the space spanned by the rows of $\mathbf{A}$, and any feasible move from $\mathbf{x}^0$ must lie within the column span (range) of $\mathbf{Z}$. Suppose we therefore make a transformation of variables

$$\mathbf{x} = \mathbf{x}^0 + \mathbf{Z}\Delta\mathbf{x}_N. \tag{2.2-11}$$

If we substitute (2.2-11) into the linear program (1.2-2) and note that for *any* $\Delta\mathbf{x}_N$,

$$\mathbf{A}(\mathbf{x}^0 + \mathbf{Z}\Delta\mathbf{x}_N) = \mathbf{A}\mathbf{x}^0 = \mathbf{b},$$

we immediately obtain the above equivalent reduced problem (2.2-9).

Let us further assume that $\mathbf{x}^0$ is *nondegenerate,* i.e., that the following condition holds

$$(\mathbf{l}_B)_i < (\mathbf{x}_B^0)_i < (\mathbf{u}_B)_i \qquad i = 1, \ldots, m. \tag{2.2-12}$$

For sufficiently small changes $\Delta\mathbf{x}_N$, the basic variables, $\mathbf{x}_B$, defined by (2.2-2) (or (2.2-11)) will continue to be strictly off the corresponding bounds. Therefore *locally,* the bound constraints on $\mathbf{x}_B$ can be dropped from the equivalent reduced problem (2.2-5) or (2.2-9), giving the *local reduced problem*

$$\begin{aligned} \text{minimize } & (\boldsymbol{\sigma}_N^0)^T\Delta\mathbf{x}_N + z^0 \\ \text{s.t. } & \mathbf{l}_N \le \mathbf{x}_N^0 + \Delta\mathbf{x}_N \le \mathbf{u}_N. \end{aligned} \tag{2.2-13}$$

$\boldsymbol{\sigma}_N^0 = \mathbf{Z}^T\mathbf{c}$ is termed the vector of *reduced costs* or the *reduced gradient.* The term *reduced* arises for the following reason: Given any smooth objective function, say $f(\mathbf{x})$, a straightforward application of the chain rule of the differential calculus shows that the gradient $\nabla f(\mathbf{x})$ transforms to $\mathbf{Z}^T\nabla f(\mathbf{x})$ under the change of variables (2.2-11). In our case, $f(\mathbf{x}) = \mathbf{c}^T\mathbf{x}$ and $\nabla f(\mathbf{x}) = \mathbf{c}$.

Let us define

$$(\boldsymbol{\pi}^0)^T \equiv \mathbf{c}_B^T(\mathbf{B}^0)^{-1}.$$

Then, from (2.2-8),

$$\boldsymbol{\sigma}_N^0 = \mathbf{c}_N - (\mathbf{N}^0)^T\boldsymbol{\pi}^0.$$

Comparing with (2.1-6), we see that $\boldsymbol{\pi}^0$ and $\boldsymbol{\sigma}_N^0$ are the multipliers corresponding to the general and the active bound constraints, respectively. Here, however, we have used a more direct derivation.

Exercise 2.2.1. Consider the vector $(\boldsymbol{\pi}^0)^T = \mathbf{c}_B^T(\mathbf{B}^0)^{-1}$ at a nondegenerate basic feasible solution $\mathbf{x}^0$, for the following linear program: Minimize $\mathbf{c}^T\mathbf{x}$ s.t. $\mathbf{Ax} = \mathbf{b}, \mathbf{x} \geq \mathbf{0}$. Show that

$$\frac{\partial z^0}{\partial b_i} = \pi_i^0 \qquad i = 1, \ldots, m,$$

where z^0 is the objective value at $\mathbf{x}^0$. For this reason, $\boldsymbol{\pi}^0$ is called the *price* vector.

Let us define

$$\sigma_j^0 \equiv c_j - \mathbf{c}_B^T(\mathbf{B}^0)^{-1}\mathbf{a}_j \qquad j \in N.$$

Then $\boldsymbol{\sigma}_N^0$ is the vector with components $\sigma_j^0, j \in N$, i.e., $(\boldsymbol{\sigma}_N^0)_k = \sigma_{m+k}^0$, $k = 1, \ldots, n-m$. We shall consistently use the index k when we refer to the sequence $1, \ldots, n-m$, and the index j when we refer to the sequence $m+1, \ldots, n$. Also, let us define $\sigma_j^0, j \in B$, in a consistent manner, as

$$\sigma_j^0 \equiv c_j - \mathbf{c}_B^T(\mathbf{B}^0)^{-1}\mathbf{a}_j \qquad j \in B.$$

Since $\mathbf{a}_j, j \in B$ is the jth column of $\mathbf{B}^0$, we have

$$\sigma_j^0 = c_j - \mathbf{c}_B^T\mathbf{e}_j = c_j - c_j = 0 \qquad j \in B.$$

The simplex method reduces the objective value of the equivalent local reduced problem, by changing just *one* component of $\Delta\mathbf{x}_N$, say $(\Delta\mathbf{x}_N)_k$. This is, of course, the same as moving along a *coordinate direction of the reduced space* defined by the $n-m$ variables $\Delta\mathbf{x}_N$. Suppose that this step is $\mathbf{e}_k\tau$ where $\mathbf{e}_k$ is a unit vector in R^{n-m} (the kth column of the $(n-m)\times(n-m)$ identity matrix) and τ is a scalar of arbitrary sign. From (2.2-11), this step, in the original space, gives the point

$$\mathbf{x} = \begin{bmatrix} \mathbf{x}_B \\ \mathbf{x}_N \end{bmatrix} = \mathbf{x}^0 + \mathbf{Z}\mathbf{e}_k\tau = \mathbf{x}^0 + \mathbf{z}_k\tau = \begin{bmatrix} \mathbf{x}_B^0 \\ \mathbf{x}_N^0 \end{bmatrix} + \begin{bmatrix} -(\mathbf{B}^0)^{-1}\mathbf{n}_k^0 \\ \mathbf{e}_k \end{bmatrix}\tau, \tag{2.2-14}$$

where $\mathbf{z}_k$ and $\mathbf{n}_k^0$ are the kth columns of $\mathbf{Z}$ and $\mathbf{N}^0$, respectively. Thus $\mathbf{x}_N = \mathbf{x}_N^0 + \mathbf{e}_k\tau$; the sign of τ must be chosen so that $(\mathbf{x}_N)_k$ satisfies $(\mathbf{l}_N)_k \leq (\mathbf{x}_N)_k \leq (\mathbf{u}_N)_k$, i.e., if $(\mathbf{x}_N^0)_k$ is at its lower (upper) bound then $\tau > 0\ (\tau < 0)$. Also, $|\tau|$ must be sufficiently small, so that $\mathbf{x}_B$ does not violate a bound. Clearly if $\mathbf{x}^0$ is nondegenerate, then $|\tau| > 0$ is possible. More will be said on this point later.

The value of the objective function at $\mathbf{x}$ is

$$z = \mathbf{c}^T\mathbf{x} = \mathbf{c}^T\mathbf{x}^0 + \mathbf{c}^T\mathbf{Z}\mathbf{e}_k\tau = z^0 + (\boldsymbol{\sigma}_N^0)_k\tau \tag{2.2-15}$$

$$\Delta z \equiv z - z^0 = (\boldsymbol{\sigma}_N^0)_k\tau. \tag{2.2-16}$$

Thus the objective can be reduced ($\Delta z < 0$) when either

1. $(\boldsymbol{\sigma}_N^0)_k < 0$ and $(\mathbf{x}_N^0)_k$ is at its lower bound ($\tau > 0$), or
2. $(\boldsymbol{\sigma}_N^0)_k > 0$ and $(\mathbf{x}_N^0)_k$ is at its upper bound ($\tau < 0$).

Variables $(\mathbf{x}_N^0)_k$ that satisfy (1) or (2) are called *improving* variables. Let C, a subset of N, denote their indices. If C is empty, then the current solution $\mathbf{x}^0$ is *optimal.* This must be true because (2.2-9) is locally equivalent to the original linear program and its objective function cannot be reduced further.

Exercise 2.2-2. Verify that the condition "C is empty" is equivalent to the optimality conditions of Theorem 1.4-1.

From (2.2-16), $(\boldsymbol{\sigma}_N^0)_k = \Delta z/\tau$ and it is thus a measure of the *rate of change* in objective value per unit change in $(\Delta\mathbf{x}_N)_k$. Any variable from C may be chosen, a common rule being to choose the improving nonbasic variable, say x_s, for which

$$|\sigma_s^0| \geq |\sigma_j^0| \qquad j \in C. \tag{2.2-17}$$

This is sometimes called the *largest-coefficient rule.* Note, from the way C is defined, that the signs of σ_s^0 and τ are correlated, i.e.,

$$\sigma_s^0 < 0 \text{ if and only if } \tau > 0 \quad \text{and} \quad \sigma_s^0 > 0 \text{ if and only if } \tau < 0.$$

Let us also define $t > 0$ by

$$\tau = -\text{sign}(\sigma_s^0)t. \tag{2.2-18}$$

Finally, from (2.2-14) with $\mathbf{n}_k^0$ corresponding to the column $\mathbf{a}_s$, we have

$$\begin{aligned} \mathbf{x}_B &= \mathbf{x}_B^0 - ((\mathbf{B}^0)^{-1}\mathbf{a}_s)\tau = \mathbf{x}_B^0 + t\,\text{sign}(\sigma_s^0)((\mathbf{B}^0)^{-1}\mathbf{a}_s) \\ x_s &= x_s^0 - \text{sign}(\sigma_s^0)t \\ x_j &= x_j^0 \qquad j \in N \quad j \neq s, \end{aligned} \tag{2.2-19}$$

and again let us define $\tilde{\mathbf{a}}_s \equiv (\mathbf{B}^0)^{-1}\mathbf{a}_s$. Expression (2.2-19) is identical to (2.1-9) and the discussion subsequent to (2.1-9) applies here for choosing the variable that leaves the basis. If τ can be increased indefinitely, without any variable becoming blocking (meeting a bound), then it follows from (2.2-16) that the linear program is unbounded from below. The corresponding direction $(\mathbf{x} - \mathbf{x}^0)$ must be an extreme direction of the polyhedral set P, as defined in Sec. 1.3. From (2.2-14), this is given by

$$\mathbf{r}_k \equiv \begin{bmatrix} -(\mathbf{B}^0)^{-1}\mathbf{a}_{m+k} \\ \mathbf{e}_k \end{bmatrix} \tag{2.2-20}$$

where $k \equiv (s - m)$.

From the correspondences pointed out throughout this section, it should be clear that the development here and the development in Sec. 2.1 are just two different perspectives on the simplex method. The convergence argument, under the nondegeneracy assumption, given in Sec. 2.1-1, applies equally well here. Convergence, when the assumption of nondegeneracy is relaxed, is discussed in Sec. 2.5.

2.2-1. Extensions

The development of the simplex method in terms of successive problem reduction permits some useful extensions of the above procedure. We introduce two such extensions here. They will be discussed in more detail in Chapter 7.

Pegged Variables. When the simplex method is initiated at $\mathbf{x}^0$, let us suppose that some of the nonbasic variables are permitted to take on values *between* their bounds, i.e., that $\mathbf{x}^0$ is not necessarily a vertex. Such variables are termed *pegged* variables and can be thought of as being temporarily fixed at feasible values, just as in the previous discussion, the nonbasics were temporarily fixed at bound values. A pegged variable can be increased or decreased without violating feasibility; thus any pegged variable, say $x_j, j \in N$, with nonzero reduced cost σ_j^0, is an improving variable with an appropriate choice of its direction of movement. The rules given just after (2.2-16) for choosing the set of improving variables, and (2.1-11) for choosing the exiting variable, can be extended in a very simple manner. Note that such extensions enable us to remove the restriction that every variable has at least one finite bound, so that the linear program being solved can then have *free variables,* with bounds of the form $-\infty \leq x_j \leq +\infty$.

Exercise 2.2-3. Extend the rules (2.2-17) and (2.1-11) of the simplex method for choosing C and p, when some nonbasic variables are pegged between their bounds.

Exercise 2.2-4. Verify that a nonbasic free variable can enter a basis, but a basic free variable can never leave its basis. Show when a linear program has more than n free variables that it will (usually) have an unbounded optimum.

Steepest-Edges. A unit step, say $\mathbf{e}_k$, in the reduced space defined by the variables $(\Delta\mathbf{x}_N)_k$, corresponds to a step $(\mathbf{x}-\mathbf{x}^0)=\mathbf{z}_k$ in the original variables, where $\mathbf{z}_k$ is the kth column of $\mathbf{Z}$ (see (2.2-7)). Thus a unit Euclidean step in the *original variables* represents a step of length $1/\|\mathbf{z}_k\|_2$ along $\mathbf{e}_k$ in the reduced space, where $\|\mathbf{z}_k\|_2 \equiv (\mathbf{z}_k^T\mathbf{z}_k)^{1/2}$. In choosing the entering variable, one might seek to compare changes in value of the objective function for steps along coordinate directions of $\Delta\mathbf{x}_N$, each of which gives a *step of unit length in the original variables.* Thus an alternative to basing choices on $(\boldsymbol{\sigma}_N^0)_k$ (which from (2.2-16) compares changes in objective value corresponding to unit steps in the *reduced* space) is to base choices on $(\sigma_N^0)_k/\|\mathbf{z}_k\|_2$. In place of (2.2-17) let

$$|(\sigma_N^0)_{m+q}|/\|\mathbf{z}_q\|_2 = \max_{1 \le k \le n-m} [|(\sigma_N^0)_{m+k}|/\|\mathbf{z}_k\|_2,\ m+k \in C]. \quad (2.2\text{-}21)$$

Then the entering variable is x_s, $s = m + q$.

Exercise 2.2-5. Show that $\mathbf{z}_k$ defines an edge (or an extreme direction) of the polyhedral set of feasible solutions.

Expression (2.2-21) is called the *steepest-edge rule,* because it follows from the foregoing exercise that $\mathbf{z}_k/\|\mathbf{z}_k\|_2$ is a unit vector along the edge $\mathbf{z}_k$, and $(\sigma_N^0)_k/\|\mathbf{z}_k\|_2 = \mathbf{c}^T(\mathbf{z}_k/\|\mathbf{z}_k\|_2)$ is the directional derivative along the edge $\mathbf{z}_k$. Efficient techniques for computing $\|\mathbf{z}_k\|_2$ will be discussed in Chapter 7.

2.3. The Simplex Method: Row-Oriented Description

Finally, we closely parallel the development in Sec. 2.1, but this time on the following (canonical) form of the problem, which we write throughout the text as follows.

$$\begin{aligned} &\text{maximize } \bar{\mathbf{c}}^T\bar{\mathbf{x}} \\ &\text{s.t. } \bar{\mathbf{A}}\bar{\mathbf{x}} \le \bar{\mathbf{b}}. \end{aligned} \quad (2.3\text{-}1)$$

$\bar{\mathbf{A}}$ is an $\bar{m} \times \bar{n}$ matrix with $\bar{m} \ge \bar{n}$ and is assumed to be of full *column* rank, $\bar{\mathbf{x}}$ and $\bar{\mathbf{c}}$ are $\bar{n}$-vectors and $\bar{\mathbf{b}}$ is an $\bar{m}$-vector. The perspective adopted in this section will yield fresh insight into the simplex method, and our development is also of direct relevance to Chapter 11. (With regard to our choice of notation, see the Notes on Sec. 1.2 at the end of Chapter 1.)

Consider a vertex $\bar{\mathbf{x}}^0$ of the polyhedral set $\bar{P} = [\bar{\mathbf{x}} \mid \bar{\mathbf{A}}\bar{\mathbf{x}} \le \bar{\mathbf{b}}]$ which is defined, in the usual manner, to be a point that satisfies all constraints of (2.3-1) and for which $\bar{n}$ linearly independent constraints are tight, i.e., $\bar{\mathbf{x}}^0$ is defined to be the solution of $\bar{n}$ *active* constraints. If more than $\bar{n}$ constraints are tight then $\bar{\mathbf{x}}^0$ is degenerate, following our now-standard definition of this term. Since the constraints of (2.3-1) can be reordered, we may assume, without loss of generality, that the first $\bar{n}$ constraints are active and indexed by $\bar{B} = \{1, \ldots, \bar{n}\}$. $\bar{N} = \{\bar{n}+1, \ldots, \bar{m}\}$ defines the *inactive constraints.* We partition $\bar{\mathbf{A}}$ and $\bar{\mathbf{b}}$ as follows:

$$\bar{\mathbf{A}} = \begin{bmatrix} \bar{\mathbf{B}}^0 \\ \bar{\mathbf{N}}^0 \end{bmatrix} \qquad \bar{\mathbf{b}} = \begin{bmatrix} \bar{\mathbf{b}}_{\bar{B}} \\ \bar{\mathbf{b}}_{\bar{N}} \end{bmatrix} \quad (2.3\text{-}2)$$

where $\bar{\mathbf{B}}^0$ is an $\bar{n} \times \bar{n}$ nonsingular matrix, $\bar{\mathbf{N}}^0$ is an $(\bar{m} - \bar{n}) \times \bar{n}$ matrix, $\bar{\mathbf{b}}_{\bar{B}}$ is an $\bar{n}$-vector and $\bar{\mathbf{b}}_{\bar{N}}$ is an $(\bar{m} - \bar{n})$-vector. In contrast to the basis matrix $\mathbf{B}^0$ in (2.1-1), which is defined in terms of *columns* of the corresponding

LP matrix $\mathbf{A}$, the matrix $\bar{\mathbf{B}}^0$ is defined in terms of *rows* of $\bar{\mathbf{A}}$. Thus

$$\begin{aligned} \bar{\mathbf{B}}^0\bar{\mathbf{x}}^0 = \bar{\mathbf{b}}_{\bar{B}} \qquad \bar{\mathbf{x}}^0 = (\bar{\mathbf{B}}^0)^{-1}\bar{\mathbf{b}}_{\bar{B}} \\ \bar{\mathbf{N}}^0\bar{\mathbf{x}}^0 \leq \bar{\mathbf{b}}_{\bar{N}}. \end{aligned} \tag{2.3-3}$$

Again following our previous terminology, $\bar{\mathbf{x}}^0$ is a *basic feasible solution.* When $\bar{\mathbf{N}}^0\bar{\mathbf{x}}^0 < \bar{\mathbf{b}}_{\bar{N}}$, then $\bar{\mathbf{x}}^0$ is *nondegenerate*. Let us assume this to be the case. Also, consistent with our earlier notation, let $\bar{\mathbf{a}}_j$ denote the jth column of $\bar{\mathbf{A}}$ and $(\bar{\mathbf{a}}^i)^T$ denote the ith row of $\bar{\mathbf{A}}$. Thus

$$\bar{\mathbf{B}}^0 = \begin{bmatrix} (\bar{\mathbf{a}}^1)^T \\ \cdot \\ \cdot \\ \cdot \\ (\bar{\mathbf{a}}^{\bar{n}})^T \end{bmatrix}.$$

To verify optimality, we compute Lagrange multipliers, say $\bar{\boldsymbol{\sigma}}^0$, associated with the active constraints with indices given by $\bar{B}$. Recall from Exercises 1.4-4 and 1.5-2 that these are found by solving

$$(\bar{\boldsymbol{\sigma}}^0)^T\bar{\mathbf{B}}^0 = \bar{\mathbf{c}}^T \quad \text{or} \quad \bar{\boldsymbol{\sigma}}^0 = (\bar{\mathbf{B}}^0)^{-T}\bar{\mathbf{c}}. \tag{2.3-4a}$$

(Note that there are no equality constraints, whose multipliers we would have denoted by $\bar{\boldsymbol{\pi}}^0$; so $\bar{\boldsymbol{\sigma}}^0$ is the appropriate notation in this context.) For our assumed ordering of constraints,

$$\bar{\mathbf{c}} = (\bar{\mathbf{B}}^0)^T\bar{\boldsymbol{\sigma}}^0 = \sum_{i=1}^{\bar{n}} (\bar{\mathbf{a}}^i)\bar{\sigma}_i^0, \tag{2.3-4b}$$

where $\bar{\mathbf{a}}^i$ corresponds to the ith constraint of (2.3-1), namely $(\bar{\mathbf{a}}^i)^T\bar{\mathbf{x}} \leq \bar{b}_i$.

To reiterate the argument of Sec. 1.5, suppose one of the active constraints is relaxed, say $(\bar{\mathbf{a}}^{\bar{s}})^T\bar{\mathbf{x}}^0 = \bar{b}_{\bar{s}}$, whose corresponding multiplier is $\bar{\sigma}_{\bar{s}}^0 < 0$, and all other active constraints remain active. Let $\bar{\mathbf{d}}$ be a feasible direction from $\bar{\mathbf{x}}^0$ (see Sec. 1.4), i.e., for $\bar{t} > 0$, $(\bar{\mathbf{a}}^i)^T(\bar{\mathbf{x}}^0 + \bar{t}\bar{\mathbf{d}}) = \bar{b}_i$, $i = 1, \ldots, (\bar{s}-1), (\bar{s}+1), \ldots, \bar{n}$ and $(\bar{\mathbf{a}}^{\bar{s}})^T(\bar{\mathbf{x}}^0 + \bar{t}\bar{\mathbf{d}}) < \bar{b}_{\bar{s}}$. Provided $\bar{t}$ is sufficiently small, all inactive constraints will continue to be satisfied at $\bar{\mathbf{x}}^0 + \bar{t}\bar{\mathbf{d}}$. These conditions, together with (2.3-3), imply that

$$(\bar{\mathbf{a}}^i)^T\bar{\mathbf{d}} = 0 \qquad i = 1, \ldots, (\bar{s}-1), (\bar{s}+1), \ldots, \bar{n} \tag{2.3-5a}$$

$$(\bar{\mathbf{a}}^{\bar{s}})^T\bar{\mathbf{d}} < 0. \tag{2.3-5b}$$

It then follows from (2.3-4b) that

$$\bar{\mathbf{c}}^T(\bar{\mathbf{x}}^0 + \bar{t}\bar{\mathbf{d}}) = \bar{\mathbf{c}}^T\bar{\mathbf{x}}^0 + \bar{t}\sum_{i=1}^{\bar{n}} ((\bar{\mathbf{a}}^i)^T\bar{\mathbf{d}})\bar{\sigma}_i^0 = \bar{\mathbf{c}}^T\bar{\mathbf{x}}^0 + \bar{t}((\bar{\mathbf{a}}^{\bar{s}})^T\bar{\mathbf{d}})\bar{\sigma}_{\bar{s}}^0$$

and

$$\Delta\bar{z} \equiv \bar{\mathbf{c}}^T(\bar{\mathbf{x}}^0 + \bar{t}\bar{\mathbf{d}}) - \bar{\mathbf{c}}^T\bar{\mathbf{x}}^0 = \bar{t}((\bar{\mathbf{a}}^{\bar{s}})^T\bar{\mathbf{d}})\bar{\sigma}_{\bar{s}}^0. \tag{2.3-6}$$

Since $((\bar{\mathbf{a}}^{\bar{s}})^T\bar{\mathbf{d}}) < 0$, the value of the objective function will increase (recall that we are maximizing the objective) when $\bar{\sigma}^0_{\bar{s}} < 0$. If $\bar{\sigma}^0_i \geq 0, i = 1, \ldots, \bar{n}$, then $\bar{\mathbf{x}}^0$ is the optimal solution. Otherwise we can choose any $\bar{s}$ for which $\bar{\sigma}^0_{\bar{s}} < 0$ to define the *outgoing constraint* and obtain the associated direction of improvement $\bar{\mathbf{d}}$ by solving (2.3-5). Since the length of $\bar{\mathbf{d}}$ is immaterial, the second relation of (2.3-5) can be replaced by $(\bar{\mathbf{a}}^{\bar{s}})^T\bar{\mathbf{d}} = \text{sign}(\bar{\sigma}^0_{\bar{s}})\mathbf{e}_{\bar{s}}$, giving the following system of equations.

$$\bar{\mathbf{B}}^0\bar{\mathbf{d}} = \text{sign}(\bar{\sigma}^0_{\bar{s}})\mathbf{e}_{\bar{s}}. \tag{2.3-7}$$

Exercise 2.3-1. Show that $\bar{\mathbf{d}}$ defines an edge of the polyhedral set of feasible solutions of (2.3-1).

Again we seek to take as large a step $\bar{t} > 0$ along $\bar{\mathbf{d}}$ as possible, because from (2.3-6), the increase in the value of the objective is directly proportional to $\bar{t}$. However, $\bar{\mathbf{x}}^0 + \bar{t}\bar{\mathbf{d}}$ must remain feasible and $\bar{t}$ must therefore satisfy

$$(\bar{\mathbf{a}}^j)^T(\bar{\mathbf{x}}^0 + \bar{t}\bar{\mathbf{d}}) \leq \bar{b}_j \qquad j = (\bar{n}+1), \ldots, \bar{m}.$$

Hence

$$\bar{t}(\bar{\mathbf{a}}^j)^T\bar{\mathbf{d}} \leq (\bar{b}_j - (\bar{\mathbf{a}}^j)^T\bar{\mathbf{x}}^0) \qquad j = (\bar{n}+1), \ldots, \bar{m}. \tag{2.3-8}$$

Note that the right-hand side of the above inequality is positive (since we have assumed that $\bar{\mathbf{x}}^0$ is nondegenerate) and that it is a measure of the amount by which the jth constraint is slack. Let

$$\bar{E} = \{j \mid (\bar{\mathbf{a}}^j)^T\bar{\mathbf{d}} > 0 \qquad j = (\bar{n}+1), \ldots, \bar{m}\}. \tag{2.3-9}$$

Then $\bar{t}$ is chosen so that

$$\bar{p} = \underset{j \in \bar{E}}{\text{argmin}} \left\{ \frac{(\bar{b}_j - (\bar{\mathbf{a}}^j)^T\bar{\mathbf{x}}^0)}{(\bar{\mathbf{a}}^j)^T\bar{\mathbf{d}}} \right. . \tag{2.3-10}$$

Similarly $\bar{\theta}_{\bar{s}}$ is defined by expression (2.3-10) with "argmin" replaced by "min," and in keeping with our notation in (2.1-11), $\bar{\theta}_{\bar{s}}$ denotes the largest possible value of $\bar{t}$ when constraint $\bar{s}$ is the outgoing constraint. When $\bar{\mathbf{x}}^0$ is nondegenerate, then $\bar{\theta}_{\bar{s}} > 0$. Note also that (2.3-10) is a *row-oriented* procedure.

If $\bar{E}$ is empty, then the optimal solution is unbounded from above. Otherwise the *incoming* constraint $(\bar{\mathbf{a}}^{\bar{p}})^T\bar{\mathbf{x}} = \bar{b}_{\bar{p}}$ becomes active. The *outgoing* constraint $(\bar{\mathbf{a}}^{\bar{s}})^T\bar{\mathbf{x}} = \bar{b}_{\bar{s}}$ is removed from the set of active constraints and is necessarily rendered slack when $\bar{\mathbf{x}}^0$ is nondegenerate. We therefore have a new basic feasible solution defined by the new set of active constraints $1, 2, \ldots, (\bar{s}-1), \bar{p}, (\bar{s}+1), \ldots, \bar{n}$. (Since $(\bar{\mathbf{a}}^{\bar{p}})^T\bar{\mathbf{d}} > 0$, it follows from (2.3-7) that constraint $\bar{p}$ is linearly independent of other constraints in this set.) The constraints can then be reordered and the procedure repeated. Convergence in a finite number of steps, when each basic feasible solution is nondegenerate, follows precisely as in the previous two sections.

The simplex method certainly enjoys its most tidy description on the alternative form (2.3-1) of this section. We have however emphasized developments on the canonical form (1.2-2) throughout this book because, from a computational standpoint, it is the more relevant of the two forms. The reader should be sure to clearly grasp the fact that Secs. 2.1, 2.2, and 2.3 simply give different *algebraic* descriptions of a *single underlying conceptual method,* namely, the *simplex method,* which we summarized in the very first paragraph of this chapter.

2.3-1. Correspondences

In Sec. 2.2 we used the terms *basic variables, nonbasic variables, entering* (*nonbasic*) *variable, exiting* (*basic*) *variable.* In Sec. 2.3 we used the terms *active constraints, inactive constraints, outgoing* (*active*) *constraint, incoming* (*inactive*) *constraint.* Section 2.1 subsumes the approaches in Secs. 2.2 and 2.3 and we now restate correspondences between these two sets of terms within the context of the canonical form (1.2-2).

A vertex or basic feasible solution of (1.2-2) is defined by n *active* (*problem*) *constraints,* where this number is determined in the usual way by the number of variables in the problem. These active constraints consist of the m general constraints (assumed to be linearly independent and always therefore active at a feasible solution) and $n-m$ linearly independent bound constraints. The remaining $(2n-(n-m))=m+n$ bounds (at most) are the *inactive constraints.* The $n-m$ active bound constraints correspond to the $n-m$ *nonbasic variables.* The remaining $m=(n-(n-m))$ variables of (1.2-2) are the *basic variables* and they define the basis matrix $\mathbf{B}^0$ (which is nonsingular because the active constraints are linearly independent). An active bound constraint, with a Lagrange multiplier of appropriate sign, is chosen to be the *outgoing* constraint (say it is $x_s=l_s$ or $x_s=u_s$). It leaves the set of active constraints and becomes inactive. The corresponding *nonbasic* variable x_s is the *entering* variable, i.e., it becomes *basic.* The *exiting* basic variable x_p meets an inactive (bound) constraint, $x_p \geq l_p$ or $x_p \leq u_p$, which then becomes active and a member of the new set of active constraints. The correspondences are thus given as follows:

active bound constraint ↔ *nonbasic variable*

outgoing active bound constraint ↔ *entering nonbasic variable*

incoming inactive bound constraint ↔ *exiting basic variable*

inactive bound constraint, tight at bound ↔ *degenerate basic variable.*

Henceforth we shall use the words "outgoing" and "incoming" in association with constraint, and "entering" and "exiting" in association with variable.

Further appreciation of the correspondences between our different derivations of the simplex method may be gained from the following two exercises.

Exercise 2.3-2. Derive the simplex method in an analogous manner to Sec. 2.3 but using the following form of the linear program.

$$\begin{aligned} &\text{maximize } \bar{\mathbf{c}}^T\bar{\mathbf{x}} \\ &\text{s.t. } \bar{\mathbf{A}}_1\bar{\mathbf{x}} = \bar{\mathbf{b}}_1 \\ &\qquad \bar{\mathbf{A}}_2\bar{\mathbf{x}} \le \bar{\mathbf{b}}_2 \end{aligned} \tag{2.3-11}$$

where $\bar{\mathbf{A}}_1$ is an $\bar{m}_1 \times \bar{n}$ matrix with $\bar{m}_1 \le \bar{n}$, $\bar{\mathbf{A}}_2$ is an $\bar{m}_2 \times \bar{n}$ matrix (with $\bar{m}_1 + \bar{m}_2 \ge \bar{n}$), and the other vectors are of matching dimensions.

Exercise 2.3-3. Suppose that the linear program (2.3-1) is put into the canonical form (1.2-2) by adding $\bar{m}$ slack variables $\bar{\mathbf{w}} = (\bar{w}_1, \ldots, \bar{w}_{\bar{m}})$ to the constraints and introducing nonnegative bounds on these slack variables. The resulting program would then take the form

$$\begin{aligned} &-\text{minimize } -\bar{\mathbf{c}}^T\bar{\mathbf{x}} \\ &\text{s.t. } \bar{\mathbf{A}}\bar{\mathbf{x}} + \bar{\mathbf{w}} = \bar{\mathbf{b}} \\ &\qquad \bar{\mathbf{w}} \ge \mathbf{0}. \end{aligned} \tag{2.3-12}$$

Let $\bar{\mathbf{x}}^0$ be a starting (nondegenerate) vertex for the linear program (2.3-1) and $(\bar{\mathbf{x}}^0, (\bar{\mathbf{b}} - \bar{\mathbf{A}}\bar{\mathbf{x}}^0))$ be the corresponding vertex for (2.3-12). Show formally that there is a one-to-one correspondence between vertices generated by the simplex method as developed in Sec. 2.3 applied to the linear program (2.3-1) and vertices generated by the simplex method as developed in Sec. 2.1, applied to the above linear program (2.3-12).

2.4. The Primal Simplex Algorithm

We have now identified (specifically in Secs. 2.1 and 2.2) the main computational steps of the simplex method for solving linear programs in the canonical form (1.2-2), and we may proceed to organize them into the mathematical algorithm known as the *primal simplex algorithm,* or more simply, as *the simplex algorithm.* Originated by G. B. Dantzig in 1947, with antecedents going back to the great mathematician J. B. J. Fourier, in the 1820s, it is the premier algorithm of linear programming.

In earlier sections, it was convenient, from a conceptual standpoint, to assume a reordering of variables (or constraints) so that at each cycle of the simplex method, the first m columns of the matrix $\mathbf{A}$ always define the basis matrix, i.e., $B = \{1, 2, \ldots, m\}$. From a computational standpoint this would not be acceptable. We now extend our notation so that B represents the indices of the set of m basic variables $\{\beta_1, \ldots, \beta_i, \ldots, \beta_m\}$, where the corresponding basis matrix is $\mathbf{B}^0 =$

$[\mathbf{a}_{\beta_1}, \ldots, \mathbf{a}_{\beta_m}]$ and is assumed to be nonsingular. Similarly, $N = \{\eta_1, \ldots, \eta_k, \ldots, \eta_{n-m}\}$ represents the indices of the $n-m$ nonbasic variables.

We shall make the same assumptions concerning $\mathbf{A}$, $\mathbf{l}$, and $\mathbf{u}$ as the beginning of Sec. 2.1. We shall assume that a starting vertex (basic feasible solution) defined by B and N is available to us, along with the initial setting $\mathbf{x}_N^0$ of the nonbasic variables, given by (finite) elements of the bound vectors $\mathbf{l}$ and $\mathbf{u}$. We shall employ the usual convention that a variable x_j unbounded from below has $l_j = -\infty$ and a variable unbounded from above has $u_j = +\infty$. For any constant $\kappa \geq 0$, $\infty \pm \kappa = \infty$, and for any constant $\kappa > 0$, $\infty . \kappa = \infty$ and $\infty/\kappa = \infty$. We denote the algorithmic operation of "assignment" by $\leftarrow$.

2.4-1. Algorithm Primal Simplex

Step P1 [*Initialize*].

$$\tilde{\mathbf{b}} \leftarrow \mathbf{b} - \sum_{k=1}^{n-m} \mathbf{a}_{\eta_k} x_{\eta_k}^0.$$

Solve the linear system $\mathbf{B}^0 \mathbf{x}_B^0 = \tilde{\mathbf{b}}$ for $\mathbf{x}_B^0$. We assume that $\mathbf{l}_B \leq \mathbf{x}_B^0 \leq \mathbf{u}_B$, where $\mathbf{l}_B$ and $\mathbf{u}_B$ are m-vectors with components $(l_{\beta_1}, \ldots, l_{\beta_m})$ and $(u_{\beta_1}, \ldots, u_{\beta_m})$ respectively.

Step P2 [*Form price vector*]. Solve the linear system

$$(\boldsymbol{\pi}^0)^T \mathbf{B}^0 = \mathbf{c}_B^T$$

for $(\boldsymbol{\pi}^0)^T$, where $\mathbf{c}_B$ is the m-vector with components $(c_{\beta_1}, \ldots, c_{\beta_m})$.

Step P3 [*Form reduced costs and choose the entering variable*]. For $\eta_k \in N$, form

$$\sigma_{\eta_k}^0 \leftarrow c_{\eta_k} - (\boldsymbol{\pi}^0)^T \mathbf{a}_{\eta_k}$$

and set

$$C \leftarrow \{\eta_k \mid (\sigma_{\eta_k}^0 < 0 \;\&\; x_{\eta_k}^0 = l_{\eta_k}) \quad \text{or} \quad (\sigma_{\eta_k}^0 > 0 \;\&\; x_{\eta_k}^0 = u_{\eta_k}) \quad \eta_k \in N\}.$$

If C is empty then stop; $\mathbf{x}^0$ is optimal. Otherwise, choose any member $\eta_q \in C$, typically by the *largest-coefficient rule*, so that $|\sigma_{\eta_q}^0| \geq |\sigma_{\eta_k}^0|$, $\eta_k \in C$. Set $s \leftarrow \eta_q$.

Step P4 [*Form vector defining the direction of improvement*]. Solve the following linear system for $\tilde{\mathbf{a}}_s$.

$$\mathbf{B}^0 \tilde{\mathbf{a}}_s = \mathbf{a}_s.$$

Step P5 [*Determine the exiting variable*]. This is defined by (2.1-11), which we express here as

$$P \leftarrow \underset{i=1,\ldots,m}{\operatorname{argmin}} \begin{cases} (x_{\beta_i}^0 - l_{\beta_i})/|(\tilde{\mathbf{a}}_s)_i| \text{ when } (\sigma_s^0)(\tilde{\mathbf{a}}_s)_i < 0 \\ (u_{\beta_i} - x_{\beta_i}^0)/|(\tilde{\mathbf{a}}_s)_i| \text{ when } (\sigma_s^0)(\tilde{\mathbf{a}}_s)_i > 0 \\ (u_s - l_s) \end{cases}$$

where P is the set of one or more indices where the minimum is attained. (If the minimum is $u_s - l_s$ then include the index s in P.) Define θ_s by an identical expression to the one above for P, but with "argmin" replaced by "min." When several variables simultaneously attain a bound, i.e., when P has more than one element, then resolve ties arbitrarily by choosing any $p \in P$.

If $\theta_s = \infty$ then stop; the solution is unbounded.

Step P6 [*Determine the new basic feasible solution and the associated basis*].

$$\tau \leftarrow -\text{sign}(\sigma_s^0)\theta_s;$$

Revise $\mathbf{x}^0 = \begin{bmatrix} \mathbf{x}_B^0 \\ \mathbf{x}_N^0 \end{bmatrix}$ as follows. ($\mathbf{e}_q$ denotes a unit vector.)

$$\mathbf{x}_B^0 \leftarrow \mathbf{x}_B^0 - \tilde{\mathbf{a}}_s \tau \qquad \mathbf{x}_N^0 \leftarrow \mathbf{x}_N^0 + \mathbf{e}_q \tau.$$

If the entering variable x_s was the first to meet its (opposite) bound, i.e., if $p = s$, then return to Step P2. Otherwise, revise $\mathbf{B}^0$, $\mathbf{N}^0$, B, and N as follows.

$$\mathbf{B}^0 \leftarrow [\mathbf{a}_{\beta_1}, \ldots, \mathbf{a}_{\beta_{p-1}}, \mathbf{a}_s, \mathbf{a}_{\beta_{p+1}}, \ldots, \mathbf{a}_{\beta_m}]$$
$$\mathbf{N}^0 \leftarrow [\mathbf{a}_{\eta_1}, \ldots, \mathbf{a}_{\eta_{q-1}}, \mathbf{a}_{\beta_p}, \mathbf{a}_{\eta_{q+1}}, \ldots, \mathbf{a}_{\eta_{n-m}}]$$
$$B \leftarrow \{\beta_1, \ldots, \beta_{p-1}, s, \beta_{p+1}, \ldots, \beta_m\}$$
$$N \leftarrow \{\eta_1, \ldots, \eta_{q-1}, \beta_p, \eta_{q+1}, \ldots, \eta_{n-m}\}.$$

Return to Step P2.

Given a starting vertex $\mathbf{x}^0$, the above algorithm is fairly complete from a mathematical standpoint, but it is very far from being an effective computational procedure. Effective implementation requires that we pay detailed attention to the following:

1. *Problem Setup,* namely, specification of the linear program and its representation as a suitable data structure.
2. *Basis Handling,* namely, solution at each cycle of the primal simplex algorithm of two systems of linear equations; see Steps P2 and P4. (A linear system must also be solved for the initial $\mathbf{x}_B^0$.) Efficient and numerically stable techniques for doing this are a key component of any practical implementation. In particular, one seeks to take advantage of the fact that the basis matrix changes in only a *single* column from one cycle of the algorithm to the next.
3. *Strategies* for choosing the entering and exiting variables.
4. *Initialization,* namely, finding a feasible solution with which to initiate the algorithm.
5. *Tactics* for the overall implementation, when the different components are put together.

Each of these points is discussed, in detail, in Part II of this book.

Exercise 2.4-1. The primal simplex algorithm simplifies considerably when general bounds $\mathbf{l} \le \mathbf{x} \le \mathbf{u}$ are replaced by nonnegative bounds $\mathbf{x} \ge \mathbf{0}$. Write out the algorithm for this case.

Exercise 2.4-2. Formulate the analogous primal simplex algorithm for the alternative form (2.3-1) used in Sec. 2.3.

2.5. Convergence and Complexity of the Primal Simplex Algorithm

The following example, due to Beale [1955], shows that the primal simplex algorithm need not converge to an optimal solution when the nondegeneracy assumption is relaxed.

Example 2.5-1. Apply the primal simplex algorithm (specifically the algorithm of Exercise 2.4-1) to the following program, with the entering variable at Step P3 being chosen by the largest-coefficient rule and with ties in the choice of exiting variable at Step P5 being arbitrarily resolved as indicated below.

$$\begin{aligned}
\text{minimize} \quad & -\tfrac{3}{4}x_4 + 20x_5 - \tfrac{1}{2}x_6 + 6x_7 \\
x_1 \quad & +\tfrac{1}{4}x_4 - 8x_5 - x_6 + 9x_7 = 0 \\
x_2 \quad & +\tfrac{1}{2}x_4 - 12x_5 - \tfrac{1}{2}x_6 + 3x_7 = 0 \\
x_3 \quad & + x_6 = 1 \\
& x_1, \ldots, x_7 \ge 0.
\end{aligned}$$

The starting feasible basis is given by $B = \{1, 2, 3\}$ and we see that it is degenerate because the basic variables x_1 and x_2 are both at their bounds. Six bound constraints are active at the initial basic feasible solution. The indices of each basis and other related quantities developed by the simplex method are as follows.

1. $B = \{1, 2, 3\}$, $N = \{4, 5, 6, 7\}$, $\eta_q = 4$, $p = 1$ (x_1 exits).
2. $B = \{4, 2, 3\}$, $N = \{1, 5, 6, 7\}$, $\eta_q = 5$, $p = 2$ (x_2 exits).
3. $B = \{4, 5, 3\}$, $N = \{1, 2, 6, 7\}$, $\eta_q = 6$, $p = 1$ (x_4 exits).
4. $B = \{6, 5, 3\}$, $N = \{1, 2, 4, 7\}$, $\eta_q = 7$, $p = 2$ (x_5 exits).
5. $B = \{6, 7, 3\}$, $N = \{1, 2, 4, 5\}$, $\eta_q = 1$, $p = 1$ (x_6 exits).
6. $B = \{1, 7, 3\}$, $N = \{6, 2, 4, 5\}$, $\eta_q = 2$, $p = 2$ (x_7 exits).
7. $B = \{1, 2, 3\}$, $N = \{6, 7, 4, 5\}$ $(=\{4, 5, 6, 7\})$.

The foregoing is a sequence of degenerate iterations during which no progress is made from the starting point $(0, 0, 1, 0, 0, 0, 0)$. The first two components of each basis are degenerate. We see that after seven iterations we have returned to the basis with which we started; the sequence will thereafter repeat itself indefinitely, when the choice of entering and exiting variables is repeated as given above. The algorithm is said to *cycle.*

Let us look at the situation in general terms. Consider the sequence of bases developed by the primal simplex algorithm, say $B_0, B_1, \ldots$. At some iteration either

1. The optimal solution is found, or
2. The solution is discovered to be unbounded, or
3. A basis, say B_k, is generated which was already encountered earlier in the sequence, say B_0. (Recall that there are only a finite number of basic solutions.)

The sequence

$$B_k = B_0 \qquad B_{k+1} = B_1, \ldots, B_{2k} = B_k$$

will then repeat itself indefinitely, provided of course the same rules for choosing the entering and exiting variables for B_j are applied to B_{k+j}.

Consider case (3) above. The value of the objective function for B_0 is obviously the same as that for B_k. Therefore there is no reduction in the objective value throughout the sequence of basis changes $B_0, \ldots, B_k$. From (2.1-10), it follows that $t = 0$, since the entering variable always has $|\sigma_s^0| \neq 0$; there is therefore no move from $\mathbf{x}^0$, the solution corresponding to B_0. $\mathbf{x}^0$ is, of course, degenerate so that more than n problem constraints are tight. n of these define the set of nonbasic variables. The surplus correspond to basic variables that are at a bound.

The situation (3) is even simpler to visualize within the setting of Sec. 2.3. (To be consistent with the notation there, we attach the symbol "bar" to each quantity.) When a cyclic sequence of bases occurs, say $\bar{B}_0, \ldots, \bar{B}_k = \bar{B}_0$, then $\bar{\mathbf{x}}^0$ is degenerate, so that a set of constraints are tight whose number exceeds $\bar{n}$. From among them, $\bar{n}$ tight constraints are linearly independent and define the basis at any one time, the remainder being out of the basis (inactive but tight). Pictorially, in two dimensions, we would have the situation shown in Figure 2.2, and the algorithm cycles indefinitely through this set of tight constraints.

From a theoretical standpoint, it is essential that there be at least one

Figure 2.2 Point where cycling can occur

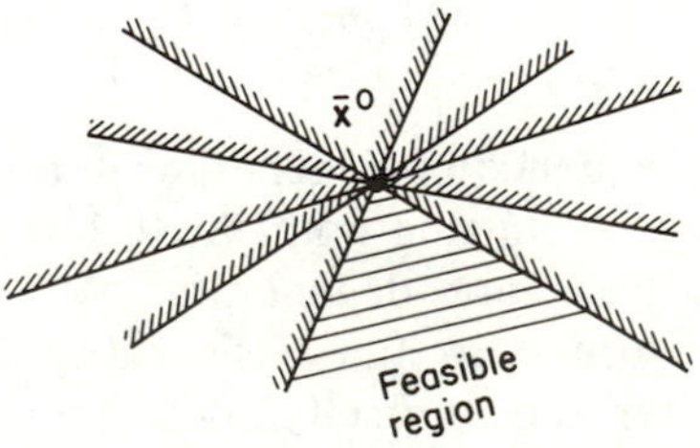

rule for choosing the entering and exiting variable at each iteration, which will guarantee that the primal simplex algorithm converges, without requiring the nondegeneracy assumption. (Whether or not this is a practical necessity will be discussed later.) The most elegant rule is due to Bland [1977], and is as follows.

Smallest-Index Rule. Assume a given ordering of the variables in the linear program (1.2-2). Whenever there is more than one choice for the entering or exiting variable, resolve the tie by always choosing the variable with the smallest index. For the linear program (2.3-1), the analogous statement is as follows. Assume a given ordering of the constraints. Whenever there is more than one choice for the outgoing or incoming constraint, resolve the tie by always choosing the constraint with the smallest index. (Recall the discussion at the end of Sec. 2.3.)

Theorem 2.5-1. Convergence of the primal simplex algorithm. The primal simplex algorithm terminates in a finite number of steps, when the entering and exiting variables (outgoing and incoming constraints) are chosen by the smallest-index rule.

Proof. The proof can be given most simply for the simplex method and algorithm on the alternative form (2.3-1). (For the canonical form (1.2-2), the proof would be identical from a conceptual standpoint, but the notation makes it technically just a little more cumbersome. It is left as Exercise 2.5-1.)

We show that if case (3) above occurs (see discussion immediately after Example 2.5-1), i.e., if we generate a cyclic sequence of bases $\bar{B}_0, \ldots, \bar{B}_k = \bar{B}_0$, then we obtain a contradiction.

Assuming that such a cyclic sequence is generated, let us define a constraint to be *fickle* if it occurs in some bases of this sequence but not in others. From among the set of fickle constraints, let the one with the *largest* index be $\bar{p}$.

Let $\bar{B}$ be a basis in this cyclic sequence for which constraint $\bar{p}$ is incoming and later in the sequence let $\bar{B}^*$ be the basis when $\bar{p}$ is outgoing. Also let constraint $\bar{s}$ be outgoing for basis $\bar{B}$.

If $\bar{\mathbf{d}}$ is the direction of improvement for $\bar{B}$, it follows directly from (2.3-5) that

$$\bar{\mathbf{d}}^T\bar{\mathbf{a}}^i = 0 \qquad i \in \bar{B} \qquad i \neq \bar{s} \quad \text{and} \quad \bar{\mathbf{d}}^T\bar{\mathbf{a}}^{\bar{s}} < 0. \tag{2.5-1}$$

Also, directly from (2.3-9), when constraint $\bar{p}$ is incoming,

$$\bar{\mathbf{d}}^T\bar{\mathbf{a}}^{\bar{p}} > 0. \tag{2.5-2}$$

From (2.3-4a) or (2.3-4b), we see that multipliers are associated with constraints defining $\bar{B}$. For convenience, let us associate zero multipliers with the remaining constraints, so that from (2.3-4b), using $\bar{\sigma}_i$ in place of $\bar{\sigma}_i^0$, we obtain

$$\bar{\mathbf{c}} = \sum_{i=1}^{\bar{m}} \bar{\mathbf{a}}^i\bar{\sigma}_i \qquad \bar{\sigma}_i = 0,\ i \notin \bar{B}. \tag{2.5-3}$$

Similarly, we associate multipliers $\bar{\sigma}_i^*$ with constraints $\bar{B}^*$, giving

$$\bar{\mathbf{c}} = \sum_{i=1}^{\bar{m}} \bar{\mathbf{a}}^i\bar{\sigma}_i^* \qquad \bar{\sigma}_i^* = 0,\ i \notin \bar{B}^*. \tag{2.5-4}$$

Equating these two expressions for $\bar{\mathbf{c}}$ and using (2.5-1), we obtain

$$(\bar{\mathbf{d}}^T\bar{\mathbf{a}}^{\bar{s}})\bar{\sigma}_{\bar{s}} = \sum_{i \in J} (\bar{\mathbf{d}}^T\bar{\mathbf{a}}^i)\bar{\sigma}_i^* + (\bar{\mathbf{d}}^T\bar{\mathbf{a}}^{\bar{p}})\bar{\sigma}_{\bar{p}}^*, \tag{2.5-5}$$

where $J = \{1, \ldots, \bar{p}-1, \bar{p}+1, \ldots, \bar{m}\}$. Since constraint $\bar{s}$ is outgoing from $\bar{B}$ and constraint $\bar{p}$ is outgoing from $\bar{B}^*$, we have

$$\bar{\sigma}_{\bar{s}} < 0 \quad \text{and} \quad \bar{\sigma}_{\bar{p}}^* < 0. \tag{2.5-6}$$

Let us write (2.5-5) as follows:

$$\sum_{j \in J} (\bar{\mathbf{d}}^T\bar{\mathbf{a}}^i)\bar{\sigma}_i^* = (\bar{\mathbf{d}}^T\bar{\mathbf{a}}^{\bar{s}})\bar{\sigma}_{\bar{s}} - (\bar{\mathbf{d}}^T\bar{\mathbf{a}}^{\bar{p}})\bar{\sigma}_{\bar{p}}^*.$$

Using (2.5-1), (2.5-2), and (2.5-6), we observe that the right-hand side of the previous expresssion is positive. Therefore there must be an index, say $\bar{r}$, on the left-hand side, for which

$$(\bar{\mathbf{d}}^T\bar{\mathbf{a}}^{\bar{r}})\bar{\sigma}_{\bar{r}}^* > 0. \tag{2.5-7}$$

Clearly $\bar{\sigma}_{\bar{r}}^* \neq 0$. Therefore $\bar{r} \in \bar{B}^*$.

Suppose $\bar{r} \notin \bar{B}$. *Then $\bar{r}$ is fickle.*

Suppose $\bar{r} \in \bar{B}$. From (2.5-7), $(\bar{\mathbf{d}}^T\bar{\mathbf{a}}^{\bar{r}}) \neq 0$. Thus (2.5-1) implies that $\bar{r} = \bar{s}$. But $\bar{s}$ is fickle. *Thus again $\bar{r}$ must be fickle.*

Now $(\bar{\mathbf{d}}^T\bar{\mathbf{a}}^{\bar{p}})\bar{\sigma}_{\bar{p}}^* < 0$, so from (2.5-7), $\bar{r} \neq \bar{p}$. Since $\bar{r}$ is fickle and $\bar{p}$ is the index of the largest fickle constraint, it follows that $\bar{r} < \bar{p}$. We cannot have $\bar{\sigma}_{\bar{r}}^* < 0$ or we would contradict the smallest index choice of outgoing constraint for $\bar{B}^*$. Therefore from (2.5-7), $(\bar{\mathbf{d}}^T\bar{\mathbf{a}}^{\bar{r}}) > 0$. But $\bar{r}$ is tight throughout the sequence of iterations, so $b_{\bar{r}} - (\bar{\mathbf{a}}^{\bar{r}})^T\bar{\mathbf{x}}^0 = 0$. Thus $\bar{r}$ is a candidate for incoming constraint for $\bar{B}$ and $\bar{r} < \bar{p}$; yet we chose $\bar{p}$. This yields the desired contradiction.

Thus we cannot have a cyclic sequence of bases, and the simplex algorithm must terminate with options (1) or (2) (see discussion immediately after Example 2.5-1). The proof is complete. ■

Corollary 2.5-1. Fundamental Theorem of Linear Programming. The statement of this corollary is identical to that of Theorem 1.3-3 and its proof is an immediate consequence of Theorem 2.5-1.

Exercise 2.5-1. Prove Theorem 2.5-1 for the canonical form (1.2-2).

Slight perturbations of the right-hand side vector **b** will usually be sufficient to move degenerate basic variables off their bounds, without significantly altering the original linear program. Indeed the rounding errors inherent in any computation in finite-precision arithmetic act as an ad hoc perturbation strategy that will stamp out degeneracy in most cases. *In practice, very many problems are degenerate, yet cycling almost never occurs,* and many implementations do not include provisions to take degeneracy explicitly into account. References to other techniques for resolving degeneracy in order to ensure theoretical convergence of the simplex algorithm are given in the notes at the end of this chapter; we shall again discuss degeneracy in a practical setting in Secs. 7.1-1, 7.3, and 9.2-7.

Finally let us consider briefly the complexity of the primal simplex algorithm in terms of the number of iterations it takes to solve a linear program. (The cost per iteration is also a factor in measuring the complexity of computation of the simplex algorithm, but this is more meaningfully discussed within the context of a specific implementation of the main cycle of the algorithm; see Part II, which follows this chapter.)

The following example shows that in certain pathological cases, the simplex algorithm can do very poorly indeed.

Example 2.5-2 (Klee and Minty [1972]). Apply the primal simplex algorithm of Sec. 2.4 (using the largest-coefficient rule to pick the entering variable) to the following problem.

$$\text{minimize} \sum_{j=1}^{m} -10^{m-j}x_j$$

$$\text{s.t.} \left(2\sum_{j=1}^{i-1} 10^{i-j}x_j\right) + x_i + z_i = 100^{i-1} \quad (i = 1, \ldots, m)$$

$$x_j \geq 0,\ z_j \geq 0 \quad (j = 1, \ldots, m).$$

Then the simplex algorithm can be shown to perform $(2^m - 1)$ iterations before finding the optimum.

The simplex method was born under a lucky star. We have just noted that the chance of cycling occurring in practice is remote. It turns out that examples like the foregoing are also largely academic in nature. These examples give a *worst-case* scenario, for which the number of iterations grows *exponentially* with m. In practice however the simplex method is remarkably efficient. The typical number of iterations increases proportionally to m (with the constant of proportionality being usually between 1 and 3), i.e., the number of iterations, on most examples, is usually a *linear* function of m. The theoretical analysis of this observed *average* case efficiency of the simplex algorithm is currently a topic of very active research.

Notes

Sec. 2.1. For a discussion of the active set approach for linear and nonlinear programs, see Gill et al. [1981].

Sec. 2.2. The reduced gradient approach was originated by Wolfe [1962]. The extension to pegged variables is given in Nazareth [1986] and is a special case of the *superbasic variables* of Murtagh and Saunders [1978]. For related approaches see also Zangwill [1969] and Solow [1984].

Sec. 2.3. The proof of Theorem 2.5–1 is a variant of the one given in Chvatal [1983]. For other anti-degeneracy techniques, in particular, *lexicographic ordering* and *epilson perturbation,* see Chvatal [1983] and references given therein. For a discussion of average-case efficiency, see Borgwardt [1982], Smale [1983], Adler and Megiddo [1985].

Part II
Practical Aspects

The development of linear programming into a practical tool closely parallels developments in the field of electronic computing. Initially, in the mid-forties, problems with 10 rows and 20 columns came close to the limits of what was solvable on first generation (vacuum tube) machines, like the SEAC at the National Bureau of Standards. Developments, however, came in rapid succession. In 1952, the early tableaux based versions were reorganized into the revised simplex algorithm. (Because tableaux have become somewhat of a historical anachronism, it is common practice nowadays to dispense with the word "revised.") By 1954–56, codes based on the simplex method could solve problems with about 100 to 256 rows and several hundred columns, on machines like the IBM 701 and IBM 704. When the era of second-generation (transistor-based) machines was ushered in at the beginning of the sixties, problems of around 1000 rows and a few thousand columns were capable of solution, for example, on the IBM 7090 or the CDC 3600. By the mid-1960s, when computers based on integrated circuits were introduced, for example, the IBM 360 series and the CDC 6400, large-scale mathematical programming systems based on the simplex method had been developed, which were capable of solving problems with several thousand rows and a virtually unlimited number of columns. The era of parallel computation is having a profound impact and is likely to lead to substantial advances in the technology of linear programming and associated reformulations of the simplex method. For a valuable survey of the history of implementation of the simplex method, see the paper of Orchard–Hays [1978].

Here, in Part II of this book, we are concerned with the body of techniques that make it possible to turn the mathematical algorithm of Sec. 2.4 (the primal simplex algorithm) and other algorithms based on the simplex method, into effective numerical procedures capable of solving the large-scale linear programs that arise in a practical setting. Such techniques rely heavily on methods of structuring data developed by computer scientists (see, in particular, the book of Knuth [1968]) and methods of numerical linear algebra (see, in particular, the classic works of Wilkinson [1963, 1965]). They are treated, in detail, in Chapters 3 through 9, following the outline given at the end of Sec. 2.4 of Part I. The final chapter of Part II, Chapter 10, gives a brief survey of large-scale Mathematical Programming (MP) Systems.

3
Problem Setup

Even the simplest linear program usually has some zero elements in the LP matrix, for example, the diet problem (1.1-1) of Chapter 1. A larger diet problem, involving $\bar{m}$ nutrients and $\bar{n}$ foods, is likely to have many zero elements, because any particular food will contain significant amounts of only a few of the $\bar{m}$ nutrients. In most practical linear programs, only a small proportion of the matrix elements are nonzero. Right from the outset, we must take this distinguishing feature into account.

Let us define the *density* and *sparsity* (expressed as a percentage) of the $\bar{m} \times \bar{n}$ linear programming matrix $\bar{\mathbf{A}}$ in (1.2-1), as follows:

Density = [(number of nonzero elements)/$(\bar{m}\bar{n})$]100.
Sparsity = (100 − density) = [(number of zero elements)/$(\bar{m}\bar{n}$]100.

Problems that are usually considered small from a practical standpoint (about 300–800 rows and 500–1200 columns) typically have a density of between 0.5 and 2%. Medium-sized problems (about 800–1500 rows and 1200–7000 columns) often have a density as low as 0.05% to 0.3%. Large problems (say in excess of 2500 rows and 30,000 columns) have a density that rarely exceeds 0.01%. Indeed, a common rule of thumb is that each column of a sizeable LP matrix has only a few nonzero elements, typically between 5 and 10. (This observation is closely linked to the way linear programs are formulated in practice, each column corresponding to an activity that involves only a few of the problem's total number of inputs and outputs.) If k denotes the typical number of nonzero elements per column of $\bar{\mathbf{A}}$, then the density is $(k\bar{n})/(\bar{m}\bar{n}))100\%$ or $(k/\bar{m})100\%$.

We see therefore that it is highly advantageous to have to explicitly specify and store *only the nonzero elements* of a linear program. We shall discuss these two aspects, namely external and internal representation, in the next two sections.

3.1. External Representation of a Linear Program

As noted in Chapter 1, specifying a linear program requires the following information.

1. A *linear objective function,* say $\bar{\mathbf{c}}^T\bar{\mathbf{x}}$, to be optimized (minimized or maximized).
2. A set of *linear constraints,* which can take any of the following three forms.

$$
\begin{aligned}
&(\bar{\mathbf{a}}^i)^T\bar{\mathbf{x}} \geq \bar{b}_i \quad \text{Greater than or equals (G) row.} \\
&(\bar{\mathbf{a}}^i)^T\bar{\mathbf{x}} \leq \bar{b}_i \quad \text{Less than or equals (L) row.} \\
&(\bar{\mathbf{a}}^i)^T\bar{\mathbf{x}} = \bar{b}_i \quad \text{Equality (E) row.}
\end{aligned}
\tag{3.1-1}
$$

Since any of these types of rows could arise quite naturally in the formulation of a linear program, it would be unreasonable to insist on a canonical description, for example, (1.2-2), when initially setting up and specifying a linear program. It is also common to have some constraints in the form of *ranges,* namely,

$$\bar{b}_i^{(1)} \leq (\bar{\mathbf{a}}^i)^T\bar{\mathbf{x}} \leq \bar{b}_i^{(2)}. \tag{3.1-2}$$

3. A set of *bounds* on the problem variables, which are of the general form,

$$\bar{l}_j \leq \bar{x}_j \leq \bar{u}_j$$

where $\bar{l}_j$ may be $-\infty$ and $\bar{u}_j$ may be $+\infty$. The most common bound on $\bar{x}_j$ is that it be nonnegative. This is usually taken to be the default and need not be specified explicitly. There are a number of special cases of the foregoing general form, which it is convenient to explicitly distinguish, namely,

$$
\begin{aligned}
\bar{x}_j &\geq \bar{l}_j \quad \text{Lower bounded (LO)} \\
\bar{x}_j &\leq \bar{u}_j \quad \text{Upper bounded (UP)} \\
\bar{l}_j = \bar{x}_j &= \bar{u}_j \quad \text{Fixed (FX)} \\
-\infty \leq \bar{x}_j &\leq +\infty \quad \text{Free (FR)} \\
\bar{x}_j &\leq +\infty \quad \text{Plus } \infty \text{ (PL)} \\
\bar{x}_j &\geq -\infty \quad \text{Minus } \infty \text{ (MI).}
\end{aligned}
\tag{3.1-3}
$$

In MPS (Mathematical Programming System) format, which has become the de facto standard, a linear program defined by the foregoing information is pictured as a tableau of numbers, in which the objective function and constraints correspond to rows and the variables and the right-hand side correspond to columns. Each row and column is given a unique name (of up to eight characters) and each *nonzero* element of the LP matrix is identified by a triple of information as follows:

$$\text{[column name] [row name] [value of element].} \tag{3.1-4}$$

The LP problem is specified by groups of information, called sections, as follows.

1. The ROWS section gives the list of all row names and their corresponding types of constraint. An objective is considered to be an unconstrained row and its row type is given by the symbol N (for "Not binding"). The other three types of row are identified by the symbols G, L, and E, as stated in (3.1-1).
2. The COLUMNS section gives the value of each nonzero element by specifying a list of the triples (3.1-4). These are grouped by column, but within a particular column, they can be provided in any order.
3. The RHS section gives the value of each nonzero right-hand side element again by specifying a list of triples (3.1-4).
4. The RANGES section provides a mechanism for specifying ranges on any constraint, of the form (3.1-2).
5. The BOUNDS section. All bounds on variables that differ from the default, are specified as

   [bound type] [bound name] [column name] [bound value] (3.1-5)

   where [bound type] consists of one of the symbols LO, UP, FX, FR, PL, MI defined in (3.1-3); [bound name] is a unique name associated with a given set of bounds; [column name] is the name associated with a variable; and [bound value] is the numerical value of the bound, if one is required.

All the previous information must follow rigid formatting conventions and be delimited by a NAME record and an ENDATA record. Precise details need not concern us at the moment (they will be given later) and an example will suffice.

Example 3.1-1. Consider the following linear program:

$$\begin{aligned} \text{minimize} \quad & \bar{x}_1 - \bar{x}_2 + \bar{x}_3 \\ \text{s.t.} \quad & 2\bar{x}_1 + 3\bar{x}_3 \leq 10 \\ & 4\bar{x}_2 + 5\bar{x}_3 \leq 20 \\ 0 \leq \bar{x}_1 \leq 100 \quad & \bar{x}_2 \geq 0 \quad \bar{x}_3 \text{ unrestricted.} \end{aligned}$$

Table 3.1 shows the general constraints of this linear program, written out as a tableau, with each row and column given a unique identifier. The numerical coefficients are treated as real numbers.

Table 3.1

	CLNAM1	CLNAM2	CLNAM3	RHS1
OBJ	1.0	−1.0	1.0	
ROW1	2.0		3.0	10.0
ROW2		4.0	5.0	20.0

Table 3.2

```
NAME
ROWS
 N   OBJ
 L   ROW1
 L   ROW2
COLUMNS
     CLNAM1     OBJ        1.0
     CLNAM1     ROW1       2.0
     CLNAM2     OBJ       -1.0
     CLNAM2     ROW2       4.0
     CLNAM3     OBJ        1.0
     CLNAM3     ROW1       3.0
     CLNAM3     ROW2       5.0
RHS
     RHS1       ROW1       10.0
     RHS1       ROW2       20.0
BOUNDS
 UP  BV1        CLNAM1     100.0
 FR  BV1        CLNAM3
ENDATA
```

The external representation in MPS format is then given by Table 3.2.

MPS format simply provides a systematic set of conventions that can serve as a standard interface to an LP code, and it is not particularly user friendly. Some further details are given in Sec. 3.3.

Exercise 3.1-1. Develop two other ways of specifying a linear program, for example, by rows and by elements.

3.2. Internal Representation of a Linear Program

To facilitate computation, a linear program must be put into a more tractable or *canonical form.* Let us consider this first and then go on to discuss how it is represented by a suitable data structure.

Given a linear program (1.2-1), we have already mentioned the necessary transformations in Sec. 1.2, and we now elaborate on them. Let us write a full identity matrix at the end of the matrix $\bar{\mathbf{A}}$ of (1.2-1) and by setting appropriate bounds on the corresponding variables, as explained below, we can distinguish between the three types of constraints. Thus (1.2-1) becomes

$$\begin{aligned} \text{minimize} \quad & \bar{\mathbf{c}}^T\bar{\mathbf{x}} \\ \text{s.t.} \quad & \bar{\mathbf{A}}\bar{\mathbf{x}} + \mathbf{I}\bar{\mathbf{y}} = \bar{\mathbf{b}} \\ & \bar{\mathbf{l}} \le \bar{\mathbf{x}} \le \bar{\mathbf{u}} \\ & 0 \le \bar{y}_i \le +\infty \text{ if corresponding row is of type } \le \\ & -\infty \le \bar{y}_i \le 0 \text{ if corresponding row is of type } \ge \\ & 0 \le \bar{y}_i \le 0 \text{ if corresponding row is of type } =. \end{aligned} \tag{3.2-1}$$

Variables $\bar{y}_i$ defined for $\le$ rows are termed *nonnegative slacks,* variables defined for $\ge$ rows are termed *nonpositive slacks,* and variables defined for = rows are sometimes called *artificial variables.* Note, however, that no real distinction is made between these three sorts of variables. They simply have different bounds that they must satisfy. Collectively, the $\bar{\mathbf{y}}$ variables are called *logical variables.* The original $\bar{\mathbf{x}}$ variables are called *structural variables.* (Henceforth, we shall discontinue use of the outmoded term "artificial variable.")

We shall associate an unrestricted *logical variable* $\bar{w}$ with the objective row and thus transform the problem to:

$$\begin{aligned} \text{minimize} \quad & -\bar{w} \\ \text{s.t.} \quad & \bar{\mathbf{c}}^T\bar{\mathbf{x}} + \bar{w} \qquad = 0 \\ & \bar{\mathbf{A}}\bar{\mathbf{x}} \qquad + \mathbf{I}\bar{\mathbf{y}} = \bar{\mathbf{b}} \\ & -\infty \le \bar{w} \le +\infty \\ & \bar{\mathbf{x}} \text{ and } \bar{\mathbf{y}} \text{ bounded as in (3.2-1).} \end{aligned} \tag{3.2-2}$$

Let us make the following definitions.

$$m = \bar{m} + 1 \qquad n = \bar{m} + \bar{n} + 1$$

$$\mathbf{y} = \begin{bmatrix} \bar{w} \\ \bar{\mathbf{y}} \end{bmatrix} \qquad \mathbf{x} = \begin{bmatrix} \bar{\mathbf{x}} \\ \mathbf{y} \end{bmatrix} = \begin{bmatrix} \bar{\mathbf{x}} \\ \bar{w} \\ \bar{\mathbf{y}} \end{bmatrix} \qquad \mathbf{b} = \begin{bmatrix} 0 \\ \bar{\mathbf{b}} \end{bmatrix}$$

$$\mathbf{A} = \begin{bmatrix} \bar{\mathbf{c}}^T & 1 & \mathbf{0} \\ \bar{\mathbf{A}} & \mathbf{0} & \mathbf{I} \end{bmatrix}$$

$$\mathbf{c}^T = [\mathbf{0}, -1, \mathbf{0}].$$

Also, let $\mathbf{l}$ and $\mathbf{u}$ be vectors of lower and upper bounds on the $\bar{\mathbf{x}}$ variables as given by $\bar{\mathbf{l}}$ and $\bar{\mathbf{u}}$ together with the bounds on $\bar{w}$ and $\bar{\mathbf{y}}$ in (3.2-1) and (3.2-2).

Finally therefore, (3.2-2) can be written in the *computational canonical form*:

$$\begin{aligned} \text{minimize } & \mathbf{c}^T\mathbf{x} \\ \text{s.t. } & \mathbf{A}\mathbf{x} = \mathbf{b} \\ & \mathbf{l} \leq \mathbf{x} \leq \mathbf{u} \end{aligned} \qquad (3.2\text{-}3)$$

where $\mathbf{A}$ is an $m \times n$ matrix, $\mathbf{b}$ is an m vector and $\mathbf{c}$, $\mathbf{l}$, and $\mathbf{u}$ are n vectors. *Note that* $\mathbf{A}$ *is full row rank,* a direct consequence of the fact that the last m columns (corresponding to the logical variables $\mathbf{y}$) define the $m \times m$ identity matrix.

Exercise 3.2-1. Extend the computational canonical form to handle the following cases.

(a) When range constraints of the form (3.1-2) are present. Show that each range constraint can be represented as a single row when appropriate bounds are placed on the corresponding logical variable.
(b) When the objective function is to be maximized.
(c) When the objective function is defined as an unrestricted row (type N in the MPS format) that could be any one of the rows of the LP matrix.

3.2-1. Terminology of Data Structures

Any collection of data usually has two important characteristics, which must be taken into account when designing a suitable internal representation. These are as follows:

1. The structural relations between individual items of data.
2. The operations that must be performed on the data.

Individual items of data are called *nodes*. Each node consists of one or more locations or *words* of computer memory, made up of a set of *fields*. The *address* of, or *pointer* to, a particular node is its first memory location. (Note that nodes are *logical records* of information that must be mapped, in turn, into *physical records*.)

A *linear list* is a set of say $k \geq 0$ nodes, $X[1], \ldots, X[k]$, whose structural properties involve only the linear (one-dimensional) relative position of nodes. If $k > 0$, $X[1]$ is the first node. When $1 < j < k$, $X[j]$ is preceded by $X[j-1]$ and followed by $X[j+1]$. $X[k]$ is the last node.

The main operations associated with a linear list are accessing a particular node, inserting a node and deleting a node. There are two principal ways of organizing information into a linear list and these have different advantages and disadvantages vis-à-vis the operations that must be performed.

Sequential Allocation. Nodes are stored consecutively in memory, with a base pointer giving the beginning of the list. If a node has a variable number of words, then the number of words associated with a node is usually stored in one of its fields. When the length of each record is the same, say w words, then the address or pointer giving access to the jth record is simply given by $\langle$base pointer$\rangle + (j-1)w$. Insertion of a node into a linear list can be expensive, because other nodes must be moved to make room for it. Deletion of nodes also requires that other nodes in the list be moved to close up gaps, unless temporary wastage of space is acceptable. This can be periodically cleaned up by a "garbage collection" operation.

Linked Allocation. Here a field within each node contains a pointer to the next node in the linear list. In contrast to sequential allocation, obtaining access to the jth node in the list requires that all preceding nodes be stepped through, so as to get at the appropriate pointer. Insertion and deletion, on the other hand, are easily accomplished by a suitable change of pointers and unused nodes can be kept in a second linked list to which nodes can be added or from which nodes can be removed. Note that the price paid for the flexibility of linked allocation is the memory needed to store pointers.

The design of an appropriate data structure for a particular application often requires a skilful blending of sequential and linked allocation.

3.2-2. Data Structures for Sparse Linear Programming Matrices

At the very least, the internal representation of an LP matrix should make it possible:

1. To store only nonzero elements.
2. To conveniently access the matrix by columns.
3. To efficiently evaluate inner products between a given vector and a given column of the matrix.

The *column list/row index packed data structure* is a compact and convenient way of achieving these goals and enjoys widespread use. It is most commonly implemented as three sequentially allocated linear lists. For the matrix shown in Table 3.3, the corresponding data structure is shown in Table 3.4.

The nonzero matrix elements, taken by columns, are stored in one list (A). A second list (HA) contains the row indices of these elements. Finally, a third list (KA) points to the start of each column of the matrix, as shown in Table 3.4. (By convention, the final element of KA points to

Table 3.3

1.0	−1.0	1.0
2.0	0.0	3.0
0.0	4.0	5.0

Table 3.4

Column Pointers (KA)	Row Indices (HA)	Matrix Elements (A)
1	1	1.0
3	2	2.0
5	1	−1.0
8	3	4.0
	1	1.0
	2	3.0
	3	5.0

the first unused location of the arrays A and HA.) Note that the jth entry in the list of column pointers must hold the pointer to the beginning of the jth column. However, entries in the other two lists corresponding to the jth column can be specified in any order, i.e., they do not necessarily have to be specified in the order in which they occur in the jth column of the tableau of Table 3.3. Columns must not, of course, be split, i.e., the entries of any column must be in a continuous group.

The foregoing data structure is most appropriate when the LP matrix is stored entirely in a computer's random-access central memory, since widely separated storage locations must be addressed, in the process of accessing the elements of any column. For example, to unpack the second column of the matrix in Table 3.4, we must access the second and third elements of the list of column pointers and use them to determine the starting location of the nonzero elements of the second column in the list of matrix elements, along with the number of elements. The rows in which these occur can be determined from the corresponding locations of the list of row indices. (Even in a paged virtual memory, the column list/row index data structure is often quite acceptable, because consecutive columns are usually accessed sequentially in linear programming.)

To insert a new column into a column list/row index data structure, say after column j, it is necessary to move the pointers in positions $j+1, j+2, \ldots$ down one position in the list of pointers. However, elements and indices of the new column can be stored at the *end* of the other two lists, i.e., entries in these lists do not have to be moved. To delete a column, say column j, entries in positions $j+1, j+2, \ldots$ in the list of column pointers must be moved up one position. Elements in the other two lists can be flagged as unused (e.g., by changing entries in the list of indices to a negative number). The lists can be periodically cleaned up by a garbage collection or compacting operation.

Note also that some "elbow-room," i.e., dummy elements that are flagged as unused, could be deliberately inserted in each column, when the column list/row index data structure is defined. These could later be used to introduce new *rows* into the packed structure. When columns include dummy elements, then $(\text{KA}(j+1) - \text{KA}(j))$ no longer gives the number of nonzero elements in column j. This information could be obtained by scanning the elements of column j or, if greater efficiency is desired, by adding one more array to the data structure whose jth element holds the number of nonzero elements in column j.

The simplicity and efficiency of the column list/row index data structure makes it attractive and widely used in implementations of the simplex method. It does however have disadvantages, in particular, it does not facilitate access by rows. When a matrix must be sequentially accessed by rows rather than by columns, then it can be packed in a *row list/column index* data structure, i.e., the nonzero elements are stored by rows along with the column index of each element and a list of pointers to the start of each row. When access is primarily by column, with occasional access by rows, then the following compromise might be acceptable. In addition to the column list/row index data structure, build the portion of the row list/column index data structure given by the list of column indices and the row pointers. (Lists of indices can often be stored in integer half-word arrays; in contrast, the list of matrix elements will often require a double precision floating-point array.) When row i is to be accessed, its column indices are available, say $j_1, \ldots, j_k$, using the partial row list/column index structure. Using the column list/row index data structure, access columns $j_1, \ldots, j_k$ in turn, and each time seek the entry with index i in the list of row indices and its corresponding value in the list of matrix elements.

Numerous alternatives and variants, including the use of linked allocation, are possible. An example is the use of doubly orthogonal linked lists, each of whose nodes consist of five fields as follows:

value of nonzero element	row index i	column index j	pointer to next nonzero in row i	pointer to next nonzero in col. j

Exercise 3.2-2. Develop the doubly orthogonal linked data structure for the example of Table 3.3.

3.3. Practical Details of Implementation

3.3-1. Further Details about MPS Format

In MPS format there are two types of input records.

1. Indicator input records, which specify the type of data that are to follow and consist primarily of a single left-adjusted word. Table 3.5 summarizes them.

Table 3.5. R—required; O—optional

NAME	R
ROWS	R
COLUMNS	R
RHS	R
RANGES	O
BOUNDS	O
ENDATA	R

2. Data input records that specify the actual input data values. They consist of up to six fields of input data and they follow the same general format, which is summarized in Table 3.6. (Note that there might be minor variations between one implementation and another in the literature.)

The MPS input must begin with a NAME record, which is then followed by an arbitrary character string giving a title to the problem. The input must be terminated with an ENDATA record.

Table 3.6. [rn]—row name; [cn]—column name; [rhsn]—right-hand side name; [rgn]—range name; [bn]—bounds name. Fields 5 and 6 are optional.

Field:	1	2	3	4	5	6
Columns:	2–3	5–12	15–22	25–36	40–47	50–61
	NAME	problem title				
	ROWS					
	N					
	G	[rn]				
	L					
	E					
	COLUMNS					
		[cn]	[rn]	[v]	[rn]	[v]
	RHS					
		[rhsn]	[rn]	[v]	[rn]	[v]
	RANGES					
		[rgn]	[rn]	[v]	[rn]	[v]
	BOUNDS					
	LO					
	UP	[bn]	[cn]	[v]		
	FX					
	FR					
	PL	[bn]	[cn]			
	MI					
	ENDATA					

Several objectives may be specified as N rows and these may be interleaved with other constraints (G, L, or E rows). For a particular run, one of these may be selected. Several objectives can be easily incorporated into the computational canonical form, by associating a different *unrestricted* logical variable with each objective row and selecting the appropriate logical variable corresponding to the particular objective to be optimized (see also Exercise 3.2-1 (c)). Several different right-hand side vectors, several different lower and upper bound vectors, and several different sets of ranges may also be specified and distinguished by unique identifiers [rhsn], [bn], and [rgn]. For a particular run, one of each would be selected. For full details of the MPS format, consult references given in the Notes at the end of this chapter.

3.3-2. Setting Up the Packed Data Structure

An LP matrix, specified in MPS format, can be internally represented in computational canonical form as a packed data structure. We assume this to be a column list/row index structure and discuss some of the details of the transformation from external to internal representation.

ROWS Section. The list of row names in the ROWS section of the input define a symbol table that is used to identify the row numbers of elements given in the COLUMNS and subsequent sections. For the example in Table 3.2, the entry "CLNAM2 ROW2 4.0" has row name ROW2 and lies in row 3 of the canonical form, since this is the third name in the list of row names. The most straightforward form of implementation is to read in the list of names in the ROWS section into a linear array (let us call it RNAME) whose ith position contains the ith name in the list. Another array can keep record of the row types. The operation of symbol table lookup for any particular row name encountered, for example, in the COLUMNS section, then consists of a linear search through the array RNAME, until the row name is matched. When RNAME has many entries, this linear search can be time consuming. *Hashing* is a technique used to save on search time during symbol table lookup. It structures the symbol table by partitioning its entries into pseudo-randomly determined classes, by means of a hashing (randomizing) function. To implement it, one sets up a "front end" to the array RNAME as follows: when the ith name in the ROWS section, say ITHROW, is read into the ith position of RNAME, a pointer to this entry is setup in another array, say KEYNAM. The element of KEYNAM that holds this pointer is determined by converting the row name, in our case ITHROW, into an address, say IA, by means of a hashing function. This function is chosen to satisfy the following two properties.

1. It efficiently converts the row name into a bit pattern used as an address.

Figure 3.1 Hashing

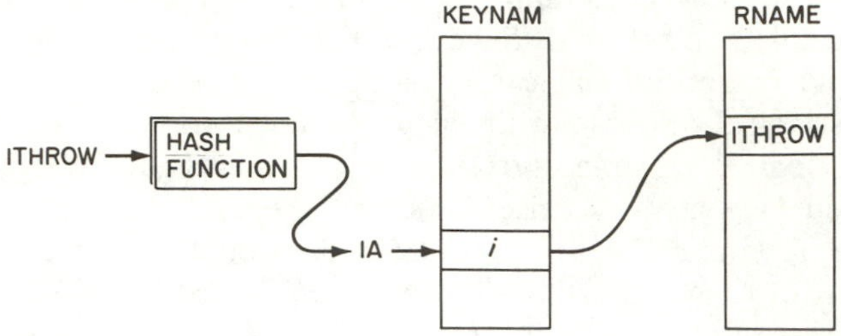

2. Names are converted into bit patterns in a relatively random manner.

Thus, following the above procedure, KEYNAM(IA) = i and RNAME(i) = ITHROW. In the absence of conflicts of addresses, each row name maps into a unique position in KEYNAM. The situation is depicted in Figure 3.1.

Now, matching a row name is much easier than carrying out a linear search. Say ITHROW is encountered in the COLUMNS section. Then we simply evaluate the hash function (this is called hashing on the name) and get a pointer. In this case, the pointer would be IA and we could go directly to address IA in KEYNAM and find out that ITHROW is row i of the matrix. In practice, no hash function is perfect and there will be conflicts when setting up KEYNAM, because two or more different names will hash to the same address in KEYNAM. In this case the entry in KEYNAM is set to point to a relatively short list of such row names, which can then be linearly searched. When a name hashes to an address in KEYNAM that has not been set, then we know that it does not occur in the list of row names, as would happen if there is an error in specifying the row identifier of an element, for example, in the COLUMNS section. (Hashing is a standard method of symbol table lookup, and much more information on it may be obtained from the references cited in the Notes at the end of this chapter.)

Effective implementation requires extensive error checking, to detect duplication of row names in the rows list, invalid row types, and so on. Note also that only after the rows section has been read in, do we know the number of rows in the LP matrix, unless this information is provided in some other manner.

COLUMNS Section. As each entry of the form [cn] [rn] [v] (or [cn] [rn] [v] [rn] [v]) is read in, the row number of the element (or row numbers of a pair of elements) is identified as just discussed and the element inserted as the next entry into the column list/row index data structure of the form discussed in Sec. 3.2. Once all elements have been read in, we know

Figure 3.2 Logical variables in packed data structure

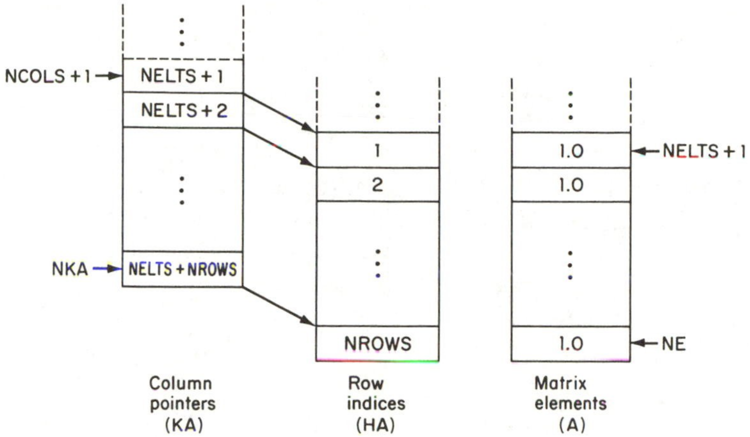

the number of columns in the matrix (say NCOLS) and the total number of elements (say NELTS). Finally, the setup of the computational canonical form requires that a packed identity matrix be included as follows: Suppose NROWS is the number of rows in the matrix as found out by reading in the ROWS section. (The number of rows was previously denoted by m, but now it is more convenient to use Fortran naming conventions.) Then the locations after NCOLS in the list of pointers and the locations after NELTS in the lists of indices and matrix elements, would be set up as shown in Figure 3.2. The total number of columns in the canonical form of the matrix is thus NKA = NCOLS + NROWS and the total number of elements is NE = NELTS + NROWS.

Again, the implementation is relatively straightforward, the main source of difficulty coming from the need for extensive error checking (invalid row names, split columns, duplicate entries, and so on) when reading in the columns of the matrix.

RHS Section. The right-hand side vector (or the one selected if several different vectors are specified in this section) can be read in just like a single column of the matrix in the COLUMNS section and packed, for example, at the end of the three arrays holding the column list/row index data structure.

RANGES Section. Ranges on a constraint of the form (3.1-2) (see also Exercise 3.2-1) are specified in this section. Either $\bar{b}_i^{(1)}$ or $\bar{b}_i^{(2)}$ (which we shall denote by b) is given in the RHS section and the lower and upper limits on the constraint in question are then defined by b and the quantity r that is specified in the RANGES section, as follows.

Row Type	Sign of r	Lower limit $\bar{b}_i^{(1)}$	Upper limit $\bar{b}_i^{(2)}$
E	+	b	$b+\|r\|$
E	−	$b-\|r\|$	b
G	+ or −	b	$b+\|r\|$
L	+ or −	$b-\|r\|$	b

BOUNDS Section. Default lower and upper bounds, typically 0 and a large positive number, say INF (defined to be the machine representation of ∞) are initially set up in arrays BL and BU. Entries are then reset, if bounds are explicitly specified in this section on the corresponding columns, as follows.

Type of Bound	Entry for BL	Entry for BU
LO	bound value	unchanged
UP	unchanged	bound value
FX	bound value	bound value
FR	−INF	+INF
PL	unchanged	+INF
MI	−INF	unchanged

The elements of BL and BU corresponding to logical variables that have not been ranged in the earlier section, are set up using the row types (given in the ROWS section) as follows.

Row Type	Entry for BL	Entry for BU
N	−INF	+INF
G	−INF	0
L	0	+INF
E	0	0

3.3-3. Communication Data Structure

We have just described the setup of the column list/row index data structure and the arrays BL and BU of lower and upper bounds that together define the internal representation of an LP problem in computational canonical form. For the Example 3.1-1 (see also Table 3.3), this packed data structure is shown in Figure 3.3. KA[NKA + 1] is set to the location of the first unused element in A and HA or if, as is shown in Figure 3.3, the right-hand side vector is also stored in these arrays, then KA[NKA + 1] points to its first element in the arrays A and HA and KA[NKA + 2] points to the first unused element in these arrays. NE gives the number of nonzeros in the packed matrix.

Figure 3.3 Communication data structure

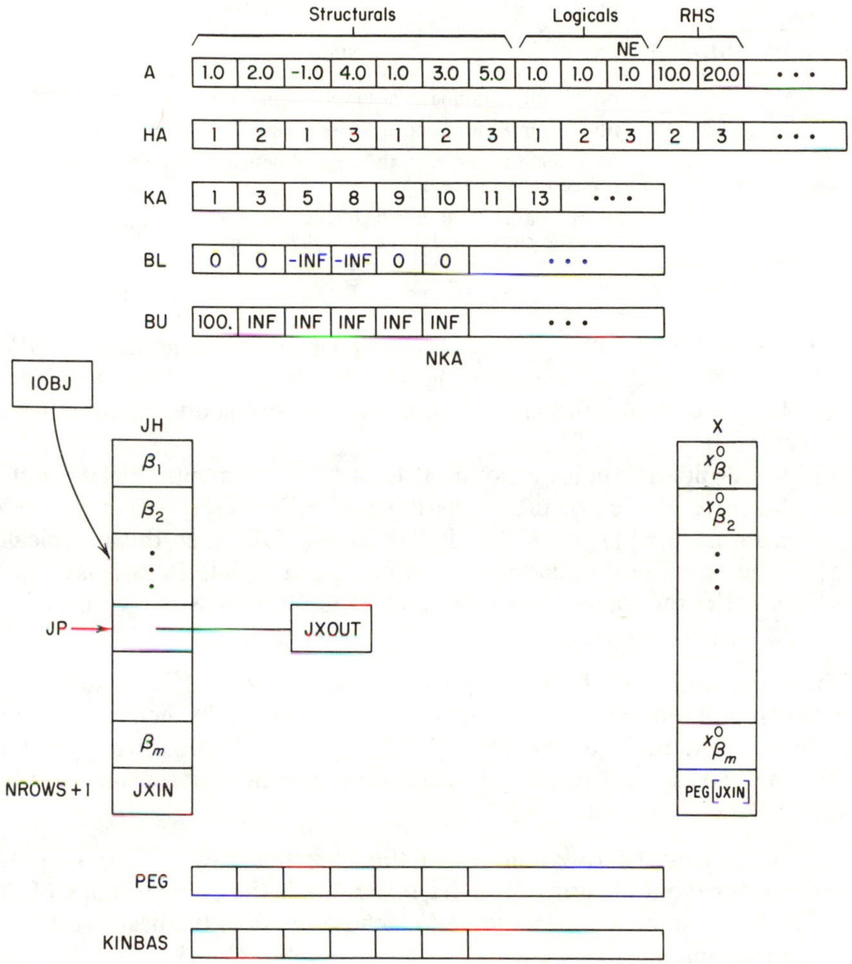

In addition, it is necessary to keep track of the status of columns (basic or nonbasic) and the values of the corresponding variables in the simplex algorithm. The following arrays, for example, would accomplish this task and again, for convenience, we use Fortran naming conventions.

JH An array (integer or half-word integer) of NROWS + 1 elements. The first NROWS elements define the index set $B = \{\beta_1, \ldots, \beta_m\}$ of basic variables. $\text{JH}[i] = \beta_i$ is a pointer into the array KA and defines the ith basic variable or column. $\text{KA}[\beta_i]$, in turn, points to the elements of this column in A and HA.

KINBAS An array (integer or half-word integer) of NKA elements that defines the status of each column of the problem. For example:

KINBAS[j]	Status
0	jth column is nonbasic at lower bound.
1	jth column is nonbasic at upper bound.
2	jth column is nonbasic and pegged between bounds (see PEG following).
3	jth column is basic, in which case this is one of the entries in JH, i.e., $j = \beta_i$ for some i.

X A floating-point array of NROWS + 1 elements. The first NROWS elements hold the current values of the basic variables. X[i] holds the value of the ith basic variable, whose corresponding column has index JH[i].

PEG A floating point array of at least NKA elements. PEG[j] holds the current value of the jth variable. If KINBAS[j] is 0 or 1, this is taken from BL[j] or BU[j]. If KINBAS[j] is 2, then this is typically some initial setting that may be off a bound. If KINBAS[j] is 3, then j = JH[i] for some i, $1 \leq i \leq m$ and PEG[JH[i]] = X[i] (see also Sec. 2.2-1, Extensions).

The foregoing set of arrays provides the means of communication between different components of the simplex algorithm, hence the name "Communication Data Structure." To complete it, the following pointers into the arrays identify the objective row and the entering and exiting variables.

IOBJ Gives the row number of the objective row in the computational canonical form. Thus JH[IOBJ] holds the column index of the corresponding logical variable (which is always in the basis, since it is a free variable).

JXIN Gives the index s of the entering column, whose first element would thus be in A[KA[JXIN]]. It is convenient, for reasons that will become apparent later, to store JXIN in position NROWS + 1 of JH and to store PEG[JXIN] in position NROWS + 1 of X.

JP Gives the index p of the exiting basic variable with JXOUT = JH[JP] identifying the corresponding column index in the packed data structure.

For Example 3.1-1, the complete communication data structure is summarized in Figure 3.3.

Exercise 3.3-1. Choose an initial basis for the linear program in Figure 3.3 and then specify the entries in the arrays JH and X (first NROWS elements), PEG, and KINBAS.

Notes

Sec. 3.0. The LP matrix statistics are given by Greenberg [1978a].

Sec. 3.2. The computational canonical form originates in Orchard–Hays [1968]. A detailed discussion of data structures is given by Knuth [1968].

Sec. 3.3. For a precise specification of MPS format and its interpretation see, for example, Murtagh and Saunders [1983]. Hashing is discussed, for example, in Wegner [1968].

4

The Basis Matrix: Fundamentals of Numerical Computation and Numerical Linear Algebra

Before turning to the solution of systems of linear equations involving the basis matrix, it is necessary to introduce the reader to a few simple principles that underlie floating-point computation on a computer and to the notion of a stable algorithm. We do this here, primarily through the use of illustrative examples.

4.1. Number Representation

Computer arithmetic is based on the floating-point number system. The binary (base 2) and hexadecimal (base 16) number systems are the most commonly used, but because the decimal number system (base 10) is the most familiar, we explain the main ideas in this system and then summarize the results for an arbitrary base.

The representation of a real number x usually requires an infinite number of digits $\mathrm{d}_1, \mathrm{d}_2, \mathrm{d}_3, \ldots$ We make the convention that the leading digit is nonzero with the (decimal) point at the left and say the number is *normalized.* The true position of the decimal point is given by a signed integer, say e, and thus

$$x = (\pm\,.\,\mathrm{d}_1\mathrm{d}_2\mathrm{d}_3\cdots)10^{\mathrm{e}} \tag{4.1-1}$$

where $1 \leq \mathrm{d}_1 \leq 9$, $0 \leq \mathrm{d}_i \leq 9$, $i = 2, 3, \ldots$. We observe that

$$|x| \geq 0.1(10^{\mathrm{e}}). \tag{4.1-2}$$

In a digital computer, only a finite number of digits, say t digits, can be stored and the range of e is restricted to

$$-m \leq \mathrm{e} \leq M \tag{4.1-3}$$

where m and M are positive integers. If the trailing digits $\mathrm{d}_{t+1}, \mathrm{d}_{t+2}, \ldots,$ are dropped, we say the resulting number, say x_c, is obtained by *chopping* x. In this case the error $|x_c - x|$ is bounded by

$$|x_c - x| \leq (\text{one unit in the last place})10^e.$$

Therefore

$$|x_c - x| \leq (10^{-t})10^e. \tag{4.1-4}$$

Rounding a number is more common and we shall only consider it, henceforth. When a number x is rounded to $fl(x)$ by adding 5 to digit $t+1$ and then chopping, the error is

$$|fl(x) - x| \leq (\tfrac{1}{2} \text{ unit in the last place})10^e.$$

Thus

$$|fl(x) - x| \leq (\tfrac{1}{2}10^{-t})10^e. \tag{4.1-5}$$

Note when x is chopped to x_c, the digits $\mathrm{d}_1, \mathrm{d}_2, \ldots, \mathrm{d}_t$ and e remain unchanged. However, when x is rounded to $fl(x)$, *all* digits $\mathrm{d}_1, \mathrm{d}_2, \ldots, \mathrm{d}_t$ and e could change. For example, $x = (0.999999\cdots)10^0$ will round to $fl(x) = (0.1)10^1$. This is why, in subsequent discussion on representable numbers, we change notation to d_i and e.

A *representable* floating-point number is, in general, given by a *signed t* digit normalized *mantissa* or *significand* $\pm d_1 d_2 d_3 \ldots d_t$ and an *exponent e* that satisfies

$$-m \leq e \leq M. \tag{4.1-6}$$

Numbers with exponents outside this range are said to *spill* and at the lower end of the range we usually say *underflow*, and at the upper end we usually say *overflow*. The number zero is represented as $(0.00\cdots 0)10^{-m}$.

For $x \neq 0$, let us define the *relative* error by

$$\delta = \frac{(fl(x) - x)}{x}. \tag{4.1-7}$$

Then from (4.1-2) and (4.1-5) we have

$$|\delta| \leq \frac{((\tfrac{1}{2}10^{-t})10^e)}{((0.1)10^e)}.$$

Thus

$$|\delta| \leq \tfrac{1}{2}10^{1-t}. \tag{4.1-8}$$

Example 4.1-1. Consider a hypothetical decimal computer with a six (decimal) digit word length. A number x is represented by $x = \pm \cdot d_1 d_2 d_3 d_4 10^e$, $-5 \leq e \leq 4$, as shown in Figure 4.1.

Figure 4.1 Six Decimal Digit Word

$\pm$	e'	d_1	d_2	d_3	d_4
sign	biased exponent	mantissa			

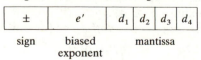

The representation of the exponent is often biased or shifted to e' by adding (in this particular case) 5 to e, so that exponents of $-5, -4, -3, \ldots, 1, 2, 3, 4$ are represented by $0, 1, 2, \ldots, 6, 7, 8, 9$. Thus only nonnegative numbers arise in representing the biased exponent and before they are used, they can be shifted back to remove the bias. Zero is represented in the above scheme by +00000 and unity by +61000.

Since $t = 4$, the relative error in representing a number x is always bounded by $\frac{1}{2}10^{-3}$, which follows from (4.1-8).

Exercise 4.1-1. Define *eps* to be the largest representable number such that $fl(1 + eps) = 1$. For the decimal machine of Example 4.1-1, show that $eps = (0.4999)10^{-3}$.

For an arbitrary base, say β, we represent a number x by

$$x = (\pm\, .\, \mathrm{d}_1\mathrm{d}_2\mathrm{d}_3 \cdots)\beta^{\mathrm{e}}. \tag{4.1-9}$$

The rounded floating-point representation of x is given by

$$fl(x) = (\pm\, .\, d_1 d_2 d_3 \cdots d_t)\beta^e \tag{4.1-10}$$

where

$$1 \le d_1 \le \beta - 1$$
$$0 \le d_i \le \beta - 1, \qquad i = 2, 3, \ldots, t$$

and

$$-m \le e \le M.$$

Now, analogously to (4.1-2) and (4.1-5)

$$|x| \ge (\beta^{-1})\beta^{\mathrm{e}} \tag{4.1-11}$$

and

$$|fl(x) - x| \ge (\tfrac{1}{2}\beta^{-t})\beta^{\mathrm{e}}. \tag{4.1-12}$$

If, as before, we define δ by (4.1-7), then

$$|\delta| \le (\tfrac{1}{2}\beta^{-t}\beta^{\mathrm{e}})/(\beta^{-1}\beta^{\mathrm{e}}) = \tfrac{1}{2}\beta^{1-t}. \tag{4.1-13}$$

Thus

$$fl(x) = x(1 + \delta) \qquad |\delta| \le \tfrac{1}{2}\beta^{1-t}. \tag{4.1-14}$$

We see that a small relative change in x is needed to represent it exactly. It is important to note that (4.1-14) holds *exactly* and is, in fact, equivalent to the definition of δ, with $|\delta|$ bounded by $\frac{1}{2}\beta^{1-t}$. We shall henceforth define the *unit of roundoff error* (or unit in the last place) by

$$ulp = \tfrac{1}{2}\beta^{1-t}. \tag{4.1-15}$$

4.2. Analysis of Basic Arithmetic Operations

Let x and y be *representable* floating-point numbers, so that without having to resort to rounding, $x = fl(x)$ and $y = fl(y)$. In general, $(x + y)$, $(x - y)$, $(x \cdot y)$, and (x/y) will *not* be representable numbers. Let $*$ stand for any of these four basic arithmetic operations $+, -, \cdot, /$. An ideal implementation of floating-point arithmetic would form $fl(x * y)$ as the representable number nearest to the *exact* result $(x * y)$, i.e., $fl(x * y)$ would be the result of rounding the quantity $(x * y)$ obtained if exact arithmetic were used. It follows directly from (4.1-14) that

$$fl(x * y) = (x * y)(1 + \delta) \tag{4.2-1}$$

with

$$|\delta| \leq \tfrac{1}{2}\beta^{1-t} \equiv ulp. \tag{4.2-2}$$

Note that δ will vary depending on the operator $*$ and the operands x and y, but the *bound* on $|\delta|$ is, in each case, the same number *ulp*, the unit of roundoff error.

4.3. Analysis of Sequences of Arithmetic Operations and Attendant Difficulties

4.3-1. Laws of Arithmetic in Finite Precision

Analysis of more complex operations is often little more than repeated application of (4.2-1) and (4.2-2). It is clear that the fundamental laws of arithmetic will only hold approximately. Let us consider the following examples.

Example 4.3-1. Associative Law. Given numbers x, y, and z

$$(x + y) + z = x + (y + z).$$

Suppose x, y, and z are representable numbers. Then

$$fl[fl(x + y) + z] = fl[(x + y)(1 + \delta_1) + z] = [(x + y)(1 + \delta_1) + z](1 + \delta_2),$$
$$fl[x + fl(y + z)] = fl[x + (y + z)(1 + \delta_3)] = [x + (y + z)(1 + \delta_3)](1 + \delta_4)$$

and

$$|\delta_i| \leq ulp \qquad i = 1, 2, 3, 4.$$

Usually δ_1, δ_2, δ_3, and δ_4 are distinct and, in general,

$$fl[fl(x + y) + z] \neq fl[x + fl(y + z)].$$

Exercise 4.3-1. Distributive Law. Given numbers x, y, and z, $x \cdot (y + z) = (x \cdot y) + (x \cdot z)$. Show that, in general, $fl[x \cdot fl(y + z)] \neq fl[fl(x \cdot y) + fl(x \cdot z)]$.

Example 4.3-2. Inner Products. Given vectors $\mathbf{x}$ and $\mathbf{y}$, which for simplicity we take to be in R^3, compute $\mathbf{x}^T\mathbf{y}$. Henceforth we shall abbreviate $x_i \cdot y_i$ by x_iy_i.

$$fl(\mathbf{x}^T\mathbf{y}) = fl[fl[fl(x_1y_1) + fl(x_2y_2)] + fl(x_3y_3)].$$

Using (4.2-1) and (4.2-2), we have

$$\begin{aligned} fl(\mathbf{x}^T\mathbf{y}) &= \{[(x_1y_1)(1+\delta_1) + (x_2y_2)(1+\delta_2)](1+\delta_3) \\ &\quad + (x_3y_3)(1+\delta_4)\}(1+\delta_5) \\ &= (x_1y_1)(1+\delta_1)(1+\delta_3)(1+\delta_5) + (x_2y_2)(1+\delta_2)(1+\delta_3)(1+\delta_5) \\ &\quad + (x_3y_3)(1+\delta_4)(1+\delta_5) \\ &= \sum_{i=1}^{3} (x_iy_i)(1+\eta_i) \end{aligned}$$

where

$$(1+\eta_1) = \prod_{i=1,3,5} (1+\delta_i)$$

$$(1+\eta_2) = \prod_{i=2,3,5} (1+\delta_i)$$

$$(1+\eta_3) = \prod_{i=4,5} (1+\delta_i)$$

and

$$|\delta_i| \leq ulp \qquad i = 1, \ldots, 5.$$

Without going into unnecessary detail, it should be apparent that

$$|\eta_i| \leq K\, ulp$$

where K is a small multiple of 3. In general, if $\mathbf{x}$ and $\mathbf{y}$ are vectors in R^n, then K would be a small multiple of n.

4.3-2. Difficulties and Remedies Associated with Exponent Spill

As we have noted earlier, exponents of representable numbers are restricted to the range $-m \leq e \leq M$. Sequences of operations involving representable numbers can produce *intermediate* results whose exponents violate these bounds, which therefore underflow or overflow, even though the final result of the computation has an exponent within range. Often a judicious reorganization of the calculation can circumvent this difficulty, as the following examples illustrate.

Example 4.3-3. Average of two positive numbers. Given x and y, it is possible when these numbers are large enough that $(x + y)$ is not representable, but $(x + y)/2$ has an exponent within the permitted range and is representable. This difficulty can be overcome by computing the average as $x + (y - x)/2$.

Example 4.3-4. Euclidean norm of a vector. Given $\mathbf{x} \in R^n$, the Euclidean norm is defined by

$$\|\mathbf{x}\|_2 = \left[\sum_{k=1}^{n} x_k^2\right]^{1/2}.$$

The (intermediate) computation of the sum of squares may overflow. To circumvent this, the definition of the norm can be reformulated as follows:

$$\mu = \max_j |x_j| \qquad \|\mathbf{x}\|_2 = \mu\left[\sum_{k=1}^{n} \left(\frac{x_k}{\mu}\right)^2\right]^{1/2}.$$

The obvious implication of examples such as the foregoing is that formulas that are mathematically equivalent are not necessarily computationally equivalent.

4.3-3. Difficulties and Remedies Associated with Cancellation of Leading Digits

Consider two numbers x and y, which are very nearly equal in magnitude, of the same sign and such that $x - y \neq 0$. If $fl(x)$ and $fl(y)$ are their machine representations, then

$$fl(x) = x(1 + \varepsilon_1), \qquad fl(y) = y(1 + \varepsilon_2) \quad \text{with} \quad |\varepsilon_1|, |\varepsilon_2| \leq ulp.$$

We seek to compute $(x - y)$, but we must, of course, be content with

$$\begin{aligned} fl[fl(x) - fl(y)] &= [x(1 + \varepsilon_1) - y(1 + \varepsilon_2)](1 + \delta) \quad \text{with} \quad |\delta| \leq ulp \\ &= (x - y)(1 + \delta) + x\varepsilon_1(1 + \delta) - y\varepsilon_2(1 + \delta). \end{aligned}$$

Let r be the relative error in the computation, given by

$$r = \frac{fl[fl(x) - fl(y)]}{(x - y)} - 1 = \delta + \frac{x\varepsilon_1(1 + \delta) - y\varepsilon_2(1 + \delta)}{(x - y)}.$$

Now x and y can be very large in comparison to $(x - y)$. Thus the relative error r can be substantially different from zero. It is important to note that the difficulty does not arise from error in the subtraction itself, although it is as a result of this operation that the calcellation of leading digits in $fl(x)$ and $fl(y)$ occurred. It is entirely possible that the subtraction is done exactly, i.e., $\delta = 0$. Cancellation, occurring as it does during the subtraction, *reveals* the presence of earlier errors ε_1 and ε_2 and it is these that get magnified and ruin the final answer, when $(x - y)$ is small.

The numbers x and y above could, of course, be the end results of extended calculations whose relative errors are again small, but sufficient nevertheless to swamp the answer, when cancellation occurs. It is also possible to have a progressive loss of leading digits, which results in a similar difficulty, as the following example illustrates.

Example 4.3-5. Series expansion. Forsythe [1970]. Compute e^x by the series expansion

$$e^x = 1 + x + \frac{x^2}{2!} + \frac{x^3}{3!} + \cdots$$

where $k! = k(k-1) \cdots 1$. Suppose that this is done on a decimal computer with a five-digit mantissa. Let $x = -5.5$. Successive terms in the series expansion for $e^{-5.5}$ are listed as follows.

+1.0000
−5.5000
+15.125
−27.730
+38.129
−41.942
+38.446
−30.208
+20.768
−12.692
+6.9803
−3.4902
+1.5997
·
·
·
+0.0026363

The above sum is stopped after 25 terms, when addition of further terms does not change it. The correct answer is approximately 0.00408677. The cancellation in the leading terms and the fact that they are not computed to a high enough precision is what causes the difficulty. The cure is to compute $e^{-5.5}$ as $1/(e^{5.5})$. All terms in the series expansion of $e^{5.5}$ are positive, cancellation cannot occur, and taking the reciprocal gives the very satisfactory answer of 0.0040865 on our five-digit decimal machine.

4.4. Analysis of Algorithms

4.4-1. Error Analysis and Stability of Algorithms

There are two main approaches to analyzing algorithms from a numerical standpoint, termed *forward error analysis* and *backward error analysis*, respectively. Suppose x is algorithmically defined in terms of data or of previously computed quantities $a_1, a_2, \ldots, a_n$, where the functional transformation that the algorithm performs is represented by

$$x = g(a_1, \ldots, a_n). \tag{4.4-1}$$

Because of rounding error, the computed quantity, which we denote by $flg(a_1, \ldots, a_n)$, differs from x. In *forward error analysis,* we seek a bound on

$$|flg(a_1, \ldots, a_n) - x|.$$

In *backward error analysis,* we try to show that $flg(a_1, \ldots, a_n)$ is the result that would have been obtained with *exact* computation starting with perturbed quantities $(a_i + \varepsilon_i)$, i.e., that

$$flg(a_1, \ldots, a_n) = g(a_1 + \varepsilon_1, \ldots, a_n + \varepsilon_n). \tag{4.4-2}$$

Furthermore, we would want $|\varepsilon_i|$ to be small relative to $|a_i|$, for the analysis to be of relevance.

Backward error analysis, which had its genesis in the works of Givens [1954] and Von Neumann and Goldstine [1947], and was brought to flower in the two classic works of Wilkinson [1963, 1965], is by far the most common form of analysis that is used in practice. It has the substantial advantage that it makes a distinction between errors that arise within an algorithm that performs the functional transformation (4.4-1) and errors that are inherent in the problem itself, in particular because of the chosen parameters $a_1, \ldots, a_n$. Thus if the quantities ε_i that emerged from the analysis involved in establishing (4.4-2) were suitably small, there is no reason to be dissatisfied with the algorithm, which is said to be *stable.* It may still, however, be the case that the computed quantity $flg(a_1, \ldots, a_n)$ differs substantially from x. This is because small changes in the problem data $a_1, \ldots, a_n$ cause *large* changes in the true answer. This is hardly the fault of the algorithm and falls within the province of perturbation theory, the study of the stability of the problem itself. We shall have more to say on this topic in Sec. 4.5.

Example 4.4-1. Basic arithmetic operations. Although we did not explicitly use the term there, the analysis of the basic arithmetic operations of Sec. 4.2 was an example of backward error analysis, albeit a rather simple one. For instance, from (4.2-1) and (4.2-2), with specific reference to addition,

$$fl(x + y) = (x + y)(1 + \delta) = x(1 + \delta) + y(1 + \delta)$$

with $|\delta| \leq ulp$.

The computed sum is the result of operating in exact arithmetic on the slightly perturbed quantities $x(1 + \delta)$ and $y(1 + \delta)$.

Example 4.4-2. Back substitution. Consider the upper triangular system $\mathbf{Ux} = \mathbf{b}$, which is to be solved for the unknown $\mathbf{x}$. For simplicity, we shall assume that $\mathbf{U}$ is a 2×2 matrix. Thus we seek to solve the nonsingular system

$$\begin{bmatrix} u_{11} & u_{12} \\ 0 & u_{22} \end{bmatrix} \begin{bmatrix} x_1 \\ x_2 \end{bmatrix} = \begin{bmatrix} b_1 \\ b_2 \end{bmatrix} \tag{4.4-3}$$

where all constants u_{ij} and b_i, $i = 1, 2$, $j = 1, 2$, are machine representable numbers. In exact arithmetic we have

$$x_2 = b_2/u_{22} \qquad x_1 = (b_1 - u_{12}x_2)/u_{11}. \tag{4.4-4}$$

In finite precision floating-point arithmetic, let us represent the computed quantities, obtained in place of x_1 and x_2, by y_1 and y_2, respectively. Thus

$$y_2 = fl(b_2/u_{22}) = (b_2/u_{22})(1 + \delta_1) \quad \text{with} \quad |\delta_1| \le ulp.$$

Let us define

$$\bar{u}_{22} = u_{22}/(1 + \delta_1).$$

Then

$$y_2 = b_2/\bar{u}_{22}. \tag{4.4-5}$$

Similarly

$$\begin{aligned} y_1 &= fl(fl[b_1 - fl(u_{12}y_2)]/u_{11}) \\ &= [[b_1 - u_{12}y_2(1 + \delta_2)](1 + \delta_3)/u_{11}](1 + \delta_4) \end{aligned}$$

where $|\delta_i| \le ulp$, $i = 2, 3, 4$. Then with the definitions

$$\bar{u}_{12} = u_{12}(1 + \delta_2) \qquad \bar{u}_{11} = u_{11}/[(1 + \delta_3)(1 + \delta_4)],$$

we have

$$y_1 = (b_1 - \bar{u}_{12}y_2)/\bar{u}_{11}. \tag{4.4-6}$$

Let us define

$$\begin{aligned} (1 + e_{11}) &= 1/[(1 + \delta_3)(1 + \delta_4)] \\ (1 + e_{12}) &= (1 + \delta_2) \\ (1 + e_{22}) &= 1/(1 + \delta_1). \end{aligned}$$

Then

$$\bar{u}_{ij} = u_{ij}(1 + e_{ij}) \qquad 1 \le i \le j \le 2, \tag{4.4-7}$$

and it is clear, without further explanation, that e_{ij} is bounded by a small multiple of *ulp*.

We therefore see from (4.4-5) and (4.4-6) that we have *exactly* solved the linear system $\bar{\mathbf{U}}\mathbf{y} = \mathbf{b}$ given by

$$\begin{bmatrix} \bar{u}_{11} & \bar{u}_{12} \\ 0 & \bar{u}_{22} \end{bmatrix} \begin{bmatrix} y_1 \\ y_2 \end{bmatrix} = \begin{bmatrix} b_1 \\ b_2 \end{bmatrix} \tag{4.4-8}$$

where from (4.4-7), each element $\bar{u}_{ij}$ is a small perturbation of the original element u_{ij}. Therefore back substitution, for this example, is an extremely stable algorithm; this statement can also be made in general, i.e., for a matrix of any given dimension. It is important to note that the errors e_{ij} are cast back into the original system *in a consistent manner,* so that the system of equations (4.4-8), involving the computed quantities, is an identity, i.e., the equality sign holds exactly and the system can be manipulated following the usual rules of algebra. We can also write (4.4-8) as

$$(\mathbf{U} + \bar{\mathbf{E}})\mathbf{y} = \mathbf{b} \tag{4.4-9}$$

where the elements $\bar{e}_{ij}$ of $\bar{\mathbf{E}}$ are given by $\bar{e}_{ij} = u_{ij}e_{ij}$. If we assume that each element $|u_{ij}|$ is bounded by some reasonable constant, (e.g., $|u_{ij}| \leq 1$) then $|\bar{e}_{ij}|$ is bounded by a small multiple of the unit of round-off error (*ulp*). Again note that $\bar{\mathbf{E}}$ is a well-defined matrix that can be manipulated like any other. It is also important to observe that the (computed) solution $\mathbf{y}$ need *not* be close to the exact solution $\mathbf{x}$, even though back substitution is stable, i.e., $(\mathbf{U} + \bar{\mathbf{E}})^{-1}\mathbf{b}$ may be very different from $\mathbf{U}^{-1}\mathbf{b}$, even when $\bar{\mathbf{E}}$ is tiny compared to $\mathbf{U}$. (We have, of course, assumed that $\mathbf{U}$ and $\mathbf{U} + \bar{\mathbf{E}}$ are invertible.) We illustrate this point by means of an example in the next section.

In the next chapter, we discuss the stable solution of the square systems of equations arising in the primal simplex algorithm and take up considerations of backward error analysis in such systems in greater generality.

4.5. Perturbation Theory and Stability of Problems

The advantage of backward error analysis over forward error analysis, and the reason that it is generally preferred, is that it distinguishes between algorithm stability and problem stability. A problem is inherently unstable, when small changes in its data result in large changes in its solution. Such problems are said to be ill conditioned.

Example 4.5-1. Consider the following 2×2 system of equations:

$$\begin{bmatrix} 1 & 0.99 \\ 0.99 & 0.98 \end{bmatrix} \begin{bmatrix} x_1 \\ x_2 \end{bmatrix} = \begin{bmatrix} 1.99 \\ 1.97 \end{bmatrix}. \tag{4.5-1}$$

The exact solution is $x_1 = 1$, $x_2 = 1$.

Let us make a small perturbation in the (2, 2) element, changing it from 0.98 to 0.9799, and seek the exact solution of the system

$$\begin{bmatrix} 1 & 0.99 \\ 0.99 & 0.9799 \end{bmatrix} \begin{bmatrix} \bar{x}_1 \\ \bar{x}_2 \end{bmatrix} = \begin{bmatrix} 1.99 \\ 1.97 \end{bmatrix}.$$

This exact solution is now $\bar{x}_1 = 1.495$, $\bar{x}_2 = 0.5$, which is substantially different from the solution of the unperturbed system. The matrix in (4.5-1) is ill conditioned.

The small perturbations in the data of a given problem may be, for example, those inescapably introduced even by a stable algorithm, when finite precision arithmetic is used. We have seen an example of this in the previous section, namely, Example 4.4-2. If the matrix **U** in that example happened to be an ill-conditioned matrix, then the components of **y** could be very different from those of **x**. The perturbation theory for a general square system of linear equations and the precise definition of an ill-conditioned matrix will be taken up in the next chapter.

4.6. Summary

We have shown that the analysis of the basic arithmetic operations in finite precision floating-point arithmetic is very simple and that the analysis of a sequence of arithmetic operations often follows, in principle, from repeated application of (4.2-1) and (4.2-2). (The technical details of such an analysis may, however, become formidable.)

We have, through examples, shown the need to reformulate calculations to avoid difficulties associated with overflow and underflow (exponent spill) and with cancellation. The aim is to organize computer calculations in a *stable* manner, whenever possible. An unstable algorithm may work well *most of the time,* but will not have the *guarantee* provided by a stable algorithm. Sometimes, of course, the price to be paid for this guarantee, namely, greater complexity of the algorithm, may be too high and there may be trade-offs that have to be taken into account, for example, between algorithm stability and ease of data management. One can introduce a posteriori tests into an algorithm, in order to detect potential instabilities, but there is generally no excuse for using an unstable algorithm when there is a stable alternative available at reasonable cost. Even a stable algorithm can generate completely erroneous results, but such an algorithm is in a position to point a finger at the problem as the real culprit, declaring it to be ill conditioned. In contrast, an unstable algorithm could produce erroneous results on a well-conditioned problem.

The ideas introduced in this chapter form part of the backdrop of any practical discussion of basis handling in linear programming. We have illustrated them here through specific examples and more general results will be quoted in subsequent chapters as and when the need arises.

Note

An excellent discussion of floating-point computation may be found in Wilkinson [1963, 1965], Forsythe [1970], Forsythe and Moler [1967].

5

The Basis Matrix: Factorizing and Solving

An elementary way to solve a nonsingular square system of linear equations that arise, in our case, within the cycle of the primal simplex algorithm, is to use *Gaussian elimination*. The reader will undoubtedly have encountered this process already, although perhaps not under this name. To review it briefly, let us assume the system to be solved is as follows:

$$
\begin{aligned}
b_{11}x_1 + b_{12}x_2 + \cdots + b_{1m}x_m &= d_1 \\
b_{21}x_1 + b_{22}x_2 + \cdots + b_{2m}x_m &= d_2 \\
&\vdots \\
b_{m1}x_1 + b_{m2}x_2 + \cdots + b_{mm}x_m &= d_m.
\end{aligned}
$$

The method proceeds as follows: First subdiagonal elements in column 1 are eliminated by subtracting suitable multiples of row 1 from all subsequent rows. This can always be done, provided $b_{11} \neq 0$. (If this condition is violated, row 1 can be interchanged with another row that has a nonzero coefficient corresponding to x_1, or column 1 can be interchanged with a subsequent column that has a nonzero coefficient in row 1.) Next, subdiagonal elements in the *new* column 2 are eliminated by subtracting multiples of the *new* row 2 from all subsequent rows. (Again, interchanges can be performed if necessary.) The process is continued in the obvious way, and at each step we can ensure that the element on the diagonal (called the *pivot element* or more simply, the *pivot*) is nonzero, by performing a suitable row (or column) interchange. After $m-1$ steps, an upper triangular system of equations has been obtained. This system can be then be solved by *back substitution*, i.e., begin with the last equation, which involves only x_m and can be trivially solved. Then the next to last equation can be solved for x_{m-1} and so on, proceeding backwards to solve successively for $x_{m-2}, x_{m-3}, \ldots, x_1$. Given that the original system is nonsingular, the above procedure is mathematically well defined.

Gaussian elimination lies at the root of most of the techniques discussed in this chapter, for solving the large sparse systems of linear equations associated with the basis matrix in the simplex method; however, substantial reformulation (in particular, in terms of matrix operations) is required, in order to ensure efficiency and numerical stability.

As we have just noted, Gaussian elimination involves the reduction of nonzero subdiagonal elements, in successive columns, to zero. In matrix terms, we shall see that this is equivalent to premultiplying the matrix $\mathbf{B} \equiv [b_{ij}]$, by a sequence of *elementary matrices*. To ensure that the pivot element is nonzero, an interchange of rows (or columns) may be necessary. This is equivalent to pre- or postmultiplication by a suitable *permutation matrix*. Finally the goal of Gaussian elimination is to reduce the matrix to a *triangular matrix*. We shall therefore first consider these three prerequisites, namely, elementary matrices, permutation matrices, and triangular matrices. Then we apply them to the reformulation of Gaussian elimination in terms of *matrix factorizations*.

5.1. Prerequisites

5.1-1. Elementary Matrices

Consider a nonsingular $m \times m$ matrix that differs from the identity in just a *single* column. If we denote this column by $\boldsymbol{\omega}$, such a matrix can be expressed as

$$(\mathbf{I} + (\boldsymbol{\omega} - \mathbf{e}_k)\mathbf{e}_k^T) = [\mathbf{e}_1, \mathbf{e}_2, \ldots, \mathbf{e}_{k-1}, \boldsymbol{\omega}, \mathbf{e}_{k+1}, \ldots, \mathbf{e}_m], \qquad (5.1\text{-}1)$$

where $\mathbf{e}_k$ denotes the kth column of the $m \times m$ identity matrix $\mathbf{I}$.

If such a matrix is nonsingular, its inverse is very easy to define. One can directly verify the following.

$$(\mathbf{I} + (\boldsymbol{\omega} - \mathbf{e}_k)\mathbf{e}_k^T)^{-1} = (\mathbf{I} - (1/\omega_k)(\boldsymbol{\omega} - \mathbf{e}_k)\mathbf{e}_k^T) = \begin{bmatrix} 1 & -\omega_1/\omega_k & \cdot \\ \cdot & \vdots & \cdot \\ \cdot & \vdots & \cdot \\ \cdot & \vdots & \cdot \\ \cdot & -\omega_{k-1}/\omega_k & \cdot \\ \cdot & 1/\omega_k & \cdot \\ \cdot & -\omega_{k+1}/\omega_k & \cdot \\ \cdot & \vdots & \cdot \\ \cdot & \vdots & \cdot \\ \cdot & \vdots & \cdot \\ \cdot & -\omega_m/\omega_k & 1 \end{bmatrix} \qquad (5.1\text{-}2)$$

(In depicting matrices, ":" denotes a nonzero element, and "." denotes a zero element.)

The assumption of nonsingularity implies that $\omega_k \neq 0$. Thus in order to find the inverse of the elementary matrix (5.1-1), we simply replace the diagonal element ω_k in column k by its reciprocal, negate all the other elements in column k, and divide them by ω_k. It is important to observe that both the elementary matrix (5.1-1) and its inverse (5.1-2) (also an elementary matrix because it differs from the identity in only a single column) have *exactly the same number of nonzero elements*. In particular, when $\boldsymbol{\omega}$ is a sparse vector, both matrices have the same *sparsity* (see beginning of Chapter 3).

Given a vector $\mathbf{w} \in R^n$, let us seek an elementary matrix of the form (5.1-1), say $\boldsymbol{\Lambda}_k$, which satisfies

$$\boldsymbol{\Lambda}_k \mathbf{w} = w_k \mathbf{e}_k = (\mathbf{e}_k^T \mathbf{w})\mathbf{e}_k, \tag{5.1-3}$$

where we assume that $w_k \neq 0$. Thus we seek a suitable vector $\boldsymbol{\omega}$ in (5.1-1), so that all elements in $\mathbf{w}$ are eliminated (zeroed) save the kth, which is left unchanged, when $\mathbf{w}$ is premultiplied by $\boldsymbol{\Lambda}_k$. If

$$(\mathbf{I} + (\boldsymbol{\omega} - \mathbf{e}_k)\mathbf{e}_k^T)\mathbf{w} = (\mathbf{e}_k^T \mathbf{w})\mathbf{e}_k,$$

then

$$\boldsymbol{\omega} = (-\mathbf{w} + 2(\mathbf{e}_k^T \mathbf{w})\mathbf{e}_k)/(\mathbf{e}_k^T \mathbf{w}).$$

Therefore

$$\boldsymbol{\Lambda}_k = (\mathbf{I} + (\mathbf{e}_k - (\mathbf{w}/w_k))\mathbf{e}_k^T) \tag{5.1-4}$$

or

$$\boldsymbol{\Lambda}_k = \begin{bmatrix} 1 & -w_1/w_k & \cdot \\ \cdot & \vdots & \cdot \\ \cdot & \vdots & \cdot \\ \cdot & \vdots & \cdot \\ \cdot & -w_{k-1}/w_k & \cdot \\ \cdot & 1 & \cdot \\ \cdot & -w_{k+1}/w_k & \cdot \\ \cdot & \vdots & \cdot \\ \cdot & \vdots & \cdot \\ \cdot & \vdots & \cdot \\ \cdot & -w_m/w_k & 1 \end{bmatrix} \tag{5.1-5}$$

From the inversion operation described just after (5.1-2), $\boldsymbol{\Lambda}_k^{-1}$ is given by the same expression as (5.1-5) but the *nondiagonal* elements of column k having their signs reversed.

A special case of (5.1-5) is the elementary matrix, say $\boldsymbol{\Gamma}_k$, which leaves $w_1, \ldots, w_k$ unchanged and eliminates the remaining elements when it is applied to $\mathbf{w}$, i.e.,

$$(\boldsymbol{\Gamma}_k\mathbf{w})_i = w_i,\ i \le k \qquad (\boldsymbol{\Gamma}_k\mathbf{w})_i = 0 \qquad i > k.$$

It is easily verified that

$$\boldsymbol{\Gamma}_k = \begin{bmatrix} 1 & 0 & \cdot \\ \vdots & \vdots & \vdots \\ & 0 & \cdot \\ \cdot & 1 & \cdot \\ \cdot & -w_{k+1}/w_k & \cdot \\ \cdot & \vdots & \cdot \\ \cdot & \vdots & \cdot \\ \cdot & \vdots & \cdot \\ \cdot & -w_m/w_k & 1 \end{bmatrix} \tag{5.1-6}$$

Exercise 5.1-1. Find the elementary matrix, say $\hat{\boldsymbol{\Gamma}}_k$, such that $\hat{\boldsymbol{\Gamma}}_k\mathbf{w} = \mathbf{e}_k$. Also find the elementary matrix, say $\boldsymbol{\Pi}_k$, for which $(\boldsymbol{\Pi}_k\mathbf{w})_i = 0,\ i < k$, $(\boldsymbol{\Pi}_k\mathbf{w})_i = w_i,\ i \ge k$.

5.1-2. Permutation Matrices

The operation of interchanging two rows, say i and j, of an $m \times m$ matrix $\mathbf{B}$, is equivalent to premultiplying $\mathbf{B}$ by an *elementary permutation matrix* $\mathbf{P}_{i,j}$, given by the $m \times m$ identity matrix, with its ith and jth rows (or columns) interchanged. ($\mathbf{P}_{i,j}$ is obviously a symmetric matrix, i.e., $\mathbf{P}_{i,j} = \mathbf{P}_{i,j}^T$.) Similarly, a permutation of several rows of $\mathbf{B}$ corresponds to premultiplying it by a *permutation matrix,* say $\mathbf{P}$, obtained by a suitable permutation of the rows (or columns) of the $m \times m$ identity matrix. (Note that $\mathbf{P}$ is *not* necessarily symmetric.) The following example illustrates the properties of such matrices.

Example 5.1-1. Consider the following permutation matrix, which is obtained from the identity matrix whose rows are taken in the order 2, 3, 1.

$$\mathbf{P} = \begin{bmatrix} 0 & 1 & 0 \\ 0 & 0 & 1 \\ 1 & 0 & 0 \end{bmatrix}.$$

Let $\mathbf{B}$ be a 3×3 matrix with columns denoted by $\mathbf{b}_i$ and rows by $(\mathbf{b}^i)^T$, $i = 1, 2, 3$. Then

$$\mathbf{PB} = \mathbf{P}\begin{bmatrix} (\mathbf{b}^1)^T \\ (\mathbf{b}^2)^T \\ (\mathbf{b}^3)^T \end{bmatrix} = \begin{bmatrix} (\mathbf{b}^2)^T \\ (\mathbf{b}^3)^T \\ (\mathbf{b}^1)^T \end{bmatrix}$$

$$\mathbf{BP} = [\mathbf{b}_1, \mathbf{b}_2, \mathbf{b}_3]\mathbf{P} = [\mathbf{b}_3, \mathbf{b}_1, \mathbf{b}_2]$$

$$\mathbf{BP}^T = [\mathbf{b}_1, \mathbf{b}_2, \mathbf{b}_3]\mathbf{P}^T = [\mathbf{b}_2, \mathbf{b}_3, \mathbf{b}_1].$$

Note that the interchange of rows brought about when **B** is premultiplied by **P** is the same as the interchange of columns brought about when **B** is postmultiplied by $\mathbf{P}^T$.

$$\mathbf{P} = \mathbf{P}_{2,3}\mathbf{P}_{1,2}$$
$$\mathbf{P}^T\mathbf{P} = \mathbf{P}_{1,2}^T\mathbf{P}_{2,3}^T\mathbf{P}_{2,3}\mathbf{P}_{1,2}.$$

Also

$$\mathbf{P}_{2,3}^T\mathbf{P}_{2,3} = \mathbf{I} \qquad \mathbf{P}_{1,2}^T\mathbf{P}_{1,2} = \mathbf{I}.$$

Hence

$$\mathbf{P}^T\mathbf{P} = \mathbf{I} \qquad \mathbf{P}^T = \mathbf{P}^{-1}.$$

We see that $\mathbf{P}^T$ will undo the interchange performed by **P**.

The results concerning permutation matrices, illustrated by the above example, can be stated more generally as follows.

Properties. (a) Any elementary permutation matrix $\mathbf{P}_{i,j}$ satisfies

$$\mathbf{P}_{i,j} = \mathbf{P}_{i,j}^T \quad \text{and} \quad \mathbf{P}_{i,j}^T\mathbf{P}_{i,j} = \mathbf{I} = \mathbf{P}_{i,j}\mathbf{P}_{i,j}^T. \tag{5.1-7}$$

(b) Any $m \times m$ permutation matrix **P** can be expressed as the product of a finite number of suitable elementary permutation matrices, namely,

$$\mathbf{P} = \mathbf{P}_{i_1,j_1}\mathbf{P}_{i_2,j_2}\cdots\mathbf{P}_{i_k,j_k}. \tag{5.1-8}$$

(c) Combining (5.1-7) and (5.1-8) gives

$$\mathbf{P}^T\mathbf{P} = \mathbf{I} = \mathbf{P}\mathbf{P}^T \qquad \mathbf{P}^T = \mathbf{P}^{-1}. \tag{5.1-9}$$

Exercise 5.1-2. Establish properties (5.1-7) through (5.1-9).

Exercise 5.1-3. Given an elementary matrix, say **E**, defined by (5.1-1) and a permutation matrix **P**, use (5.1-8) to show that ($\mathbf{PEP}^T$) is also an elementary matrix.

5.1-3. Triangular Matrices

A lower (or upper) triangular matrix is one whose super (or sub) diagonal elements are zero. It is readily verified that the product of lower (or upper) triangular matrices is lower (or upper) triangular. Furthermore, a lower triangular matrix **L**, namely,

$$\mathbf{L} = \begin{bmatrix} l_{11} & \cdot & & \cdot \\ l_{21} & l_{22} & & \cdot \\ \vdots & \vdots & & \cdot \\ \vdots & \vdots & & \cdot \\ \vdots & \vdots & & \cdot \\ l_{m1} & l_{m2} & & l_{mm} \end{bmatrix}$$

may be expressed in *product form* as follows.

$$\mathbf{L} = \begin{bmatrix} l_{11} & \cdot & & \cdot \\ l_{21} & 1 & & \cdot \\ \vdots & \cdot & & \cdot \\ \vdots & \cdot & & \cdot \\ \vdots & \cdot & & \cdot \\ l_{m1} & \cdot & & 1 \end{bmatrix} \begin{bmatrix} 1 & \cdot & & \cdot \\ \cdot & l_{22} & & \cdot \\ \cdot & \vdots & & \cdot \\ \cdot & \vdots & & \cdot \\ \cdot & \vdots & & \cdot \\ \cdot & l_{m2} & & 1 \end{bmatrix} \cdots \begin{bmatrix} 1 & \cdot & & \cdot \\ \cdot & 1 & & \cdot \\ \cdot & \cdot & & \cdot \\ \cdot & \cdot & & \cdot \\ \cdot & \cdot & & \cdot \\ \cdot & \cdot & & l_{mm} \end{bmatrix}$$

If we denote successive matrices in the above product by $\mathbf{L}_i$, $i = 1, \ldots, m$, where each matrix $\mathbf{L}_i$ is an elementary (lower triangular) matrix as defined in Sec. 5.1-1, then

$$\mathbf{L} = \mathbf{L}_1\mathbf{L}_2 \cdots \mathbf{L}_m. \tag{5.1-10}$$

Analogous statements can be made about upper triangular matrices as follows.

$$\mathbf{U} = \begin{bmatrix} u_{11} & u_{12} & & u_{1m} \\ \cdot & u_{22} & & u_{2m} \\ \cdot & \cdot & & \vdots \\ \cdot & \cdot & & \vdots \\ \cdot & \cdot & & \vdots \\ \cdot & \cdot & & u_{mm} \end{bmatrix}$$

and

$$\mathbf{U} = \begin{bmatrix} 1 & & \cdot & u_{1m} \\ \cdot & & \cdot & u_{2m} \\ \cdot & & \cdot & \vdots \\ \cdot & & \cdot & \vdots \\ \cdot & & 1 & u_{m-1,m} \\ \cdot & & \cdot & u_{mm} \end{bmatrix} \begin{bmatrix} 1 & & u_{1,m-1} & \cdot \\ \cdot & & u_{2,m-1} & \cdot \\ \cdot & & \vdots & \cdot \\ \cdot & & \vdots & \cdot \\ \cdot & & u_{m-1,m-1} & \cdot \\ \cdot & & \cdot & 1 \end{bmatrix} \cdots \begin{bmatrix} u_{11} & & \cdot & \cdot \\ \cdot & & \cdot & \cdot \\ \cdot & & \cdot & \cdot \\ \cdot & & \cdot & \cdot \\ \cdot & & 1 & \cdot \\ \cdot & & \cdot & 1 \end{bmatrix}$$

If we denote successive terms in the above product by $\mathbf{U}_j$, $j = m, m-1, \ldots, 1$, where again each matrix $\mathbf{U}_j$ is an elementary matrix as defined in Sec. 5.1-1, then

$$\mathbf{U} = \mathbf{U}_m\mathbf{U}_{m-1} \cdots \mathbf{U}_1. \tag{5.1-11}$$

Assume that $l_{ii} \neq 0$ and $u_{ii} \neq 0$, $1 \leq i \leq m$. As we have noted in Sec. 5.1-1, each elementary matrix in (5.1-10) and (5.1-11) is readily invertible. (See, in particular, the paragraph after (5.1-2).) We can therefore express $\mathbf{L}^{-1}$ and $\mathbf{U}^{-1}$ in *product form* as follows:

$$\mathbf{L}^{-1} = \mathbf{L}_m^{-1}\mathbf{L}_{m-1}^{-1} \cdots \mathbf{L}_1^{-1} \tag{5.1-12}$$

$$\mathbf{U}^{-1} = \mathbf{U}_1^{-1}\mathbf{U}_2^{-1} \cdots \mathbf{U}_m^{-1}. \tag{5.1-13}$$

It is important to observe when **L** and **U** are sparse matrices that $\mathbf{L}^{-1}$ and $\mathbf{U}^{-1}$ *can be expressed as the product of sparse elementary matrices.* This follows from (5.1-12) and (5.1-13) and the fact that the jth columns of $\mathbf{L}_j^{-1}$ and $\mathbf{U}_j^{-1}$ have the same number of nonzero elements as the jth columns of $\mathbf{L}_j$ and $\mathbf{U}_j$, respectively. However $\mathbf{L}^{-1}$ and $\mathbf{U}^{-1}$ *may not be sparse.* This is illustrated by the following example.

Example 5.1-2. Consider the matrix **U** and its product form (5.1-11).

$$\mathbf{U}=\begin{bmatrix} 1/1 & -1/1 & . & & . \\ . & 1/2 & -1/2 & & . \\ . & . & 1/3 & & . \\ . & . & . & & . \\ . & . & . & & . \\ . & . & . & & -1/(\mathrm{m}-1) \\ . & . & . & & 1/\mathrm{m} \end{bmatrix}$$

$$=\begin{bmatrix} 1 & . & . \\ . & 1 & . \\ . & . & . \\ . & . & . \\ . & . & . \\ . & . & -1/(\mathrm{m}-1) \\ . & . & 1/\mathrm{m} \end{bmatrix} \cdots \begin{bmatrix} 1/1 & . & . \\ . & 1 & . \\ . & . & . \\ . & . & . \\ . & . & . \\ . & . & . \\ . & . & 1 \end{bmatrix}$$

It is easily verified that

$$\mathbf{U}^{-1}=\begin{bmatrix} 1 & 2 & 3 & m \\ . & 2 & 3 & m \\ . & . & 3 & m \\ . & . & . & \vdots \\ . & . & . & \vdots \\ . & . & . & m \\ . & . & . & m \end{bmatrix}=\begin{bmatrix} 1/1 & . & . \\ . & 1 & . \\ . & . & . \\ . & . & . \\ . & . & . \\ . & . & . \\ . & . & 1 \end{bmatrix}\cdots\begin{bmatrix} 1 & . & . \\ . & 1 & . \\ . & . & . \\ . & . & . \\ . & . & . \\ . & . & m/(m-1) \\ . & . & m \end{bmatrix}$$

Note that **U** is a *sparse* upper triangular matrix, but $\mathbf{U}^{-1}$ is a *dense* upper triangular matrix that may be expressed as the product of *sparse* factors.

Throughout this book we use inverses as mathematical operators, e.g., $\mathbf{U}^{-1}$, but *never as computational operators,* e.g., $fl(\mathbf{U}^{-1})$, for reasons we shall explain in due course.

Much of what follows in this chapter relies on the simple ideas introduced in this section; readers should therefore be sure to have these prerequisites at their fingertips.

5.2. The LU Factorization

LU factorization is a reformulation, in matrix terms, of Gaussian elimination. It rests on the following theorem.

Theorem 5.2-1. Consider a matrix **B** with *nonsingular* leading principal submatrices. (The submatrix of **B** formed from its first k rows and its first k columns, $1 \le k \le m$, is called the leading principal submatrix of order k.) There exists a *unique unit lower triangular matrix* **L** (i.e., $l_{ii} = 1$, $1 \le i \le m$) and a *unique upper triangular matrix* **U** such that

$$\mathbf{B} = \mathbf{LU}. \tag{5.2-1}$$

Proof. The proof will follow from constructive reformulation of Gaussian elimination.

Let us attach a superscript to **B** to indicate that it is our starting matrix, i.e., $\mathbf{B}^{(1)} \equiv \mathbf{B}$. The elements of $\mathbf{B}^{(1)}$ are denoted by $b_{ij}^{(1)}$.

$\mathbf{B}^{(1)}$ will be progressively transformed into an upper triangular matrix, by premultiplying it by a sequence of elementary matrices of the form (5.1-6). The elementary matrix, say $\mathbf{\Gamma}_1$, which transforms $\mathbf{b}_1^{(1)}$ (the first column of $\mathbf{B}^{(1)}$) into $b_{11}^{(1)}\mathbf{e}_1$, is given by (5.1-6) with $k = 1$ and $\mathbf{w} = \mathbf{b}_1^{(1)}$, namely,

$$\mathbf{\Gamma}_1 = \begin{bmatrix} 1 & \cdot & & \cdot \\ -b_{21}^{(1)}/b_{11}^{(1)} & 1 & & \cdot \\ \vdots & \cdot & & \cdot \\ \vdots & \cdot & & \cdot \\ \vdots & \cdot & & \cdot \\ -b_{m1}^{(1)}/b_{11}^{(1)} & \cdot & & 1 \end{bmatrix}$$

where the first leading principal submatrix of $\mathbf{B}^{(1)}$, namely $b_{11}^{(1)}$, is nonzero by assumption.

$$\mathbf{B}^{(2)} \equiv \mathbf{\Gamma}_1\mathbf{B}^{(1)} = \begin{bmatrix} b_{11}^{(1)} & b_{12}^{(1)} & & b_{1m}^{(1)} \\ 0 & b_{22}^{(2)} & & b_{2m}^{(2)} \\ \cdot & \vdots & & \vdots \\ \cdot & \vdots & & \vdots \\ \cdot & \vdots & & \vdots \\ 0 & b_{m2}^{(2)} & & b_{mm}^{(2)} \end{bmatrix}$$

Row j of $\mathbf{B}^{(2)}$ is, by construction, a linear combination of rows 1 and j of $\mathbf{B}^{(1)}$. If $b_{22}^{(2)}$ was zero, then there would be a linear combination of the rows of the 2×2 leading principal submatrix of $\mathbf{B}$ $(=\mathbf{B}^{(1)})$ whose elements are all zero, contradicting our assumption that the leading principal submatrices of $\mathbf{B}$ are nonsingular. Therefore $b_{22}^{(2)}\neq 0$.

Define $\mathbf{\Gamma}_2$ to be the elementary matrix that transforms the second column $\mathbf{b}_2^{(2)}$ of the matrix $\mathbf{B}^{(2)}$, so that all subdiagonal elements are zero and other elements of the column are unchanged. This is given by (5.1-6) with $k=2$ and $\mathbf{w}=\mathbf{b}_2^{(2)}$. Thus

$$\mathbf{B}^{(3)}\equiv\mathbf{\Gamma}_2\mathbf{B}^{(2)}=\mathbf{\Gamma}_2\mathbf{\Gamma}_1\mathbf{B}^{(1)}.$$

The process can be continued in this fashion, $b_{kk}^{(k)}\neq 0$ at each step. (If it were zero, by an argument analogous to that used above, the $k\times k$ leading principal submatrix of $\mathbf{B}$ would have linearly dependent rows.) At the beginning of the kth step,

$$\mathbf{B}^{(k)}=\mathbf{\Gamma}_{k-1}\cdots\mathbf{\Gamma}_2\mathbf{\Gamma}_1\mathbf{B}^{(1)}. \tag{5.2-2}$$

Hence

$$\mathbf{B}^{(k)}=\begin{bmatrix} b_{11}^{(1)} & b_{12}^{(1)} & \vdots & \vdots & b_{1m}^{(1)} \\ \cdot & b_{22}^{(2)} & \vdots & \vdots & \vdots \\ \cdot & \cdot & \vdots & \vdots & \vdots \\ \cdot & \cdot & \vdots & \vdots & \vdots \\ \cdot & \cdot & b_{k-1,k-1}^{(k-1)} & \vdots & \vdots \\ \cdot & \cdot & \cdot & b_{kk}^{(k)} & b_{km}^{(k)} \\ \cdot & \cdot & \cdot & \vdots & \vdots \\ \cdot & \cdot & \cdot & \vdots & \vdots \\ \cdot & \cdot & \cdot & b_{mk}^{(k)} & b_{mm}^{(k)} \end{bmatrix}$$

At the completion of step k we have

$$\mathbf{B}^{(k+1)}\equiv\mathbf{\Gamma}_k\mathbf{B}^{(k)}, \tag{5.2-3}$$

where

$$\mathbf{\Gamma}_k=\begin{bmatrix} 1 & \cdot & \cdot \\ \cdot & \cdot & \cdot \\ \cdot & \cdot & \cdot \\ \cdot & \cdot & \cdot \\ \cdot & 1 & \cdot \\ \cdot & -b_{k+1,k}^{(k)}/b_{kk}^{(k)} & \cdot \\ \cdot & \vdots & \cdot \\ \cdot & \vdots & \cdot \\ \cdot & \vdots & \cdot \\ \cdot & -b_{mk}^{(k)}/b_{kk}^{(k)} & 1 \end{bmatrix}$$

Thus

$$\boldsymbol{\Gamma}_k = \mathbf{I} + (\boldsymbol{\omega} - \mathbf{e}_k)\mathbf{e}_k^T \tag{5.2-4}$$

where

$$\boldsymbol{\omega} = [0, \ldots, 1, -b_{k+1,k}^{(k)}/b_{kk}^{(k)}, \ldots, -b_{mk}^{(k)}/b_{kk}^{(k)}]^T.$$

The final matrix $\mathbf{B}^{(m)}$ takes the form

$$\mathbf{B}^{(m)} = \boldsymbol{\Gamma}_{m-1}\boldsymbol{\Gamma}_{m-2} \cdots \boldsymbol{\Gamma}_2\boldsymbol{\Gamma}_1\mathbf{B}^{(1)}. \tag{5.2-5}$$

This is an upper triangular matrix, explicitly given by

$$\mathbf{B}^{(m)} = \begin{bmatrix} b_{11}^{(1)} & b_{12}^{(1)} & & b_{1m}^{(1)} \\ \cdot & b_{22}^{(2)} & & b_{2m}^{(2)} \\ \cdot & \cdot & & \vdots \\ \cdot & \cdot & & \vdots \\ \cdot & \cdot & & \vdots \\ \cdot & \cdot & & b_{mm}^{(m)} \end{bmatrix} \equiv \mathbf{U}.$$

Hence

$$\mathbf{B} = \boldsymbol{\Gamma}_1^{-1}\boldsymbol{\Gamma}_2^{-1} \cdots \boldsymbol{\Gamma}_{m-1}^{-1}\mathbf{U}.$$

Let us define $\mathbf{L}_k \equiv \boldsymbol{\Gamma}_k^{-1}$, $k = 1, \ldots, m-1$. From Sec. 5.1-1, we know that each elementary matrix $\mathbf{L}_k$ is obtained from (5.2-4), or the expression preceding it, simply by reversing the sign of the subdiagonal elements in column k. Using (5.1-10), we see also that $\mathbf{L}_1, \ldots, \mathbf{L}_{m-1}$ is the product form of the following unit lower triangular matrix,

$$\mathbf{L} = \mathbf{L}_1\mathbf{L}_2 \cdots \mathbf{L}_{m-1} = \begin{bmatrix} 1 & \cdot & \cdot & & \cdot \\ b_{21}^{(1)}/b_{11}^{(1)} & 1 & \cdot & & \cdot \\ b_{31}^{(1)}/b_{11}^{(1)} & b_{32}^{(2)}/b_{22}^{(2)} & 1 & & \cdot \\ \vdots & \vdots & \vdots & & \cdot \\ & & & & \cdot \\ b_{m1}^{(1)}/b_{11}^{(1)} & b_{m2}^{(2)}/b_{22}^{(2)} & \vdots & & 1 \end{bmatrix} \tag{5.2-6}$$

Combining (5.2.5) and (5.2-6) gives

$$\mathbf{B} = \mathbf{LU} \tag{5.2-7}$$

as we had set out to prove. Note also that (5.2-7) can be put in product form, using (5.1-11), namely,

$$\mathbf{B} = \mathbf{L}_1\mathbf{L}_2 \cdots \mathbf{L}_{m-1}\mathbf{U}_m\mathbf{U}_{m-1} \cdots \mathbf{U}_1. \tag{5.2-8}$$

It remains to establish uniqueness of the LU factorization. Suppose $\mathbf{B} = \mathbf{L}_A\mathbf{U}_A = \mathbf{L}_B\mathbf{U}_B$, where $\mathbf{L}_A$ and $\mathbf{L}_B$ are unit lower triangular matrices and $\mathbf{U}_A$ and $\mathbf{U}_B$ are upper triangular matrices. Clearly all these matrices are nonsingular. Therefore

$$\mathbf{L}_B^{-1}\mathbf{L}_A = \mathbf{U}_B\mathbf{U}_A^{-1}. \tag{5.2-9}$$

The left-hand side is a unit lower triangular matrix and the right-hand side is an upper triangular matrix. Therefore each side of (5.2-9) must equal the identity matrix. Then from (5.2-9), $\mathbf{L}_A = \mathbf{L}_B$ and $\mathbf{U}_A = \mathbf{U}_B$. Hence the LU factorization is unique and the theorem is proved. ■

We can summarize the recurrence relations (5.2-3) for carrying out the LU factorization as follows. For $k = 1, 2, \ldots, m-1$:

$$b_{ij}^{(k+1)} = b_{ij}^{(k)} \qquad i \le k \qquad 1 \le j \le m$$

$$b_{ij}^{(k+1)} = b_{ij}^{(k)} - \frac{b_{ik}^{(k)} b_{kj}^{(k)}}{b_{kk}^{(k)}} \qquad k < i \le m \qquad k \le j \le m. \tag{5.2-10}$$

Other elements of $\mathbf{B}^{(k+1)}$ equal zero.

It is also useful to note that the LU factorization can be somewhat differently organized. What we have just described is a *row-oriented* procedure, with rows of $\mathbf{U}$ being determined in succession. We could also proceed in a *column-oriented* manner as follows. For $k = 1, 2, \ldots, m$ and with the convention $\boldsymbol{\Gamma}_0 = \boldsymbol{\Gamma}_m = \mathbf{I}$:

1. Select column k of $\mathbf{B}$.
2. Form column k of $\mathbf{B}^{(k)}$ as

   $$\mathbf{b}_k^{(k)} = \boldsymbol{\Gamma}_{k-1} \cdots \boldsymbol{\Gamma}_0 \mathbf{b}_k,$$

   or by solving

   $$(\mathbf{L}_0 \cdots \mathbf{L}_{k-1})\mathbf{b}_k^{(k)} = \mathbf{b}_k.$$

   (See also Sec. 5.7.) This gives the kth column of $\mathbf{U}$ by setting

   $$u_{ik} = (\mathbf{b}_k^{(k)})_i \qquad i \le k \quad \text{and} \quad u_{ik} = 0 \qquad i > k.$$

3. Form $\boldsymbol{\Gamma}_k$ or $\mathbf{L}_k$ from the subdiagonal elements of $\mathbf{b}_k^{(k)}$, using (5.2-4).

Since LU factorization of $\mathbf{B}$ is a reformulation of Gaussian elimination, we shall use the terminology of the latter process to call the diagonal element $b_{kk}^{(k)}$ the pivot element or *pivot*. Also, given the LU factors of $\mathbf{B}$, it is a simple matter to solve the system $\mathbf{Bx} = \mathbf{d}$ as $\mathbf{L}(\mathbf{Ux}) = \mathbf{d}$, namely, to solve in turn, $\mathbf{Ly} = \mathbf{d}$ for $\mathbf{y}$ and $\mathbf{Ux} = \mathbf{y}$ for $\mathbf{x}$. We shall use the term *solving by back substitution* for both these cases. (The former is often called *forward elimination,* because one solves first for y_1, next for y_2, and so on.) More will be said on this topic in Sec. 5.7.

Exercise 5.2-1. Give the LU factorization of a general (i.e., not necessarily unit) lower triangular matrix.

Exercise 5.2-2. Show that the number of floating-point multiplications or divisions (flops) in (5.2-10) is $m(m^2 - 1)/3$.

5.3. The LU Factorization with Row and/or Column Interchanges

As we noted at the beginning of this chapter, Gaussian elimination requires row (or column) interchanges whenever the current candidate for pivot element on the diagonal is zero. Equivalently, LU factorization requires occasional pre (or post) multiplications by suitable elementary permutation matrices. It relies on the following result, which holds without placing any restrictions on the matrix **B**.

Theorem 5.3-1. Given any $m \times m$ matrix **B**, there exist permutation matrices **P** and **Q**, a unit lower triangular matrix **L** and an upper triangular matrix **U** such that

$$\mathbf{PBQ} = \mathbf{LU}. \tag{5.3-1}$$

Proof. Again, we give a constructive proof of this result.

Suppose at the start of the kth step of the LU factorization described in the previous section, the candidate for pivot, namely, $b_{kk}^{(k)}$ is zero. Suppose further that $b_{k'k''}^{(k)} \neq 0$, for some particular pair of indices k' and k'', $k \leq k' \leq m$, $k \leq k'' \leq m$. We can bring element (k', k'') to position (k, k) on the diagonal using the elementary permutation matrices $\mathbf{P}_{k,k'}$ and $\mathbf{P}_{k,k''}$ to give

$$\tilde{\mathbf{B}}^{(k)} = \mathbf{P}_{k,k'}\mathbf{B}^{(k)}\mathbf{P}_{k,k''} \qquad \text{with} \qquad \tilde{b}_{kk}^{(k)} \neq 0.$$

In effect, we make element (k', k'') the candidate for pivot element at step k. The factorization can then be continued in the usual way to give

$$\mathbf{B}^{(k+1)} = \tilde{\mathbf{\Gamma}}_k\tilde{\mathbf{B}}^{(k)} = \tilde{\mathbf{\Gamma}}_k\mathbf{P}_{k,k'}\mathbf{B}^{(k)}\mathbf{P}_{k,k''},$$

where $\tilde{\mathbf{\Gamma}}_k$ is defined in an analogous way to (5.2-4), with elements of $\tilde{\mathbf{B}}^{(k)}$ replacing those of $\mathbf{B}^{(k)}$. (As a general rule of thumb regarding notation, we employ a "tilde" or "wiggle" when there is a permutation (wiggling) of the matrix.)

Since permutations may be carried out at any stage, we have

$$\mathbf{B}^{(k+1)} = \tilde{\mathbf{\Gamma}}_k\mathbf{P}_{k,k'}\tilde{\mathbf{\Gamma}}_{k-1}\mathbf{P}_{k-1,(k-1)'} \cdots \tilde{\mathbf{\Gamma}}_2\mathbf{P}_{2,2'}\tilde{\mathbf{\Gamma}}_1\mathbf{P}_{1,1'}\mathbf{B}\mathbf{P}_{1,1''} \cdots \mathbf{P}_{k,k''} \tag{5.3-2}$$

and at the end of the process, we have an upper triangular matrix **U** given by

$$\mathbf{B}^{(m)} = \tilde{\mathbf{\Gamma}}_{m-1}\mathbf{P}_{m-1,(m-1)'}\tilde{\mathbf{\Gamma}}_{m-2}\mathbf{P}_{m-2,(m-2)'} \cdots \tilde{\mathbf{\Gamma}}_1\mathbf{P}_{1,1'}\mathbf{B}\mathbf{P}_{1,1''} \cdots \mathbf{P}_{m-1,(m-1)''} \equiv \mathbf{U}. \tag{5.3-3}$$

Let us define permutation matrices $\boldsymbol{P}_k$ recursively by

$$\boldsymbol{P}_m = \mathbf{I}$$

and for $k = m-1, m-2, \ldots, 1$

$$\boldsymbol{P}_k = \mathbf{P}_{k,k'}\boldsymbol{P}_{k+1}. \tag{5.3-4}$$

Recall that $P_k P_k^T = \mathbf{I}$ and that we can therefore write (5.3-3) as

$$P_m^T \tilde{\mathbf{\Gamma}}_{m-1}(P_m P_m^T)\mathbf{P}_{m-1,(m-1)'}\tilde{\mathbf{\Gamma}}_{m-2}(P_{m-1}P_{m-1}^T)\mathbf{P}_{m-2,(m-2)'}\tilde{\mathbf{\Gamma}}_{m-3} \cdots \tilde{\mathbf{\Gamma}}_1(P_2 P_2^T)\mathbf{P}_{1,1'}\mathbf{B}\mathbf{P}_{1,1''}\cdots \mathbf{P}_{m-1,(m-1)''} = \mathbf{U}.$$

Now using (5.3-4) and regrouping terms,

$$(P_m^T \tilde{\mathbf{\Gamma}}_{m-1} P_m)(P_{m-1}^T \tilde{\mathbf{\Gamma}}_{m-2} P_{m-1})(P_{m-2}^T \tilde{\mathbf{\Gamma}}_{m-3} P_{m-2}) \cdots (P_2^T \tilde{\mathbf{\Gamma}}_1 P_2) P_1^T \mathbf{B}\mathbf{P}_{1,1''}\cdots \mathbf{P}_{m-1,(m-1)''} = \mathbf{U}. \quad (5.3\text{-}5)$$

Let us define

$$\mathbf{Q} \equiv \mathbf{P}_{1,1''}\mathbf{P}_{2,2''}\cdots \mathbf{P}_{m-1,(m-1)''} \quad (5.3\text{-}6)$$

$$\mathbf{P} \equiv P_1^T = \mathbf{P}_{m-1,(m-1)'}\cdots \mathbf{P}_{2,2'}\mathbf{P}_{1,1'} \quad (5.3\text{-}7)$$

and

$$\mathbf{\Gamma}_k \equiv P_{k+1}^T \tilde{\mathbf{\Gamma}}_k P_{k+1} \qquad k = 1, \ldots, m-1. \quad (5.3\text{-}8)$$

It can be easily verified that $\mathbf{\Gamma}_k$ is also a *unit lower triangular matrix*. Thus

$$\mathbf{\Gamma}_{m-1}\mathbf{\Gamma}_{m-2}\cdots \mathbf{\Gamma}_1 \mathbf{PBQ} = \mathbf{U}. \quad (5.3\text{-}9)$$

As before, let $\mathbf{L}_k = \mathbf{\Gamma}_k^{-1}$, $k = 1, \ldots, m-1$ and $\mathbf{L} \equiv \mathbf{L}_1 \cdots \mathbf{L}_{m-1}$. The final factorization then takes the desired form,

$$\tilde{\mathbf{B}} \equiv \mathbf{PBQ} = \mathbf{LU}. \quad ■ \quad (5.3\text{-}10)$$

Two clarifications concerning the above proof will be helpful. First, we have assumed when $b_{kk}^{(k)} = 0$ that there is a nonzero subdiagonal element in column k of $\mathbf{B}^{(k)}$ to be eliminated. Otherwise we could proceed directly to step $k+1$ with $\tilde{\mathbf{\Gamma}}_k \equiv \mathbf{I}$. Second, we have assumed that the process continues to the final column m. If it terminates prematurely with some index, say $\bar{m}$ and $b_{ij}^{(\bar{m})} = 0$, $\bar{m} \le i \le m$, $\bar{m} \le j \le m$, then replace m by $\bar{m}$ from equation (5.3-3) to the end of the proof. Of course, in this case, we are dealing with a singular matrix $\mathbf{B}$.

It is also useful to note that column permutations provide greater flexibility but are not essential to the above result and proof, i.e., we can always set $\mathbf{P}_{k,k''} = \mathbf{I}$ and hence $\mathbf{Q} = \mathbf{I}$, giving the factorization $\mathbf{PB} = \mathbf{LU}$. However, column interchanges are of great practical importance as we shall soon see, and we therefore prefer to develop the above result (5.3-10) in its full generality.

Exercise 5.3-1. Develop a column-oriented version of the LU factorization with interchanges analogous to the procedure at the end of Sec. 5.2.

Exercise 5.3-2. Show that the factorization (5.3-10) can be expressed in the form $\mathbf{B} = \tilde{\mathbf{L}}\tilde{\mathbf{U}}$ where $\tilde{\mathbf{L}}$ is a symmetrically permuted (identical permutation of rows and columns) unit lower triangular matrix and $\tilde{\mathbf{U}}$ is a permuted upper triangular matrix.

Given any *nonsingular* matrix **B**, there is always a way to permute its rows and columns to give a new matrix $\tilde{\mathbf{B}} = \mathbf{PBQ}$ (indeed, as just mentioned, permutation of rows is all that is necessary) so that LU factorization with pivoting down the diagonal of $\tilde{\mathbf{B}}$ will not break down. With **P** and **Q** fixed, this will give *unique* factors **L** and **U**, by Theorem 5.2-1. There is, of course, much flexibility in the choice of **P** and **Q**, i.e., there is nothing unique about them, but once **P** and **Q** and hence $\tilde{\mathbf{B}}$ are decided on, then the resulting factors are unique. As far as implementation is concerned, once (k', k'') are chosen there is no need to explicitly carry out the permutation to bring the pivot to the (k, k) position. Instead, we could do it *implicitly* by eliminating the nonzero elements in column k'' in the rows that have, so far, not contained a pivot element, by adding to them suitable multiples of row k'. We shall call this *pivoting in place* and more will be said on it later, in particular, in the final section of this chapter.

We now return to ways of choosing **P** and **Q** in order to ensure numerical stability and preserve sparseness.

5.4. Pivoting Strategies to Ensure Stability of the LU Factorization

Zero pivots are disastrous for the LU factorization—it simply cannot proceed. Near zero pivots are equally disastrous from a numerical standpoint. Consider the following example.

Example 5.4-1. Forsythe and Moler [1967]. In three-decimal floating-point arithmetic, let us compute the LU factors of the following 2×2 matrix,

$$\mathbf{B} = \begin{bmatrix} 0.0001 & 1. \\ 1. & 1. \end{bmatrix}$$

and let us use them to solve the system $\mathbf{Bx} = \begin{bmatrix} 1. \\ 2. \end{bmatrix}$. The true solution is

$$x_1 = 10{,}000/9999(\approx 1.00010) \qquad x_2 = 9998/9999(\approx 0.99990).$$

The $(1, 1)$ element is close to zero, but nevertheless let us pivot on it to obtain the LU factors

$$\mathbf{L} = \begin{bmatrix} 1. & 0. \\ 10{,}000. & 1. \end{bmatrix} \qquad \mathbf{U} = \begin{bmatrix} 0.0001 & 1. \\ 0. & -10{,}000. \end{bmatrix}$$

If we now solve $\mathbf{Ly} = \begin{bmatrix} 1. \\ 2. \end{bmatrix}$ in three-decimal floating-point arithmetic, we obtain

$$\mathbf{y} = \begin{bmatrix} 1. \\ -10{,}000. \end{bmatrix}.$$

Then solving $\mathbf{Ux} = \mathbf{y}$, we obtain the completely erroneous solution $x_1 = 0.$, $x_2 = 1$.

As the foregoing example illustrates, the problem is that near zero pivot elements lead to *uncontrolled growth* in the elements of $\mathbf{L}$ and $\mathbf{U}$. This, in turn, results in large numerical errors. The solution is to choose pivot elements suitably (hence suitable permutations in the LU factorization of Sec. 5.3) so as to prevent such element growth. Two common strategies are as follows.

Complete Pivoting. At the kth step of the LU factorization with permutations (Sec. 5.3), choose as the pivot the element in rows k through m and columns k through m of $\mathbf{B}^{(k)}$ that is *largest in absolute value*. The pivot element is therefore given by (k', k''), where

$$|b_{k'k''}^{(k)}| \equiv \max_{i,j} [|b_{ij}^{(k)}| \mid k \leq i \leq m \qquad k \leq j \leq m]. \tag{5.4-1}$$

Partial Pivoting. At the kth step of the LU factorization with permutations (Sec. 5.3), choose as the pivot the element in column k and rows k through m that is *largest in absolute value*. The pivot element is therefore given by (k', k), where

$$|b_{k'k}^{(k)}| = \max_{i} [|b_{ik}^{(k)}| \mid k \leq i \leq m]. \tag{5.4-2}$$

Note with partial pivoting, we exercise the option mentioned in Sec. 5.3, namely, $\mathbf{Q} \equiv \mathbf{I}$.

From (5.2-4), or the alternative expression preceding it, we see that the foregoing strategies ensure that the elements of $\mathbf{L}$ satisfy $|l_{ij}| \leq 1$ for all i, j, and from (5.2-10) we see also that an element of $\mathbf{B}^{(k+1)}$ can exceed the corresponding element of $\mathbf{B}^{(k)}$ by a factor of *at most* 2. Thus if $|b_{ij}^{(1)}| \leq 1$ for all i, j then $|u_{ij}| \leq 2^m$ for all i, j. This is a *worst-case bound* and it is strict only on very artificial examples. In practice, element growth in $\mathbf{U}$, by a factor of at most 16, is the norm. (See Wilkinson [1965].) Although complete pivoting is more successful in keeping down element growth, it requires more computational effort to search for the largest element (in absolute value) in $\mathbf{B}^{(k)}$ in the appropriate submatrix, rather than the largest subdiagonal element in the pivot column. A detailed error analysis and extensive numerical experience show that partial pivoting is entirely satisfactory in practice and it is obviously easier to implement.

The numerical stability of LU factorization with partial pivoting is expressed by the following result, which we quote without proof.

Theorem 5.4-1. Consider the LU factorization of a given matrix $\mathbf{B}$ with partial pivoting (5.4-2). Let the factorization in exact arithmetic be

$$\tilde{\mathbf{B}} = \mathbf{PB} = \mathbf{LU},$$

and let $\mathbf{L}_c$ and $\mathbf{U}_c$ be the *computed* factors of $\tilde{\mathbf{B}}$ using finite precision floating-point arithmetic, with unit round-off error given by *ulp* (see Chapter 4, (4.2-2)). Then

$$(\tilde{\mathbf{B}} + \tilde{\mathbf{E}}) = \mathbf{L}_c\mathbf{U}_c \tag{5.4-3}$$

where

$$\|\tilde{\mathbf{E}}\|_\infty \leq m^2\left[\max_{i,j,k} |b_{ij}^{(k)}|\right] ulp. \tag{5.4-4}$$

($\|\tilde{\mathbf{E}}\|_\infty$ denotes the "infinity" norm, namely, $\max_{1\leq i\leq m} \sum_{j=1}^m |\tilde{e}_{ij}|$, where $\tilde{e}_{ij}$ denotes the (i, j)th element of $\tilde{\mathbf{E}}$.)

Proof. This may be found in Forsythe and Moler [1967], p. 103. ■

The partial pivoting strategy is equivalent to permuting the rows of **B** by the permutation matrix **P** and then pivoting down the diagonal of $\tilde{\mathbf{B}}$. The LU factorization of $\tilde{\mathbf{B}}$ is *stable,* because the *computed* factors $\mathbf{L}_c$ and $\mathbf{U}_c$ are the *true* factors of a *slight perturbation* of $\tilde{\mathbf{B}}$.

Exercise 5.4-1. Consider LU factorization with partial *column* pivoting, namely, pick the pivot element (k, k'') at the kth step by

$$|b_{kk''}^{(k)}| = \max_j \left[|b_{kj}^{(k)}| \mid k \leq j \leq m\right]. \tag{5.4-5}$$

Show that this will control element growth in **U** but *not* in **L**. Is the resulting factorization stable?

5.5. Pivoting Strategies to Control Fill-in during LU Factorization

So far we have neglected sparsity in the matrix $\mathbf{B}^{(k)}$. Taking it into account is important for two reasons. Zero subdiagonal elements in $\mathbf{B}^{(k)}$ in the relevant column do not need to be eliminated and this reduces the operation count. (Exercise 5.2-2 now only provides an upper bound.) Furthermore, when $\mathbf{B}^{(k)}$ is sparse, the factors $\boldsymbol{\Gamma}_k$ (hence $\mathbf{L}_k$ and **L**) and **U** will also have many zero elements. Use of a packed storage scheme to hold these factors will result in significant savings in storage.

We turn, therefore, to pivoting strategies that seek to preserve as much sparsity in $\mathbf{B}^{(k)}$ as possible. Let us first consider an example.

Example 5.5-1. Suppose that **B** $(=\mathbf{B}^{(1)})$ is the following matrix, whose nonzeros are denoted by x.

$$\mathbf{B} = \begin{bmatrix} x & x & x & x & x \\ x & x & 0 & 0 & 0 \\ x & 0 & x & 0 & 0 \\ x & 0 & 0 & x & 0 \\ x & 0 & 0 & 0 & x \end{bmatrix} \tag{5.5-1}$$

If we pivot on the element (1, 1) at the first step of the LU factorization of **B**, thereby eliminating the elements in positions (2, 1), . . . , (5, 1), we obtain the matrix

$$\mathbf{B}^{(2)} = \mathbf{\Gamma}_1\mathbf{B}^{(1)} = \begin{bmatrix} x & x & x & x & x \\ 0 & x & x & x & x \\ 0 & x & x & x & x \\ 0 & x & x & x & x \\ 0 & x & x & x & x \end{bmatrix}$$

We observe that all zero elements have filled in. The operation count (floating-point multiplications and divisions) is therefore given by Exercise 5.2-2 with $m = 5$, when (5.5-1) is factored by pivoting down the diagonal. (We assume, of course, that the factorization does not break down due to a zero pivot.) Furthermore, the LU factors *will be completely dense,* i.e., each will contain $m(m-1)/2$ nonzero off-diagonal elements, with $m = 5$.

Suppose, on the other hand, we interchange rows 1 and 5 and columns 1 and 5 at the first step of the factorization. We do this even though element (1, 1) could be used as a pivot without causing instability. Thus we define $\tilde{\mathbf{B}}^{(1)} = \mathbf{P}_{1,5}\mathbf{B}^{(1)}\mathbf{P}_{1,5}$ and

$$\tilde{\mathbf{B}}^{(1)} = \begin{bmatrix} x & 0 & 0 & 0 & x \\ 0 & x & 0 & 0 & x \\ 0 & 0 & x & 0 & x \\ 0 & 0 & 0 & x & x \\ x & x & x & x & x \end{bmatrix}$$

If we now pivot on the (1, 1) element of $\tilde{\mathbf{B}}^{(1)}$, we obtain

$$\mathbf{B}^{(2)} = \begin{bmatrix} x & 0 & 0 & 0 & x \\ 0 & x & 0 & 0 & x \\ 0 & 0 & x & 0 & x \\ 0 & 0 & 0 & x & x \\ 0 & x & x & x & x \end{bmatrix}$$

No fill-in whatsoever has occurred. We can continue in this fashion and observe that at each step *only one element* must be eliminated. The computational cost is therefore much smaller than that of factorizing (5.5-1). Also the LU factors will be much more sparse than those of (5.5-1). Each of the factors **L** and **U** will now only contain $(m-1)$ off-diagonal elements with $m = 5$.

As the foregoing example illustrates, the choice of pivot plays a crucial role in preserving sparsity of the LU factors. Let us now consider strategies for choosing the permutation matrices at each step (or equivalently the pivot element) that seek to reduce, as far as possible, the creation of new nonzero elements during the LU factorization. It is here that interchange of columns, namely, the matrix **Q** of (5.3-1), becomes as significant as interchange of rows.

5.5-1. Dynamic Strategies

Suppose that we are at the kth step of LU factorization and for convenience of exposition, let us assume that we have pivoted down the diagonal so far, or alternatively have explicitly permuted rows and columns at earlier steps of the factorization, so that we are dealing with the matrix $\mathbf{B}^{(k)}$ of (5.2-2).

Suppose now that element (k', k'') is chosen as the candidate for pivot, where k' and k'' are indices in the range $k \le k' \le m$ and $k \le k'' \le m$. Let us pivot in place on element (k', k''); see the end of Sec. 5.3 for an explanation of this term. We would then be adding multiples of row k' to all rows i such that $k \le i \le m$, $i \ne k'$ and $b_{ik''}^{(k)} \ne 0$. For each such row i, nonzero elements in row k' would introduce nonzero elements in the corresponding columns of row i, which may previously have contained zeros in these positions, i.e., there is a potential fill-in. We ignore the rare event that cancellation of elements could occur, leading to new zero elements in row i.

Let $r_{k'}^{(k)}$ be the number of nonzero elements in row k', where we may observe that these can only occur in columns $k, k+1, \ldots, m$ because of eliminations in previous columns. Let $c_{k''}^{(k)}$ be the number of nonzero elements in column k'', where we *restrict counting* to only rows k, $k+1, \ldots, m$. Then an upper bound on the fill-in, when we pivot on (k', k''), is

$$f_{k',k''}^{(k)} = (r_{k'}^{(k)} - 1)(c_{k''}^{(k)} - 1). \tag{5.5-2}$$

The *Markowitz strategy* chooses the particular element (k', k'') in the permissible range of indices, $k \le k' \le m$, $k \le k'' \le m$ so that $f_{k',k''}^{(k)}$ is *least* and for which a pivot is possible, i.e., $b_{k'k''}^{(k)} \ne 0$. Formally, it chooses the pair given by

$$\underset{(k',k'')}{\operatorname{argmin}} \left[(r_{k'}^{(k)} - 1)(c_{k''}^{(k)} - 1) \mid k \le k' \le m \qquad k \le k'' \le m \qquad b_{k'k''}^{(k)} \ne 0 \right]. \tag{5.5-3}$$

This element is used as the pivot at step k; then the row and column counts are revised and the LU factorization proceeds to the next step. The Markowitz strategy is a *greedy* or myopic strategy because at each step it does what is locally best. It works surprisingly well in practice. Let us illustrate this strategy for the following example.

Example 5.5-2. Lower triangular matrix with a bump. Consider the matrix of Figure 5.1. The nonzero elements are denoted by x. **A** and **F**

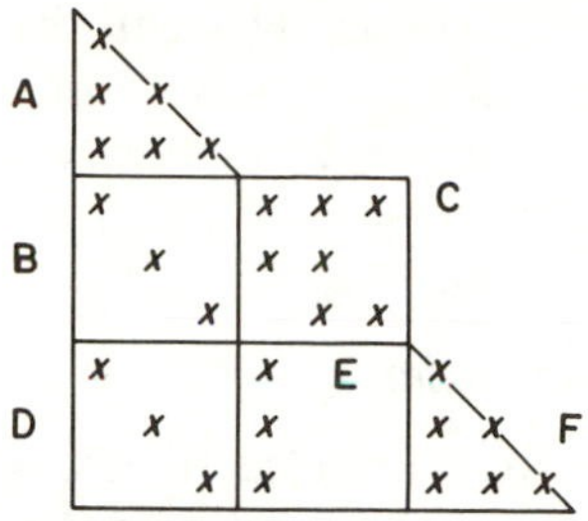

Figure 5.1 Lower triangular matrix with bump

are lower triangular matrices. Suppose we were to factorize this matrix with pivoting down the diagonal. The first three pivots down the diagonal elements of **A** are essentially trivial (recall Exercise 5.2-1), create no fill-in, and the associated elementary matrices $\mathbf{\Gamma}_1$, $\mathbf{\Gamma}_2$, and $\mathbf{\Gamma}_3$ have the same zero structure as columns 1, 2, and 3 of the matrix of Figure 5.1. However, when the factorization continues into the **C** submatrix (the so-called *bump*) there will be fill-in, in particular in the elements of the **E** submatrix. As we pivot down the diagonal of **C**, **E** will fill in completely.

Suppose however that we reorder the matrix to give the matrix of Figure 5.2. This is obtained from the matrix of Figure 5.1 by first interchanging $[\mathbf{B}\ \ \mathbf{C}\ \ \mathbf{0}]$ and $[\mathbf{D}\ \ \mathbf{E}\ \ \mathbf{F}]$, where $\mathbf{0}$ denotes the zero matrix of appropriate dimension, then interchanging $\begin{bmatrix}\mathbf{E}\\ \mathbf{C}\end{bmatrix}$ and $\begin{bmatrix}\mathbf{F}\\ \mathbf{0}\end{bmatrix}$, reversing the rows in the $[\mathbf{D}\ \ \mathbf{F}\ \ \mathbf{E}]$ portion giving $[\tilde{\mathbf{D}}\ \ \hat{\mathbf{F}}\ \ \tilde{\mathbf{E}}]$ and finally reversing the columns in the $\hat{\mathbf{F}}$ portion to give $\tilde{\mathbf{F}}$. Now we see that the first six pivots down the diagonal are essentially trivial, *no* fill-in occurs in $\tilde{\mathbf{E}}$ and one need *only* further factorize **C**.

The remainder of the example is left to the following exercise.

Exercise 5.5-1. Show that the Markowitz strategy (5.5-3) applied to the matrix of Figure 5.1 accomplishes precisely the above reordering, leading to the matrix of Figure 5.2 in the foregoing example.

Because the decision about pivot choice at step k depends on $\mathbf{B}^{(\kappa)}$, i.e., the current state of the factorization, the Markowitz strategy is called a

Figure 5.2 Reordered matrix

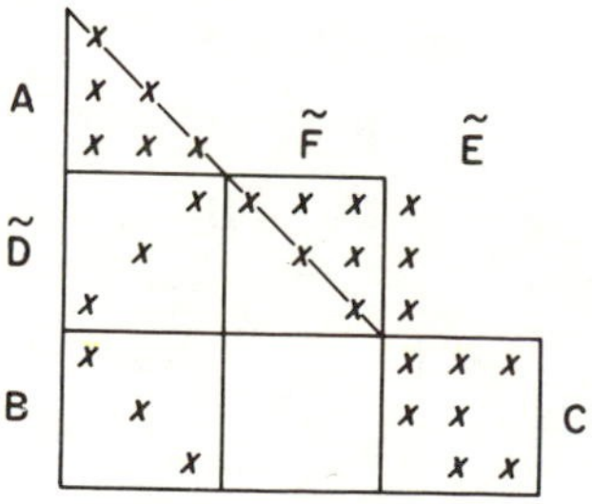

dynamic strategy. Variants on it are given in the references in the notes at the end of this chapter. We turn now to a priori or *static* strategies, which determine a pivot ordering from the zero structure of the *initial* matrix **B**.

5.5-2. Static (a priori) Strategies

When portions of the matrix **B** are already triangular, the task of finding the LU factors is made easier. Example 5.5-2 has illustrated this. Note that in the leading portion of the matrix of Figure 5.1 in this example, the first row consists of a singleton, i.e., the row count is 1. When the first column, which contains this (row) singleton is ignored, the same comment applies to the second row of the matrix and so on. Similar remarks can be made with regard to the column singletons and column counts in the trailing triangular portion of the matrix.

Exercise 5.5-2. Give a simple procedure based on the foregoing remarks, for permuting a matrix **B** into a leading and a trailing triangular part separated by a "bump," as illustrated by Figure 5.1.

The "bump" may itself have further structure, i.e., it may consist of several smaller bumps. It is therefore natural to seek a reordering of the rows and columns of **B** that would give a block lower (or upper) triangular matrix of the form illustrated by Figure 5.3.

Figure 5.3 Block lower triangular matrix

$$\begin{array}{lllllll}
x \\
x & x & x & x \\
x & x & x & x \\
x & x & x & x \\
x & x & x & x & x & x \\
x & x & x & x & x & x \\
x & x & x & x & x & x & x
\end{array}$$

We describe here an elegant, yet simple, algorithm due to Sargent and Westerberg [1964], for achieving such a reordering, using symmetric permutation of rows and columns of **B**. A symmetric permutation of **B** reorders rows and columns in the same way. Other more sophisticated versions of the algorithm are surveyed in Chapter 10. Specifically, we seek a permutation matrix **P** such that

$$\mathbf{PBP}^T = \begin{bmatrix} \mathbf{B}_{11} & \mathbf{0} & \mathbf{0} \\ \mathbf{B}_{21} & \mathbf{B}_{22} & . \\ \vdots & \vdots & . \\ \vdots & \vdots & \mathbf{0} \\ \mathbf{B}_{M1} & \mathbf{B}_{M2} & \mathbf{B}_{MM} \end{bmatrix} \tag{5.5-4}$$

where the blocks $\mathbf{B}_{ii}$ are square and cannot themselves be further symmetrically permuted to block lower triangular form. During the course of the discussion we also introduce the reader to an interesting and widely used technique for describing a sparse matrix in terms of a *directed graph.*

We associate with the $m \times m$ matrix $\mathbf{B}$, a *directed graph* that consists of a set of *nodes* $1, 2, \ldots, m$ and a set of *edges,* each of which is an ordered pair of nodes (i, j) corresponding to a nonzero off-diagonal element b_{ij} of $\mathbf{B}$. (Note that (i, j) and (j, i) are different edges and the latter would only be defined if $b_{ji} \neq 0$.) A symmetric permutation of $\mathbf{B}$ corresponds to a relabeling of the nodes of the directed graph of $\mathbf{B}$, as illustrated in Figure 5.4. A *path* is a sequence of edges $(i, j), (j, k), (k, l), \ldots$ and a *cycle* is a path from a node i that leads back to itself, i.e., the final edge in the cycle is of the form (l, i). A *subgraph* consists of a subset of nodes and all edges that are pairs of nodes belong to the subset. A subgraph is *strongly*

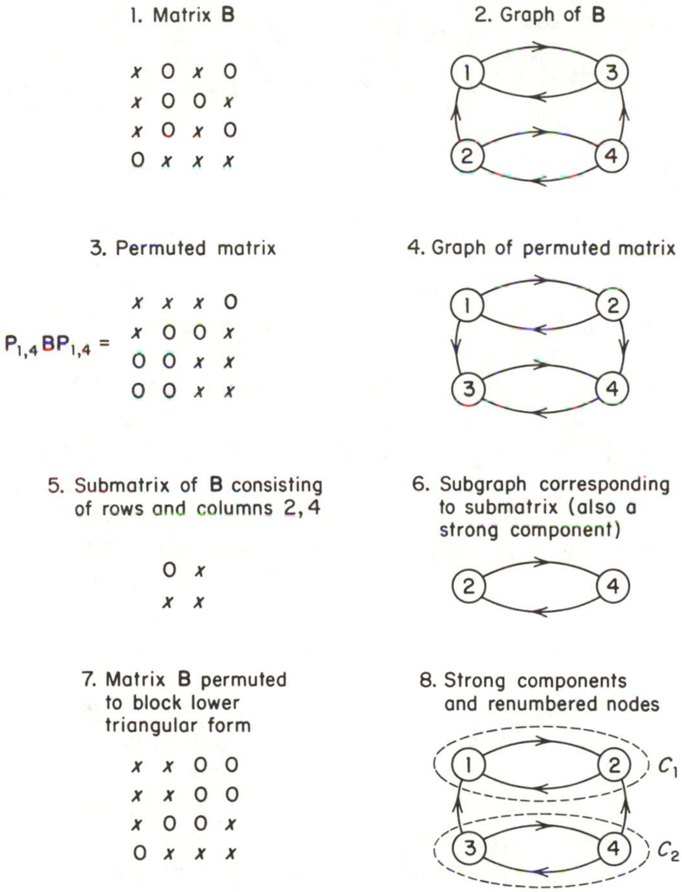

Figure 5.4 Illustration of directed graph representation

connected if there is a path from any of its nodes to any other. A subgraph is a *strong component* if it is strongly connected and cannot be enlarged further without losing this property. All these definitions are illustrated in Fig. 5.4.

The key idea behind the algorithm is to *identify and order the strong components* and it is based on the following easily established observations:

1. Each node belongs to at most one strong component.
2. Nodes in a cycle lie within the same strong component.
3. There is at least one strong component, say C_1, with the property that no constituent node has an edge that leads to a node *not* in C_1, i.e., there is no path out of C_1. Similarly, if the subgraph C_1 and the edges leading to it are removed from the graph of **B**, then the remaining subgraph will have the same property, i.e., there will be a strong component, say C_2, with no path leading out of it, and so on.
4. If we reorder the strong components as $C_1, C_2, \ldots$ and relabel their nodes so that nodes of C_1 precede those of C_2 and so on, then the associated matrix (which is a symmetric permutation of **B**) is block lower triangular.

Exercise 5.5-3. Establish properties (1) through (4) above.

The following description and flowchart (Figure 5.5) of the algorithm is due to Duff and Reid [1978]. The algorithm traces paths in the graph associated with **B** and in modifications of the graph. Starting from any node, a path is followed through the graph until a cycle is found (identified by encountering the same node twice) or a node is encountered with no edges leaving it. All nodes in a cycle must belong in the same strong component and when one is found, the graph is modified by collapsing all the nodes in the cycle into a single "composite node." Edges between nodes that constitute a composite node are ignored and edges entering or leaving constituent nodes from elsewhere in the graph are regarded as entering or leaving the new composite node. If a node (or composite node) is encountered with no edges leaving it, then it must correspond to a strong component of the original graph having no edge to any other strong component. This, therefore, corresponds to the first block of the required triangular form. The composite node and its associated edges are then deleted from the graph to leave a graph corresponding to the permuted matrix **B** with its first *block* row and column removed. The algorithm continues from the last node on the remaining path, or starts from another node if the path becomes empty. In this way the blocks of the required form are obtained successively.

Elaborate strategies have been devised for further a priori ordering of columns and rows *within* a bump (or block) of the block lower triangular

Figure 5.5 Flowchart of Sargent and Westerberg's algorithm (Duff and Reid 1978)

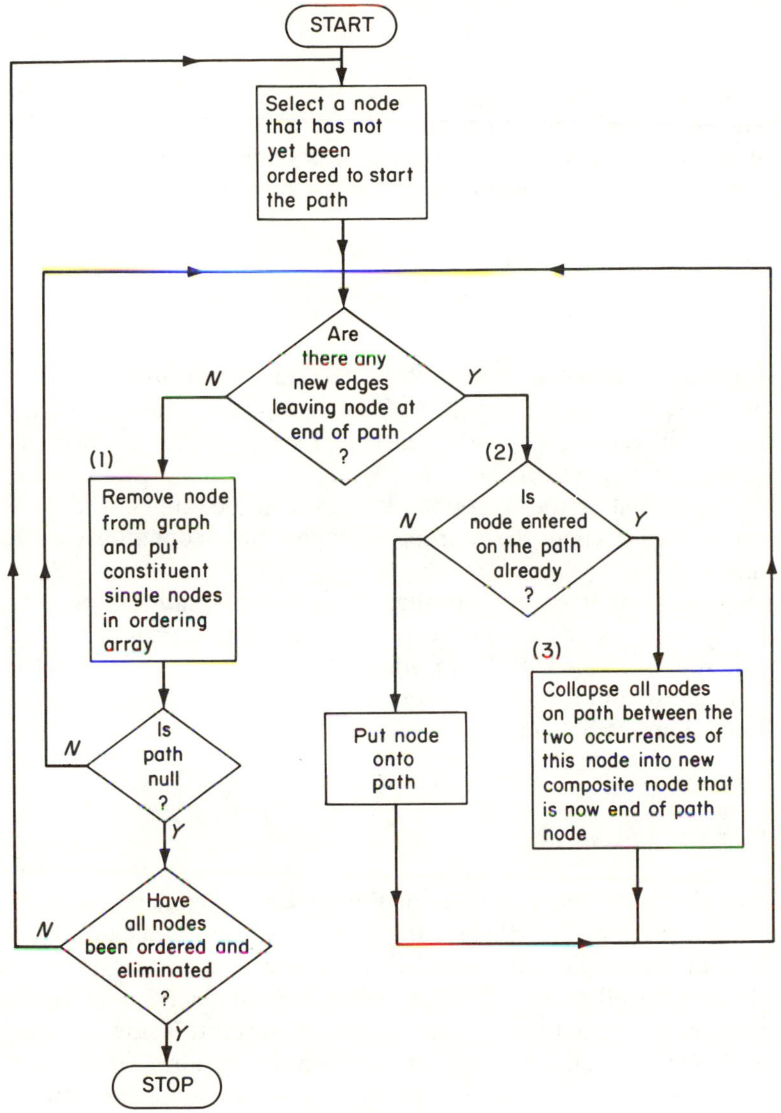

form, so as to further reduce fill-in during the LU factorization. It would take us too far afield to discuss these extensively here, but they are briefly surveyed in Chapter 10. An example is the *merit ordering* of Tomlin [1972a]. Assume that **B** has been ordered into leading and trailing triangular portions separated by a bump, see Exercise 5.5-2. Let us now confine attention to the bump, say $\bar{\mathbf{B}}$. For each row of $\bar{\mathbf{B}}$ compute the row count $\bar{r}_i$ as the number of nonzeros in row i (restricting the count, of course, just to columns of the bump). For each column j of $\bar{\mathbf{B}}$ compute a *merit count* as follows:

$$M_j = \sum_{[i \mid \bar{b}_{ij} \neq 0]} (\bar{r}_i - 1). \tag{5.5-5}$$

Then arrange the columns of $\bar{\mathbf{B}}$ in increasing order of M_j. Suppose that $\bar{c}_j$ were the associated column count of $\bar{\mathbf{B}}$ for column j. Then $(\bar{r}_i - 1)\bar{c}_j$ is an indication of the number of nonzeros associated with a pivot on $\bar{b}_{ij}$, counting here also the elements in the pivot row. Thus

$$\frac{\left[\sum_{[i \mid \bar{b}_{ij} \neq 0]} (\bar{r}_i - 1)\bar{c}_j\right]}{\bar{c}_j} \tag{5.5-6}$$

is the *average* number of nonzeros associated with a pivot operation in column j and this is, of course, M_j. Sorting M_j and ordering columns on the basis of increasing M_j is a much better strategy than ordering based just on the column counts $\bar{c}_j$.

It is clear that a merit count strategy can be applied to a block triangular matrix, where the reordering of columns is done for each block in turn.

Finally, it is important to note that the *numerical* values of elements in $\mathbf{B}$ are not used, so static (a priori) ordering schemes can work with a Boolean representation of $\mathbf{B}$, in which nonzeros are represented by 1. Such schemes could thus operate on a bit array, rather than on an array holding floating-point numbers.

5.6. Combined Strategies

We have discussed the two ends of the spectrum, namely, pivoting for stability and pivoting for sparsity. In practice, we want a balance between the two. For example, consider LU factorization with partial pivoting (5.4-2). Among all members of the pivot column in relevant rows, we chose the pivot element to be the largest in absolute value. Instead we could introduce an added degree of freedom by requiring that the pivot element be at least as large, in absolute value, as some fraction, say u (e.g., $u = 0.1$) of the largest element. We will now usually have a set of candidates for pivot element and we can choose from among them the one that gives the least fill-in. This can be easily incorporated into the Markowitz strategy (5.5-3) by modifying it to

$$\underset{(k',k'')}{\operatorname{argmin}} \left[(r_{k'}^{(k)} - 1)(c_{k''}^{(k)} - 1) \mid k \leq k' \leq m,\ k \leq k'' \leq m \right.$$
$$\left. \text{and} \quad |b_{k'k''}^{(k)}| \geq u\left(\max_{k \leq i \leq m} |b_{ik''}^{(k)}|\right)\right]. \tag{5.6-1}$$

In the merit count strategy (5.5-5), reorder the columns based on M_j. Traverse the columns in the new order $1, 2, \ldots, k, \ldots$ When pivoting at the kth step, pick the row index as

$$\operatorname*{argmin}_{k \le k' \le m} \left[r_{k'}^{(k)} \mid |b_{k'k}^{(k)}| \ge u \left[\max_{k \le i \le m} |b_{ik}^{(k)}| \right] \right] \tag{5.6-2}$$

where $r_{k'}^{(k)}$ denotes the row count for row k', and where we make the simplifying assumption that pivots were on the diagonal prior to step k.

5.7. Solving Systems of Equations

We have developed the LU factorization of a given matrix **B**, with and without permutations. Let us now use the factorization to solve systems of linear equations of the form $\mathbf{Bx} = \mathbf{d}$, where **B** is nonsingular. This takes the following two basic forms.

1. Given $\mathbf{B} = \mathbf{LU}$ as in (5.2-1), solve as follows.

$$\mathbf{Ly} = \mathbf{d} \quad \text{for} \quad \mathbf{y} \quad \text{and} \quad \mathbf{Ux} = \mathbf{y} \quad \text{for} \quad \mathbf{x}. \tag{5.7-1}$$

2. Given $\tilde{\mathbf{B}} = \mathbf{PBQ} = \mathbf{LU}$ as in (5.3-1), solve as follows.

$$\begin{aligned} &\text{Form } \tilde{\mathbf{d}} = \mathbf{Pd}. \\ &\text{Solve } \mathbf{Ly} = \tilde{\mathbf{d}} \quad \text{for} \quad \mathbf{y} \quad \text{and} \quad \mathbf{Uz} = \mathbf{y} \quad \text{for} \quad \mathbf{z}. \\ &\text{Form } \mathbf{x} = \mathbf{Qz}. \end{aligned} \tag{5.7-2}$$

In either case we see that the solution of triangular systems is involved. We call this solving by *back substitution*. (Sometimes the term *forward elimination* is used when the system being solved is lower triangular and back substitution when it is upper triangular. However, the solution processes are essentially the same, the only difference between them being that one starts at opposite ends of the system of equations. We therefore prefer to use back substitution as a generic term to cover both cases.) Let us begin by considering the upper triangular system $\mathbf{Ux} = \mathbf{d}$, where **U** is nonsingular, i.e., $u_{ii} \neq 0$, $1 \le i \le m$.

5.7-1. Row-oriented Back Substitution

We have seen a particular example of this in Chapter 4, Example 4.4-2. The general recurrence relations are straightforward and are given as follows.

$$x_m = d_m / u_{mm}.$$

For $i = m-1, m-2, \ldots, 1$

$$x_i = (d_i - u_{i,i+1}x_{i+1} - \cdots - u_{im}x_{im}) / u_{ii}. \tag{5.7-3}$$

Exercise 5.7-1. Show that the operation count for back substitution based on (5.7-3) is $m(m+1)/2$ flops, when sparsity in $\mathbf{U}$ is not taken into consideration.

One potential advantage of the recurrence relation (5.7-3) is the ability to *accumulate the inner products in double the working precision* and only round them to working precision when defining x_i.

The divisors u_{ii} in (5.7-3) can be arbitrarily small, so it would appear at first sight that the above recurrence is potentially unstable. However, as demonstrated by Example 4.4-2 and the analysis given there, computational errors can be cast back in a consistent manner as small perturbations of the original system, and the same argument can be made in general to show that backsubstitution is *very satisfactory from a numerical standpoint.* This takes the form of the following theorem.

Theorem 5.7-1. The vector, say $\mathbf{x}_c$, computed using (5.7-3) in finite precision floating-point arithmetic, is the *exact* solution of a perturbed triangular system $(\mathbf{U}+\boldsymbol{\delta}\mathbf{U})\mathbf{x}_c = \mathbf{d}$ where

$$\|\boldsymbol{\delta}\mathbf{U}\|_\infty \le (m(m+1)/2)1.01 \max_{i,j} |u_{ij}|\ ulp \tag{5.7-4}$$

Proof. The proof is essentially a generalization of the argument given under Example 4.4-2 and full details may be found in Forsythe and Moler [1967], p. 105. (The infinity norm $\|.\|_\infty$ was defined in Theorem 5.4-1.) ■

Note that even though $\|\boldsymbol{\delta}\mathbf{U}\|_\infty$ is small relative to $\|\mathbf{U}\|_\infty$, the computed solution $\mathbf{x}_c$ could be quite different from the true solution $\mathbf{x}$, when $\mathbf{U}$ is ill conditioned. We discuss this possibility in Sec. 5.8.

5.7-2. Column-oriented Back Substitution

As noted in Sec. 5.1-3, in particular (5.1-11), $\mathbf{U}$ can be expressed in the product form

$$\mathbf{U} = \mathbf{U}_m\mathbf{U}_{m-1}\cdots\mathbf{U}_1. \tag{5.7-5}$$

In particular, this may be the representation of choice when the LU factorization is carried out in a column-oriented way as discussed at the end of Sec. 5.2. With the above product representation of $\mathbf{U}$, we could solve $\mathbf{Ux}=\mathbf{d}$ as follows.

$$\begin{aligned} &\text{Solve } \mathbf{U}_m\mathbf{y}_m = \mathbf{d} \quad \text{for} \quad \mathbf{y}_m.\\ &\text{Solve } \mathbf{U}_k\mathbf{y}_k = \mathbf{y}_{k+1} \quad \text{for} \quad \mathbf{y}_k,\ k = m-1, m-2, \ldots, 1.\\ &\text{Set } \mathbf{x} = \mathbf{y}_1. \end{aligned} \tag{5.7-6}$$

Since each $\mathbf{U}_k$, $m \ge k \ge 1$ is an (elementary) upper triangular matrix, each step of (5.7-6) involves an (elementary) back substitution of the form (5.7-3). It follows from repeated use of Theorem 5.7-1 that the foregoing process is also numerically stable. Equation (5.7-6) does not permit accumulation of inner products in double the working precision, but if this option is not needed, then it is no different from (5.7-3), in terms of the final solution $\mathbf{x}$. *However from the standpoint of data management, the two processes are quite different.*

Exercise 5.7-2. Show that (5.7-3) and (5.7-6) yield identical final solutions when the same precision is used throughout.

Exercise 5.7-3. Develop analogues of (5.7-3) and (5.7-6) for solving $\mathbf{Lx} = \mathbf{d}$ where $\mathbf{L}$ is a unit lower triangular matrix.

5.7-3. Successive Back Substitution

Solving a system of equations $\mathbf{Bx} = \mathbf{d}$ given the LU factors of $\mathbf{B}$, simply involves repeated back substitution, as we have already discussed at the beginning of this section, see (5.7-1) and (5.7-2). Let us now examine the effects of finite precision arithmetic on these processes.

Let $\mathbf{L}_c$ and $\mathbf{U}_c$ be the *computed* LU factors of $\mathbf{B}$ using partial pivoting. From Theorem 5.4-1,

$$(\tilde{\mathbf{B}} + \tilde{\mathbf{E}}) = \mathbf{L}_c\mathbf{U}_c \tag{5.7-7}$$

where $\tilde{\mathbf{B}} \equiv \mathbf{PB} = \mathbf{LU}$ is the true LU factorization. It is the *computed* factors that we now use in (5.7-2). Therefore we must solve

$$\mathbf{L}_c\mathbf{y} = \tilde{\mathbf{d}} \qquad \text{where} \quad \tilde{\mathbf{d}} = \mathbf{Pd}. \tag{5.7-8}$$

The computed solution of this system, in finite precision floating-point arithmetic, is, say $\mathbf{y}_c$, and now using Theorem 5.7-1, in its obvious application to a lower triangular system, we see that $\mathbf{y}_c$ is the exact solution of

$$(\mathbf{L}_c + \boldsymbol{\delta}\mathbf{L}_c)\mathbf{y}_c = \tilde{\mathbf{d}}. \tag{5.7-9}$$

Finally we must solve $\mathbf{U}_c\mathbf{z} = \mathbf{y}_c$ and its computed solution, say $\mathbf{z}_c \equiv \mathbf{x}_c$ (recall we are taking $\mathbf{Q} \equiv \mathbf{I}$ in (5.7-2)), is the exact solution of

$$(\mathbf{U}_c + \boldsymbol{\delta}\mathbf{U}_c)\mathbf{x}_c = \mathbf{y}_c. \tag{5.7-10}$$

We can now combine these expressions (5.7-9) and (5.7-10) to give

$$(\mathbf{L}_c + \boldsymbol{\delta}\mathbf{L}_c)(\mathbf{U}_c + \boldsymbol{\delta}\mathbf{U}_c)\mathbf{x}_c = \tilde{\mathbf{d}}.$$

Then using (5.7-7), we obtain

$$(\tilde{\mathbf{B}} + \boldsymbol{\delta}\tilde{\mathbf{B}})\mathbf{x}_c = \tilde{\mathbf{d}} \tag{5.7-11}$$

where

$$\boldsymbol{\delta}\tilde{\mathbf{B}} = (\tilde{\mathbf{E}} + \boldsymbol{\delta}\mathbf{L}_c\mathbf{U}_c + \mathbf{L}_c\boldsymbol{\delta}\mathbf{U}_c + \boldsymbol{\delta}\mathbf{L}_c\boldsymbol{\delta}\mathbf{U}_c). \tag{5.7-12}$$

Finally, using the error bounds given by Theorems 5.4-1 and 5.7-1, it can be shown by a straightforward argument that

$$\|\boldsymbol{\delta}\tilde{\mathbf{B}}\|_\infty \leq 1.01(m^3 + 3m^2) \max_{i,j,k} |b_{ij}^{(k)}| \, ulp. \tag{5.7-13}$$

(For details see Forsythe and Moler [1967], p. 106.) It is usually the case that

$$\|\boldsymbol{\delta}\tilde{\mathbf{B}}\|_\infty \leq m(ulp)\, \|\mathbf{B}\|_\infty.$$

In summary, we have obtained the exact solution $\mathbf{x}_c$ of a slightly perturbed initial system $(\tilde{\mathbf{B}} + \boldsymbol{\delta}\tilde{\mathbf{B}})$. Thus LU factorization with back substitution is a stable algorithm for computing the solution of a given system of linear equations.

As a practical matter, there are numerous variants on the basic processes (5.7-1) and (5.7-2) that we have studied here. They derive from the specific way in which one chooses to develop and maintain the elementary and permutation matrices that define **L** and **U**, and on whether the back substitution is organized in row- or column-oriented fashion. For example, with partial pivoting, the factorization might take the form (5.3-3) with $\mathbf{P}_{i,i''} = \mathbf{I}$ for all i, namely,

$$\tilde{\boldsymbol{\Gamma}}_{m-1}\mathbf{P}_{m-1,(m-1)'}\tilde{\boldsymbol{\Gamma}}_{m-2} \cdots \tilde{\boldsymbol{\Gamma}}_1\mathbf{P}_{1,1'}\mathbf{B} = \mathbf{U}.$$

The system $\mathbf{Bx} = \mathbf{d}$ could then be solved as follows.

$$\begin{aligned} &\text{Set } \mathbf{d}_0 = \mathbf{d}. \\ &\text{Form } \mathbf{d}_k = \tilde{\boldsymbol{\Gamma}}_k\mathbf{P}_{k,k'}\mathbf{d}_{k-1}, \; k = 1, 2, \ldots, m-1. \\ &\text{Solve } \mathbf{Ux} = \mathbf{d}_{m-1}. \end{aligned} \tag{5.7-14}$$

Depending on how **U** is represented, we could carry out the last step, in turn, in a row-oriented or in a column-oriented manner, as discussed in Sec. 5.7-1 and Sec. 5.7-2, respectively. A specific implementation will be discussed, in detail, in Sec. 5.10.

Exercise 5.7-4. Given the LU factorization of **B**, show how to solve the system $\mathbf{B}^T\mathbf{x} = \mathbf{d}$ for the two cases $\mathbf{B} = \mathbf{LU}$ and $\tilde{\mathbf{B}} \equiv \mathbf{PBQ} = \mathbf{LU}$.

Exercise 5.7-5. Given the product from representation of a *unit* lower triangular matrix $\mathbf{L} = \mathbf{L}_1\mathbf{L}_2 \cdots \mathbf{L}_m$ and $\mathbf{L}^{-1} = \boldsymbol{\Gamma}_m\boldsymbol{\Gamma}_{m-1} \cdots \boldsymbol{\Gamma}_1$ (with $\mathbf{L}_i^{-1} \equiv \boldsymbol{\Gamma}_i$), show that even in finite precision arithmetic, the solution of $(\mathbf{L}_1 \cdots \mathbf{L}_m)\mathbf{x} = \mathbf{d}$, by a sequence of back substitutions, is *identical* to that given by forming the product $(\boldsymbol{\Gamma}_m(\boldsymbol{\Gamma}_{m-1}(\cdots(\boldsymbol{\Gamma}_1\mathbf{d})\cdots)$. Would this be true if **L** is lower triangular but not necessarily *unit* lower triangular?

5.7-4. Some Comments on Inversion

Although we have frequently used the inverse of matrices as a mathematical operator, for example, $\mathbf{B}^{-1}$, it is a maxim of numerical analysis that one avoids *computing* the inverse whenever possible. For example, to solve $\mathbf{Bx} = \mathbf{d}$, one does not normally form the matrix $\mathbf{B}^{-1}$ and then develop $\mathbf{x} = \mathbf{B}^{-1}\mathbf{d}$. There are several reasons for this.

1. In order to obtain $\mathbf{B}^{-1}$ as an $m \times m$ matrix, it is necessary to solve m systems of the form $\mathbf{Bz}_i = \mathbf{e}_i$, $i = 1, \ldots, m$, using the LU factors of $\mathbf{B}$. (As usual, $\mathbf{e}_i$ denotes the ith column of the identity matrix.) This gives $\mathbf{B}^{-1} \equiv [\mathbf{z}_1, \ldots, \mathbf{z}_m]$ and $\mathbf{x} = \mathbf{B}^{-1}\mathbf{d}$. For dense matrices, the cost (including that of factorizing $\mathbf{B}$) is approximately $(m^3 + m^2)$ flops. This is roughly three times as expensive as factorizing and solving by back substitution.
2. Because more operations are involved, as just noted, it is numerically less satisfactory to form the inverse of $\mathbf{B}$ and then apply it to $\mathbf{d}$, rather than to solve directly.
3. For large sparse matrices it is normally out of the question to invert $\mathbf{B}$, because $\mathbf{B}^{-1}$ can be completely dense even though $\mathbf{B}$ and its LU factors are sparse. Example 5.1-2 illustrates this for the case of an upper triangular matrix.

There remains the alternative of maintaining $\mathbf{B}^{-1}$ implicitly in a *product form,* namely,

$$\mathbf{B}^{-1} = \mathbf{U}^{-1}\mathbf{L}^{-1} = \mathbf{U}_1^{-1} \cdots \mathbf{U}_m^{-1}\mathbf{L}_m^{-1} \cdots \mathbf{L}_1^{-1}. \tag{5.7-15}$$

(See (5.1-12) and (5.1-13).) Then $\mathbf{Bx} = \mathbf{d}$ can be solved as follows.

$$\mathbf{x} = (\mathbf{U}_1^{-1}(\cdots(\mathbf{U}_m^{-1}(\mathbf{L}_m^{-1} \cdots (\mathbf{L}_1^{-1}\mathbf{d}) \cdots). \tag{5.7-16}$$

However, we could equally well define the product form of $\mathbf{B}$ as

$$\mathbf{B} = \mathbf{LU} = \mathbf{L}_1 \cdots \mathbf{L}_m\mathbf{U}_m \cdots \mathbf{U}_1, \tag{5.7-17}$$

and solve $\mathbf{Bx} = \mathbf{d}$ by successive column-oriented back substitution, as discussed in Secs. 5.7-2. and 5.7-3. Indeed, with a suitable representation of the elementary matrices and organization of calculations, the two processes just outlined *can be made identical in terms of both intermediate and final results, even in finite precision arithmetic*; see also Exercise 5.7-5. We shall therefore make it a practice, whenever possible, to compute with representations of $\mathbf{B}$ and back substitution, and reserve $\mathbf{B}^{-1}$ for use purely as a mathematical operator.

5.8. The Accuracy of the Computed Solution

Although the computed solution $\mathbf{x}_c$ is the exact solution of $(\tilde{\mathbf{B}} + \boldsymbol{\delta}\tilde{\mathbf{B}})\mathbf{x}_c = \tilde{\mathbf{d}}$ (cf. (5.7-11) through (5.7-13)), this does *not* imply that $\mathbf{x}_c$ is close to the true solution $\mathbf{x}$ of $\tilde{\mathbf{B}}\mathbf{x} = \tilde{\mathbf{d}}$. Conditions under which it is close are revealed through the following *perturbation analysis.*

Let $\boldsymbol{\delta}\mathbf{x} \equiv \mathbf{x}_c - \mathbf{x}$. Then $(\tilde{\mathbf{B}} + \boldsymbol{\delta}\tilde{\mathbf{B}})(\mathbf{x} + \boldsymbol{\delta}\mathbf{x}) = \tilde{\mathbf{d}}$ and hence $(\tilde{\mathbf{B}} + \boldsymbol{\delta}\tilde{\mathbf{B}})\boldsymbol{\delta}\mathbf{x} = -(\boldsymbol{\delta}\tilde{\mathbf{B}})\mathbf{x}$. Let us assume that $(\tilde{\mathbf{B}} + \boldsymbol{\delta}\tilde{\mathbf{B}})$ is also nonsingular. For example, this would be true when $\|\tilde{\mathbf{B}}^{-1}\boldsymbol{\delta}\tilde{\mathbf{B}}\|_\infty < 1$. (For convenience, we shall work throughout with the infinity norm, but any other matrix norm could be substituted in its place.)

$$\begin{aligned}\boldsymbol{\delta}\mathbf{x} &= -(\tilde{\mathbf{B}} + \boldsymbol{\delta}\tilde{\mathbf{B}})^{-1}(\boldsymbol{\delta}\tilde{\mathbf{B}})\mathbf{x}\\ &= -(\mathbf{I} + \tilde{\mathbf{B}}^{-1}\boldsymbol{\delta}\tilde{\mathbf{B}})^{-1}\tilde{\mathbf{B}}^{-1}(\boldsymbol{\delta}\tilde{\mathbf{B}})\mathbf{x}.\end{aligned}$$

Then

$$\begin{aligned}\|\boldsymbol{\delta}\mathbf{x}\|_\infty &= \|(\mathbf{I} + \tilde{\mathbf{B}}^{-1}\boldsymbol{\delta}\tilde{\mathbf{B}})^{-1}\tilde{\mathbf{B}}^{-1}(\boldsymbol{\delta}\tilde{\mathbf{B}})\mathbf{x}\|_\infty\\ &\le \|(\mathbf{I} + \tilde{\mathbf{B}}^{-1}\boldsymbol{\delta}\tilde{\mathbf{B}})^{-1}\|_\infty\|\tilde{\mathbf{B}}^{-1}\|_\infty\|\boldsymbol{\delta}\tilde{\mathbf{B}}\|_\infty\|\mathbf{x}\|_\infty\\ &\le (1/(1 - \|\tilde{\mathbf{B}}^{-1}\boldsymbol{\delta}\tilde{\mathbf{B}}\|_\infty)\|\tilde{\mathbf{B}}^{-1}\|_\infty\|\boldsymbol{\delta}\tilde{\mathbf{B}}\|_\infty\|\mathbf{x}\|_\infty\\ &\le (\|\tilde{\mathbf{B}}^{-1}\|_\infty\|\boldsymbol{\delta}\tilde{\mathbf{B}}\|_\infty\|\mathbf{x}\|_\infty)/(1 - \|\tilde{\mathbf{B}}^{-1}\|_\infty\|\boldsymbol{\delta}\tilde{\mathbf{B}}\|_\infty).\end{aligned}$$

Hence

$$\frac{\|\boldsymbol{\delta}\mathbf{x}\|_\infty}{\|\mathbf{x}\|_\infty} \le \frac{\|\tilde{\mathbf{B}}\|_\infty\|\tilde{\mathbf{B}}^{-1}\|_\infty(\|\boldsymbol{\delta}\tilde{\mathbf{B}}\|_\infty/\|\tilde{\mathbf{B}}\|_\infty)}{(1 - \|\tilde{\mathbf{B}}\|_\infty\|\tilde{\mathbf{B}}^{-1}\|_\infty(\|\boldsymbol{\delta}\tilde{\mathbf{B}}\|_\infty/\|\tilde{\mathbf{B}}\|_\infty)} \tag{5.8-1}$$

The factor $\kappa \equiv \|\tilde{\mathbf{B}}\|_\infty\|\tilde{\mathbf{B}}^{-1}\|_\infty$ determines whether or not $\|\boldsymbol{\delta}\mathbf{x}\|_\infty$ is small relative to $\|\mathbf{x}\|_\infty$, and it is termed *the condition number with respect to inversion*. When κ is large, then $\mathbf{B}$ is said to be *ill conditioned*. It is, however, important to note that (5.8-1) only provides an upper bound, so that the condition number could be unduly pessimistic. This is particularly the case when we are dealing with triangular systems. For example, when solving $\mathbf{Lx} = \mathbf{d}$, the fact that $\|\mathbf{L}\|_\infty\|\mathbf{L}^{-1}\|_\infty$ is large could be misleading, i.e., the computed solution could still be very accurate. (See notes at the end of this chapter.)

During the course of the LU factorization and back substitution to solve $\mathbf{Bx} = \mathbf{d}$, the following are sufficient conditions to indicate ill conditioning.

1. A "large" residual vector $\mathbf{r} \equiv \mathbf{Bx}_c - \mathbf{d}$ or $\tilde{\mathbf{r}} \equiv \tilde{\mathbf{B}}\mathbf{x}_c - \tilde{\mathbf{d}}$.
2. A "large" computed solution $\mathbf{x}_c$.
3. A "small" pivot element.

We use quotes above to imply that the terms are not precisely defined.

It is however possible for the system to be ill conditioned and for the solution computed from it to be completely inaccurate, yet for *none* of the above symptoms to appear. Consider, for example, the residual vector $\tilde{\mathbf{r}}$. From the stability analysis, $(\tilde{\mathbf{B}} + \boldsymbol{\delta}\tilde{\mathbf{B}})\mathbf{x}_c = \tilde{\mathbf{d}}$. Thus

$$\tilde{\mathbf{r}} = -(\boldsymbol{\delta}\tilde{\mathbf{B}})\mathbf{x}_c.$$

From (5.7-13),

$$\|\tilde{\mathbf{r}}\|_\infty \le \|\boldsymbol{\delta}\tilde{\mathbf{B}}\|_\infty\|\mathbf{x}_c\|_\infty \le 1.01(m^3 + 3m^2)\max_{i,j,k}|b_{ij}^{(k)}|\,(ulp)\,\|\mathbf{x}_c\|_\infty. \tag{5.8-2}$$

Suppose now that $\|\mathbf{x}_c\|_\infty$ is of order unity *but completely inaccurate.* $\|\tilde{\mathbf{r}}\|_\infty$ would then be of order $\|\boldsymbol{\delta}\tilde{\mathbf{B}}\|_\infty$. As noted just after (5.7-13), this is generally of order $m(ulp)\,\|\mathbf{B}\|_\infty$ and therefore small, unless $\|\mathbf{B}\|_\infty$ is unusually large. In a similar vein, examples can be constructed of ill-conditioned matrices $\mathbf{B}$ for which all pivots are of order unity when performing the LU factorization. The above conditions (1) through (3) are sufficient but not *necessary* conditions for ill conditioning. A more detailed discussion may be found in Wilkinson [1965].

There is no simple cure for ill conditioning. However scaling of the rows of $\mathbf{B}$ and the right-hand side vector, and scaling of the columns of $\mathbf{B}$, i.e., of the associated variables, can make a significant difference. Thus if $\mathbf{D}_1 \equiv \mathrm{diag}[s_1^{(1)}, \ldots, s_m^{(1)}] > 0$ and $\mathbf{D}_2 \equiv \mathrm{diag}[s_1^{(2)}, \ldots, s_m^{(2)}] > 0$ and we make the transformations

$$\mathbf{d} = \mathbf{D}_1\bar{\mathbf{d}}$$

$$\mathbf{x} = \mathbf{D}_2\bar{\mathbf{x}}$$

then the system of equations $\mathbf{Bx} = \mathbf{d}$ becomes

$$(\mathbf{D}_1^{-1}\mathbf{B}\mathbf{D}_2)\bar{\mathbf{x}} = \bar{\mathbf{d}}.$$

The diagonal matrices are chosen so as to make the condition number of $(\mathbf{D}_1^{-1}\mathbf{B}\mathbf{D}_2)$ less than that of $\mathbf{B}$, and indeed, as small as possible. A common approach is to *equilibrate* $\mathbf{B}$ by choosing scale factors to make its rows and columns roughly of the same length, in some appropriate norm. A full discussion of scaling techniques may be found in Forsythe and Moler [1967] and in Wilkinson [1965]. See also Tomlin [1975b].

5.9. Variants on LU Factorization

5.9-1. Gauss–Jordan Elimination

This is a variant on the Gaussian elimination procedure reviewed at the beginning of this chapter. At each step both subdiagonal and super-diagonal elements in a column are eliminated, i.e., at step k, elements in column k, other than the one on the diagonal, are reduced to zero by adding suitable multiples of row k to all other rows. After m steps, Gauss–Jordan elimination gives a diagonal system of equations, which can then be trivially solved.

Let us now reformulate this in terms of matrix operations. The process is precisely the same as that of LU factorization in Sec. 5.2 with elementary matrices of the form (5.1-5) substituted for those of the form (5.1-6). It is defined as follows.

$$\mathbf{B}^{(1)} \equiv \mathbf{B}.$$

For $k = 1, 2, \ldots, m$

$$\mathbf{B}^{(k+1)} = \boldsymbol{\Lambda}_k\mathbf{B}^{(k)}, \qquad (5.9\text{-}1a)$$

where

$$\mathbf{B}^{(k)} = \begin{bmatrix} b_{11}^{(1)} & . & . & b_{1k}^{(k)} & b_{1m}^{(k)} \\ . & b_{22}^{(2)} & . & \vdots & \vdots \\ . & . & . & \vdots & \vdots \\ . & . & . & \vdots & \vdots \\ . & . & . & \vdots & \vdots \\ . & . & b_{k-1,k-1}^{(k-1)} & b_{k-1,k}^{(k)} & b_{k-1,m}^{(k)} \\ . & . & . & b_{kk}^{(k)} & b_{km}^{(k)} \\ . & . & . & \vdots & \vdots \\ . & . & . & \vdots & \vdots \\ . & . & . & \vdots & \vdots \\ . & . & . & b_{mk}^{(k)} & b_{mm}^{(k)} \end{bmatrix} \tag{5.9-1b}$$

$$\mathbf{\Lambda}_k = \begin{bmatrix} 1 & . & . & -b_{1k}^{(k)}/b_{kk}^{(k)} & . \\ . & 1 & . & \vdots & . \\ . & . & . & \vdots & . \\ . & . & . & \vdots & . \\ . & . & . & \vdots & . \\ . & . & 1 & -b_{k-1,k}^{(k)}/b_{kk}^{(k)} & . \\ . & . & . & 1 & . \\ . & . & . & \vdots & . \\ . & . & . & \vdots & . \\ . & . & . & \vdots & . \\ . & . & . & -b_{mk}^{(k)}/b_{kk}^{(k)} & 1 \end{bmatrix} \tag{5.9-1c}$$

and we assume that $b_{kk}^{(k)} \neq 0,\ k = 1, \ldots, m$.

We can summarize the recurrence relations for carrying out the Gauss–Jordan elimination process (5.9-1) as follows:

$$\begin{aligned} b_{ii}^{(k+1)} &= b_{ii}^{(k)} = b_{ii}^{(i)} \qquad i \leq k \\ b_{ij}^{(k+1)} &= b_{ij}^{(k)} - \frac{b_{ik}^{(k)} b_{kj}^{(k)}}{b_{kk}^{(k)}} \qquad j > k \qquad i \neq k \\ b_{ij}^{(k+1)} &= 0 \qquad j \leq k,\ j \neq i. \end{aligned} \tag{5.9-2}$$

After m steps we have the factorization

$$\mathbf{\Lambda}_m \mathbf{\Lambda}_{m-1} \cdots \mathbf{\Lambda}_1 \mathbf{B} = \mathbf{D} = \operatorname{diag}[b_{11}^{(1)}, \ldots, b_{mm}^{(m)}].$$

Let us define

$$\hat{\mathbf{E}}_j \equiv \mathbf{\Lambda}_j^{-1} \quad \text{and} \quad \mathbf{D}_k = \text{diag}[1, \ldots, 1, b_{kk}^{(k)}, 1, \ldots, 1] \tag{5.9-3}$$

Then

$$\mathbf{B} = \hat{\mathbf{E}}_1\hat{\mathbf{E}}_2 \cdots \hat{\mathbf{E}}_m\mathbf{D} = (\hat{\mathbf{E}}_1\mathbf{D}_1)(\hat{\mathbf{E}}_2\mathbf{D}_2) \cdots (\hat{\mathbf{E}}_m\mathbf{D}_m).$$

Define $\mathbf{E}_k \equiv (\hat{\mathbf{E}}_k\mathbf{D}_k)$, $k = 1, \ldots, m$. From (5.9-1c) and our simple rule for inverting an elementary matrix given in Sec. 5.1-1,

$$\mathbf{E}_k = \begin{bmatrix} 1 & \cdot & b_{1k}^{(k)} & \cdot & \cdot \\ \cdot & \cdot & \vdots & \cdot & \cdot \\ \cdot & \cdot & \vdots & \cdot & \cdot \\ \cdot & \cdot & \vdots & \cdot & \cdot \\ \cdot & 1 & b_{k-1,k}^{(k)} & \cdot & \cdot \\ \cdot & \cdot & b_{kk}^{(k)} & \cdot & \cdot \\ \cdot & \cdot & \vdots & 1 & \cdot \\ \cdot & \cdot & \vdots & \cdot & \cdot \\ \cdot & \cdot & \vdots & \cdot & \cdot \\ \cdot & & b_{mk}^{(k)} & \cdot & 1 \end{bmatrix} \tag{5.9-4}$$

Therefore

$$\mathbf{B} = \mathbf{E}_1\mathbf{E}_2 \cdots \mathbf{E}_m. \tag{5.9-5}$$

This is often called *the product form* of $\mathbf{B}$.

Now we can solve $\mathbf{Bx} = \mathbf{d}$ as follows.

$$\begin{aligned} &\text{Solve } \mathbf{E}_1\mathbf{y}_1 = \mathbf{d} \text{ for } \mathbf{y}_1. \\ &\text{Solve } \mathbf{E}_k\mathbf{y}_k = \mathbf{y}_{k-1},\ k = 2, \ldots, m \text{ for } \mathbf{y}_k. \\ &\text{Set } \mathbf{x} = \mathbf{y}_m. \end{aligned} \tag{5.9-6}$$

Each elementary operation $\mathbf{E}_k\mathbf{y}_k = \mathbf{y}_{k-1}$ can be solved as follows.

$$(\mathbf{P}_{1,k}\mathbf{E}_k\mathbf{P}_{1,k})(\mathbf{P}_{1,k}\mathbf{y}_k) = (\mathbf{P}_{1,k}\mathbf{y}_{k-1}) \tag{5.9-7}$$

where $\mathbf{P}_{1,k}$ is the elementary permutation which interchanges rows (columns) 1 and k. $(\mathbf{P}_{1,k}\mathbf{E}_k\mathbf{P}_{1,k})$ is an elementary lower triangular matrix and the above equation (5.9-7) can thus be solved by back substitution. Of course, the permutation is not really needed. From the form of $\mathbf{E}_k$ it is clear that the solution of $\mathbf{E}_k\bar{\mathbf{y}} = \mathbf{y}$ is given by $\bar{y}_k = y_k/b_{kk}^{(k)}$ and then $\bar{y}_i = y_i - b_{ik}^{(k)}\bar{y}_k$ for $i \neq k$.

Exercise 5.9-1. In analogy to the column-oriented LU factorization of Sec. 5.2, develop a column-oriented version of Gauss–Jordan elimination.

Exercise 5.9-2. Give the version of Gauss–Jordan elimination in which off-diagonal elements of $\mathbf{B}$ are eliminated and, in addition, elements on the diagonal are transformed to unity. Show that this procedure transforms

(a) $[\mathbf{B}^0 \quad \mathbf{I}]$ to $[\mathbf{I} \quad (\mathbf{B}^0)^{-1}]$.
(b) $[\mathbf{B}^0 \quad \mathbf{N}^0]$ to $[\mathbf{I} \quad (\mathbf{B}^0)^{-1}\mathbf{N}^0]$.
(c)

$$\begin{bmatrix} 1 & \mathbf{c}_B^T & \mathbf{c}_N^T & 0 \\ \mathbf{0} & \mathbf{B}^0 & \mathbf{N}^0 & \mathbf{b} \end{bmatrix}$$

to

$$\begin{bmatrix} 1 & 0 & (\boldsymbol{\sigma}_N^0)^T & z^0 \\ \mathbf{0} & \mathbf{I} & (\mathbf{B}^0)^{-1}\mathbf{N}^0 & (\mathbf{B}^0)^{-1}\mathbf{b} \end{bmatrix}$$

where $\mathbf{B}^0$ is an $m \times m$ nonsingular basis matrix, $\mathbf{N}^0$ is an $m \times (n-m)$ matrix of nonbasic columns and $\mathbf{c}_B$, $\mathbf{c}_N$, $\boldsymbol{\sigma}_N^0$, and z^0 denote the usual quantities in the cycle of the simplex method; see, in particular, Sec. 2.2.

(d) Use (c) above to develop an algorithm for carrying out the cycle of the simplex method on a tableau.

5.9-2. Relationship between LU and Gauss–Jordan Product Forms and Relative Merits of the Two Factorizations

Suppose that the matrix $\mathbf{B}$ has nonsingular leading principal submatrices and is factorized as $\mathbf{B} = \mathbf{LU}$. Then $\mathbf{U}^{-1}$ is also nonsingular and upper triangular. Using (5.1-10) and analogously to (5.1-11), let us express $\mathbf{L}$ and $\mathbf{U}^{-1}$ in product forms as follows:

$$\mathbf{L} = \mathbf{L}_1\mathbf{L}_2 \cdots \mathbf{L}_m \tag{5.9-8a}$$

$$\mathbf{U}^{-1} = [\mathbf{U}^{-1}]_m[\mathbf{U}^{-1}]_{m-1} \cdots [\mathbf{U}^{-1}]_1. \tag{5.9-8b}$$

Note that $[\mathbf{U}^{-1}]_k$ is *not* the same as $\mathbf{U}_k^{-1}$, where $\mathbf{U}_k$ is the kth element in the product form of $\mathbf{U} = \mathbf{U}_m \cdots \mathbf{U}_1$.

Lemma 5.9-1. Suppose that the foregoing matrix $\mathbf{B}$ is alternatively factorized as in (5.2-7) and (5.9-5) with $\mathbf{L}$ and $\mathbf{U}^{-1}$ expressed in the product forms (5.9-8). Then

$$\mathbf{E}_k = \mathbf{L}_k([\mathbf{U}^{-1}]_k)^{-1}.$$

Proof. $\mathbf{B} = \mathbf{LU}$ implies that $\mathbf{U}^{-1}\mathbf{L}^{-1}\mathbf{B} = \mathbf{I}$. Thus, from (5.9-8),

$$[\mathbf{U}^{-1}]_m \cdots [\mathbf{U}^{-1}]_1\mathbf{L}_m^{-1} \cdots \mathbf{L}_1^{-1}\mathbf{B} = \mathbf{I}. \tag{5.9-9}$$

Now

$$[\mathbf{U}^{-1}]_i\mathbf{L}_j^{-1} = \mathbf{L}_j^{-1}[\mathbf{U}^{-1}]_i, \; i < j.$$

Therefore (5.9-9) can be expressed as

$$([\mathbf{U}^{-1}]_m\mathbf{L}_m^{-1})([\mathbf{U}^{-1}]_{m-1}\mathbf{L}_{m-1}^{-1}) \cdots ([\mathbf{U}^{-1}]_1\mathbf{L}_1^{-1})\mathbf{B} = \mathbf{I},$$

where each term in the above product, namely, $([\mathbf{U}^{-1}]_k \mathbf{L}_k^{-1})$ is an elementary matrix. We can therefore identify this term with $\mathbf{E}_k^{-1}$ in (5.9-5). Furthermore if they were not equal, we could reverse the above argument to show that the LU factorization of **B** is not unique, leading to a contradiction. Therefore $\mathbf{E}_k = \mathbf{L}_k([\mathbf{U}^{-1}]_k)^{-1}$ as required. This completes the proof. ■

Gauss–Jordan factorization has little to recommend it over LU factorization. First, the operation count can easily be shown to compare unfavorably with that of LU factorization with back substitution. Second, when the pivot element $b_{kk}^{(k)}$ in Gauss–Jordan factorization is zero or close to zero, a row and/or column interchange can bring a more suitable element, $b_{k'k''}^{(k)}$, $k \le k' \le m$, $k \le k'' \le m$ to the pivot position, just as in LU factorization. However, no interchange of row k with *earlier* rows is possible because this would destroy the off-diagonal zero structure of $\mathbf{B}^{(k)}$ in columns 1 through k. Since $\mathbf{\Lambda}_k$ (see (5.9-1c)) involves elements of the form $-b_{ik}^{(k)}/b_{kk}^{(k)}$, $i<k$, there is no guarantee that these are bounded in magnitude even with the above pivoting strategy. Thus Gauss–Jordan factorization is potentially unstable. Third, from Lemma 5.9-1, the sparsity of the Gauss–Jordan product form is comparable to the sparsity of the product forms of **L** and $\mathbf{U}^{-1}$. Since $\mathbf{U}^{-1}$ will generally be less sparse than **U** (cf Example 5.1-2), the sparsity of the product form (5.9-5) compares unfavorably with the sparsity of the LU factors.

5.10. Practical Details of Implementation

In this concluding section, we describe implementation techniques for sparse LU factorization and the solution of associated systems of linear equations. Our aim is to highlight some of the practical considerations when implementing methods introduced earlier in this chapter. (Obviously a complete account cannot be given here; for full details the reader should consult Reid [1976].) The implementation is based on a *Markowitz pivoting strategy* that balances considerations of *stability and sparsity* (Sec. 5.6) and performs the eliminations by *row-oriented pivoting in place* (see end of Sec. 5.3) on a suitable *packed* representation of the matrix.

When pivoting in place with the following pivot sequence $(1', 1'')$, $(2', 2''), \ldots, ((m-1)', (m-1)'')$, the LU factorization takes the form

$$\mathbf{B}^{(1)} = \mathbf{B}.$$

For $k = 1, \ldots, m-1$,

$$\mathbf{B}^{(k+1)} = \mathbf{\Pi}_k \mathbf{B}^{(k)} \tag{5.10-1}$$

where each $\mathbf{\Pi}_k$ is an elementary matrix. Typically, $\mathbf{\Pi}_k$ would be of the form

$$\boldsymbol{\Pi}_k = \begin{bmatrix} 1 & . & 0 & . & . \\ . & . & x & . & . \\ . & . & 0 & . & . \\ . & 1 & x & . & . \\ . & . & 1 & . & . \\ . & . & 0 & 1 & . \\ . & . & 0 & . & . \\ . & . & x & . & 1 \end{bmatrix} \tag{5.10-2}$$

where column k' in (5.10-2) is the distinguished column. The zero elements in column k' are in the rows where a pivot has already occurred at a previous step, namely, rows $1', 2', \ldots, (k-1)'$. The remaining off-diagonal elements in column k' are given by

$$(\boldsymbol{\Pi}_k)_{ik'} = -b_{ik''}^{(k)}/b_{k'k''}^{(k)}. \tag{5.10-3}$$

After m steps we have $\mathbf{B}^{(m)} \equiv \tilde{\mathbf{U}}$. It should be clear that $\tilde{\mathbf{U}}$ is a *permuted* upper triangular matrix. With the definitions

$$\mathbf{P} \equiv \mathbf{P}_{(m-1),(m-1)'} \cdots \mathbf{P}_{1,1'}$$
$$\mathbf{Q} \equiv \mathbf{P}_{1,1''} \cdots \mathbf{P}_{(m-1),(m-1)''},$$

we obtain

$$\mathbf{U} = \mathbf{P}\tilde{\mathbf{U}}\mathbf{Q} \tag{5.10-4}$$

where $\mathbf{U}$ is an upper triangular matrix. Thus

$$\mathbf{P}\boldsymbol{\Pi}_{m-1} \cdots \boldsymbol{\Pi}_1 \mathbf{B}\mathbf{Q} = \mathbf{P}\tilde{\mathbf{U}}\mathbf{Q} = \mathbf{U}. \tag{5.10-5}$$

From (5.3-3) and (5.3-9),

$$\tilde{\boldsymbol{\Gamma}}_{m-1}\mathbf{P}_{(m-1),(m-1)'} \cdots \tilde{\boldsymbol{\Gamma}}_1 \mathbf{P}_{1,1'}\mathbf{B}\mathbf{Q} = \mathbf{U} \tag{5.10-6a}$$

and

$$\boldsymbol{\Gamma}_{m-1} \cdots \boldsymbol{\Gamma}_1 \mathbf{P}\mathbf{B}\mathbf{Q} = \mathbf{U}. \tag{5.10-6b}$$

The reader can easily verify the connections between $\boldsymbol{\Pi}_k$, $\tilde{\boldsymbol{\Gamma}}_k$ and $\boldsymbol{\Gamma}_k$, which we leave to the following exercise.

Exercise 5.10-1. Show that

$$\boldsymbol{\Pi}_k = \mathbf{P}_{1,1'} \cdots \mathbf{P}_{k,k'}\tilde{\boldsymbol{\Gamma}}_k \mathbf{P}_{k,k'} \cdots \mathbf{P}_{1,1'} = \mathbf{P}^T \boldsymbol{\Gamma}_k \mathbf{P}. \tag{5.10-7}$$

Let us define

$$\tilde{\mathbf{L}}^{-1} \equiv \boldsymbol{\Pi}_{m-1} \cdots \boldsymbol{\Pi}_1. \tag{5.10-8}$$

From (5.10-7), we then have

$$\tilde{\mathbf{L}}^{-1} = \mathbf{P}^T(\mathbf{\Gamma}_{m-1} \cdots \mathbf{\Gamma}_1)\mathbf{P} = \mathbf{P}^T\mathbf{L}^{-1}\mathbf{P},$$

$$\mathbf{L} = \mathbf{P}\tilde{\mathbf{L}}\mathbf{P}^T, \tag{5.10-9}$$

and recall that $\mathbf{L}$ is a *unit* lower triangular matrix. Thus after a factorization, $\tilde{\mathbf{L}}$ is a *symmetrically permuted* unit lower triangular matrix and $\tilde{\mathbf{U}}$ a permuted upper triangular matrix with $\mathbf{B} = \tilde{\mathbf{L}}\tilde{\mathbf{U}}$ (cf. also Exercise 5.3-2).

5.10-1. The FACTOR Module

The LU factorization is defined in terms of the permutations $\mathbf{P}$ and $\mathbf{Q}$, the elementary matrices $\mathbf{\Pi}_1, \ldots, \mathbf{\Pi}_{m-1}$, and the partially factorized matrices $\mathbf{B}^{(k)}$ that are transformed through successive eliminations, into the permuted upper triangular matrix $\tilde{\mathbf{U}}$.

Let us consider each of these items in turn.

Permutations. $\mathbf{P}$ and $\mathbf{Q}$ are simply defined by the following two arrays.

PMAP	QMAP
$1'$	$1''$
$2'$	$2''$
k'	k''
$(m-1)'$	$(m-1)''$

(5.10-10)

Thus element k of PMAP gives the row number k' of $\tilde{\mathbf{U}}$ (the permuted upper triangular matrix) corresponding to the kth row of $\mathbf{U}$ (the true upper triangular matrix). Element k of QMAP gives the column number k'' of $\tilde{\mathbf{U}}$ corresponding to the kth column of $\mathbf{U}$.

Sparse Elementary Matrices and the* $\mathbf{\Gamma}$ *File. Column k' of $\mathbf{\Pi}_k$ in (5.10-2) is sparse for two reasons. First, there are zeros in positions corresponding to previous pivot rows $1', \ldots, (k-1)'$. Second, many elements are zero because the corresponding element $b_{ik''}^{(k)}$ in (5.10-3) is zero, the result of propagated sparsity of the original matrix $\mathbf{B}$. Suppose that the (relatively few) nonzero elements of $\mathbf{\Pi}_k$ are in rows $i_1, \ldots, i_{n_k}$. We can then express $\mathbf{\Pi}_k$ as

$$\mathbf{\Pi}_k = \mathbf{\Pi}_k^{i_{n_k}}\mathbf{\Pi}_k^{i_{n_k-1}} \cdots \mathbf{\Pi}_k^{i_1}, \tag{5.10-11}$$

or an arbitrary permutation of these matrices, where $\mathbf{\Pi}_k^{i_l}$ is defined as follows.

$$\mathbf{\Pi}_k^{i_l} = \begin{bmatrix} 1 & \cdot & \cdot & \cdot \\ \cdot & \cdot & \cdot & \cdot \\ \cdot & \cdot & \cdot & \cdot \\ \cdot & 1 & v & \cdot \\ \cdot & \cdot & \cdot & \cdot \\ \cdot & \cdot & \cdot & \cdot \\ \cdot & \cdot & 1 & \cdot \\ \cdot & \cdot & \cdot & \cdot \\ \vdots & \cdot & \cdot & \cdot \\ \cdot & \cdot & \cdot & \cdot \\ \cdot & \cdot & \cdot & 1 \end{bmatrix} \qquad (5.10\text{-}12)$$

The distinguished column has index k', v is in position (i_l, k') with $v \equiv -b_{i_lk''}^{(k)}/b_{k'k''}^{(k)}$, and $\mathbf{\Pi}_k^{i_l}$ is stored as the operator

$$\textit{Add row } i_l \textit{ to } v \textit{ times row } k'. \qquad (5.10\text{-}13)$$

This requires storage of one floating-point number v and two integer indices. $(\mathbf{\Pi}_k^{i_l})^T$ is defined by the same information and an analogous operator.

The complete sequence of elementary matrices is maintained as a sequentially allocated linear list (see Chapter 3) called the $\mathbf{\Gamma}$ file. (Note that the inverse operators $\mathbf{\Pi}_k^{i_l}$ are stored, so we refer to this as the $\mathbf{\Gamma}$ file rather than the $\mathbf{L}$ file.) This must be accessed in a forward and in a backward direction as we shall discuss in Sec. 5.10-2. As new elementary matrices are generated during the elimination process, operators of the form (5.10-13) are simple added to the end of the $\mathbf{\Gamma}$ file.

Data Structure for B and B$^{(k)}$. The implementation being row oriented, the $m \times m$ sparse matrix $\mathbf{B}$ is stored internally as a row list/column index packed data structure (Sec. 3.2-2), i.e., three arrays hold row pointers, column indices of rows and numerical values of rows, respectively. Entries *within* a row are in any order. The column structure is also needed during the elimination process (i.e., the row indices of each column) so that a *partial* column list/row index data structure is maintained, consisting of column pointers and row indices but *not* numerical values. (A similar idea was discussed at the end of Sec. 3.2 for holding a column list/row index data structure and a partial row list/column index data structure, i.e., the reverse situation to the one we have here.) Normally the number of nonzero elements in a row or column can be computed from the corresponding arrays of pointers.

However because elimination and fill-in occurs during the course of the factorization, the arrays holding indices are permitted to have a certain amount of unused space or "elbow room." Such entries can be flagged, for example, by storing a zero entry in place of an index. Two additional arrays are therefore used to hold the number of nonzeros in each row and in each column, respectively. Also, the partial column list/row index structure is maintained only for the *candidate* rows and columns of $\mathbf{B}^{(k)}$, by which we mean rows and columns that have not (yet) contained a pivot element.

Markowitz Structure. To implement the Markowitz strategy, it is necessary to find from among the candidate rows and columns (i.e., the relevant submatrix of $\mathbf{B}^{(k)}$) the nonzero element for which the product of number of other nonzeros in its row and number of other nonzeros in its column is least. This can be efficiently implemented by maintaining doubly linked lists of all candidate rows having the same number of nonzeros and all candidate columns having the same number of nonzeros. The kth position of another array holds the row index of the first candidate row with k nonzeros and this serves as a pointer into the corresponding doubly linked list of rows with k nonzeros elements. A similar array is defined for the first candidate column with k nonzeros elements. (The latter two arrays could have up to m elements, the dimension of $\mathbf{B}$, but generally this number will be much smaller than m, since $\mathbf{B}$ is sparse.)

Note that the doubly linked lists hold only candidate rows and columns. The foregoing structure makes it easy to search efficiently for columns with one nonzero, then rows with one nonzero, then columns with two nonzeros and so on, in a rather straightforward procedure to find the element with least Markowitz cost. Then the stability criterion is verified. Reid [1976] checks that the potential pivot element is at least as large, in magnitude, as 0.1 times the largest element *in the same row and the candidate columns.* This is convenient to implement within the above data structure, but it does not bound the size of elements in the $\mathbf{\Gamma}$ file. (See Exercise 5.4-1.) Instead, at a slight additional cost, one could implement partial pivoting with rows, by using the partial column list/row index structure to access all elements in candidate rows and the (potential) pivot column. (The technique for accessing these elements is analogous to that discussed at the end of Sec. 3.2.) If the pivot is not suitable, then another element with the same Markowitz cost can be found using the doubly linked lists and, if none exists, the element with the next best Markowitz cost. The stability criterion can again be checked, and so on.

We see that the above structure makes it possible to define efficiently the candidate rows and columns and to find the most suitable pivot (k', k''). These indices can be inserted into the arrays PMAP and QMAP holding the permutations.

At the end of each elimination step, i.e., after $\mathbf{B}^{(k+1)}$ has been formed from $\mathbf{B}^{(k)}$ as will be described next, the Markowitz structure is revised to take account of fill-in and to remove row k' and column k'', since they are no longer candidates.

The Elimination Step. To carry out the elimination, we must pivot in place on (k', k''). We discover from the partial column list/row index data structure, the candidate rows that contain an element to be eliminated. These rows and their elements can then be accessed from the row list/column index data structure. An elementary row operation for each such row, i.e., the appropriate linear combination of it and the pivot row, can then be efficiently implemented. Because the row containing an element to be eliminated will change after the row operation, a new entry may be opened for it at the end of the row list/column index structure and the old space temporarily wasted. (The structure is periodically compacted using a "garbage collection" routine.) As each element $\mathbf{\Pi}_k^{i_l}$ in (5.10-13) is formed it is simply added to the end of the $\mathbf{\Gamma}$ file.

When fill-in occurs, the partial column list/row index structure must also be updated suitably. Recall that it only keeps track of the candidate rows and columns. If cancellation during a row operation leads to an element falling below a small threshold level that defines it to be zero, then the corresponding entry in the column structure is flagged as being elbow room and moved to the end of the column. If fill-in occurs and there is insufficient elbow room in the column, then a new entry is opened for the column at the end of the partial column list/row index data structure and the old space temporarily wasted. It too is periodically compacted by the garbage collection.

The Permuted Upper Triangular Matrix $\tilde{U}$. At the end of the previous elimination process, $\mathbf{B}^{(m)}$ is held in the row list/column index data structure and defined to be $\tilde{\mathbf{U}}$. The column list/row index data structure for $\tilde{\mathbf{U}}$ can then be reconstructed from it. (It will not of itself be available, since $\mathbf{B}^{(m)}$ has no candidate rows and columns.)

Exercise 5.10-2. For any convenient sparse matrix $\mathbf{B}$ of small dimension, depict the complete data structure described above and carry out the first step of the elimination process.

5.10-2. Solving Systems of Equations: *FTRAN and BTRAN Modules*

At the end of the factorization, the $\mathbf{\Gamma}$ file will hold the list of elementary matrices, namely,

$$\tilde{\mathbf{L}}^{-1} = (\mathbf{\Pi}_{m-1}^{i_{n_{m-1}}} \cdots \mathbf{\Pi}_{m-1}^{i_1}) \cdots (\mathbf{\Pi}_1^{i_{n_1}} \cdots \mathbf{\Pi}_1^{i_1}). \tag{5.10-14}$$

We solve a system of equations $\mathbf{Bx} = \mathbf{d}$ by first solving

$$\tilde{\mathbf{U}}\mathbf{x} = (\mathbf{\Pi}_{m-1}^{i_{n_{m-1}}} \cdots \mathbf{\Pi}_1^{i_1})\mathbf{d}.$$

Thus the sequence of elementary operators in the $\mathbf{\Gamma}$ file must first be applied directly to $\mathbf{d}$ in a forward direction (hence the name "forward transformation" or *FTRAN*) giving, say, $\mathbf{y}$. $\tilde{\mathbf{U}}\mathbf{x} = \mathbf{y}$ can then be solved by a row-oriented back substitution (5.7-3), provided we access rows and columns of $\tilde{\mathbf{U}}$ through one level of indirection defined by the arrays PMAP and QMAP, i.e., replace row index i by the contents of PMAP[i] (say p_i, for convenience) and column index j by the contents of QMAP[j] (say q_j). Then the back substitution is defined by

$$x_{q_i} = \left(y_{p_i} - \sum_{j \neq q_i} \tilde{u}_{p_i j} x_j\right) \Big/ \tilde{u}_{p_i q_i} \qquad i = m, m-1, \ldots, 1.$$

Reid [1976] gives some useful devices for enhancing program efficiency, when $\mathbf{y}$ is sparse.

Similarly $\mathbf{B}^T\mathbf{x} = \mathbf{d}$ is solved as follows. First solve $\tilde{\mathbf{U}}^T\mathbf{y} = \mathbf{d}$, namely,

$$y_{p_i} = \left(d_{q_i} - \sum_{j \neq p_i} \tilde{u}_{j q_i} y_j\right) \Big/ \tilde{u}_{p_i q_i} \qquad i = 1, 2, \ldots, m.$$

This too can be programmed very efficiently as described in Reid [1976]. Then form

$$\mathbf{x} = ((\mathbf{\Pi}_1^{i_1})^T \cdots (\mathbf{\Pi}_1^{i_{n_1}})^T) \cdots ((\mathbf{\Pi}_{m-1}^{i_1})^T \cdots (\mathbf{\Pi}_{m-1}^{i_{n_{m-1}}})^T)\mathbf{y},$$

by applying the transposed operators in the $\mathbf{\Gamma}$ file traversed in a backward direction (hence the name "backward transformation" or *BTRAN*).

Exercise 5.10-3. Design a version of the entire implementation just described, when $\mathbf{B}$ (and $\mathbf{B}^{(k)}$) is stored as a *column list/row index* data structure and a *partial row list/column index* data structure (the reverse of the one used above) and column-oriented LU factorization is used (see end of Sec. 5.2 and Exercise 5.3-1). Note how much easier it is to perform partial pivoting by rows in this setting.

Notes

Secs. 5.1–5.4. See also Wilkinson [1965], Forsythe and Moler [1967].

Sec. 5.5. See also George and Ng [1984]. Example 5.5-2 is a variant of one given in Tomlin [1972a]. Generalizations of the Markowitz strategy are given in Osterby and Zlatev [1983].

Sec. 5.7. Operation counts may be found in Isaacson and Keller [1966].

Sec. 5.8. The perturbation analysis given here and the accuracy of computed solutions of triangular systems are fully discussed in Wilkinson [1965].

Sec. 5.9. The tableau version of the simplex method is discussed in numerous texts, for example, Bazaraa and Jarvis [1977] and Chvatal [1983].

Sec. 5.10. Reid [1976] gives a number of programming details to enhance efficiency. A column-oriented version of the procedure of this section is given by Gill et al. [1986].

6

The Basis Matrix: Updating and Solving

We have just seen how to develop a suitable representation or factorization of a basis matrix, which permits convenient and efficient solution of associated systems of linear equations in the simplex method. We now turn to the question of *updating* a factorization, as the columns of the basis matrix are replaced, one at a time, in successive cycles of the simplex method. We discuss first updating techniques when a *single* column is replaced. Here, for simplicity of notation, we shall assume that the nonsingular basis matrix is defined by the first m columns of the LP matrix $\mathbf{A}$ and denote it as usual by $\mathbf{B}^0 = [\mathbf{a}_1, \ldots, \mathbf{a}_m]$. We shall assume that column $\mathbf{a}_p$, for some index p, $1 \leq p \leq m$, is replaced by $\mathbf{a}_s$, $m < s \leq n$, giving a new nonsingular basis matrix

$$\mathbf{B} = [\mathbf{a}_1, \ldots, \mathbf{a}_{p-1}, \mathbf{a}_s, \mathbf{a}_{p+1}, \ldots, \mathbf{a}_m]. \tag{6.0-1}$$

We shall also assume that a factored representation of $\mathbf{B}^0$ is available, which permits convenient solution of $\mathbf{B}^0\tilde{\mathbf{a}}_s = \mathbf{a}_s$ for $\tilde{\mathbf{a}}_s$. Within this context, we discuss updating techniques in Secs. 6.1 through 6.5. Then in Sec. 6.6 we discuss sequences of updates as successive columns of $\mathbf{B}^0$ are replaced, and the solution of associated systems of linear equations. Finally in Sec. 6.7 we consider details of a practical nature.

6.1. The Product Form Update

Replacing column p of $\mathbf{B}^0$ by $\mathbf{a}_s$ to give $\mathbf{B}$ is equivalent to

$$\mathbf{B} = \mathbf{B}^0 - \mathbf{a}_p\mathbf{e}_p^T + \mathbf{a}_s\mathbf{e}_p^T \tag{6.1-1a}$$

where $\mathbf{e}_p$ is the pth column of the identity matrix. Since $\mathbf{B}^0\tilde{\mathbf{a}}_s = \mathbf{a}_s$ and $\mathbf{a}_p = \mathbf{B}^0\mathbf{e}_p$,

$$\mathbf{B} = \mathbf{B}^0 + (\mathbf{B}^0\tilde{\mathbf{a}}_s - \mathbf{B}^0\mathbf{e}_p)\mathbf{e}_p^T = \mathbf{B}^0(\mathbf{I} + (\tilde{\mathbf{a}}_s - \mathbf{e}_p)\mathbf{e}_p^T). \tag{6.1-1b}$$

Thus

$$\mathbf{B} = \mathbf{B}^0\mathbf{E} \tag{6.1-2}$$

where $\mathbf{E} \equiv (\mathbf{I} + (\tilde{\mathbf{a}}_s - \mathbf{e}_p)\mathbf{e}_p^T)$ is an elementary matrix (Sec. 5.1-1).

Exercise 6.1-1. Given $\mathbf{B}$ defined by (6.0-1) and $(\mathbf{B}^0)^{-1}$, show that $(\mathbf{B}^0)^{-1}\mathbf{B}$ is an elementary matrix. Use this to give an alternative derivation of (6.1-2).

Exercise 6.1-2. Show that the Gauss–Jordan factorization (5.9-5) of the matrix $\mathbf{B}^0$ can be developed by beginning with the identity matrix, replacing its columns, one at a time, by columns of $\mathbf{B}^0$ and applying (6.1-2) after each replacement.

Expression (6.1-2) is known as *the product form update.* Historically the update was developed in the form $\mathbf{B}^{-1} = \mathbf{E}^{-1}(\mathbf{B}^0)^{-1}$ and known as the product form of the *inverse.* However, it is just as easy to develop the update in the form (6.1-2) and, in keeping with our practice of not using the inverse in computational procedures, we shall solve $\mathbf{Bx} = \mathbf{d}$ as follows:

1. Use the factorization of $\mathbf{B}^0$ to solve $\mathbf{B}^0\mathbf{y} = \mathbf{d}$ for $\mathbf{y}$.
2. Solve $\mathbf{Ex} = \mathbf{y}$ by back substitution, as in (5.9-7).

A system of the form $\mathbf{B}^T\mathbf{x} = \mathbf{d}$ can be solved in an analogous way.

It should be clear that the update (6.1-2) corresponds to a single step of Gauss–Jordan factorization (see also Exercise 6.1-2) and it suffers from the disadvantages of the latter, as was discussed in Sec. 5.9-2.

First, there is the potential for numerical instability when $|(\tilde{\mathbf{a}}_s)_p|$ is small relative to other elements of $\tilde{\mathbf{a}}_s$ taken in modulus, i.e., when $\mathbf{E}$ is an ill-conditioned matrix. This means, of course, that $\mathbf{a}_s$, the column replacing $\mathbf{a}_p$, is almost linearly dependent on other columns of $\mathbf{B}^0$, so that $\mathbf{B}$ is itself ill conditioned. We cannot then expect to accurately solve $\mathbf{Bx} = \mathbf{d}$. However at a later cycle of the simplex method, $\mathbf{B}$ may, in turn, lead to a well-conditioned matrix; then *retaining an ill-conditioned* elementary matrix $\mathbf{E}$ in the product form will have serious numerical consequences. This will become clearer in Sec. 6.6 when we discuss sequences of updates.

Second, $\tilde{\mathbf{a}}_s$ can be quite dense, for reasons that are identical to those given after Lemma 5.9-1, i.e., the fill-in characteristics of the product form update may be less desirable than those of an update of the LU factors, which we discuss next.

6.2. Updating the LU Factors

Suppose $\mathbf{P}^0\mathbf{B}^0 = \mathbf{L}^0\mathbf{U}^0$ is a stable factorization of the nonsingular matrix $\mathbf{B}^0$ obtained by partial pivoting (see Theorem 5.4-1). In what follows it will be convenient for purposes of exposition, to employ the mathematical operator $(\mathbf{L}^0)^{-1}$, but note that in keeping with our usual practice, all computations involving it will be carried out by back substitution, as in Sec. 5.7.

Let **B** as usual be the updated nonsingular basis matrix (6.0-1), namely,

$$\mathbf{B} = \mathbf{B}^0 + (\mathbf{a}_s - \mathbf{a}_p)\mathbf{e}_p^T = [\mathbf{a}_1, \ldots, \mathbf{a}_{p-1}, \mathbf{a}_s, \mathbf{a}_{p+1}, \ldots, \mathbf{a}_m].$$

Then

$$\begin{aligned}(\mathbf{L}^0)^{-1}\mathbf{P}^0\mathbf{B} &= (\mathbf{L}^0)^{-1}[\mathbf{P}^0\mathbf{a}_1, \ldots, \mathbf{P}^0\mathbf{a}_{p-1}, \mathbf{P}^0\mathbf{a}_s, \mathbf{P}^0\mathbf{a}_{p+1}, \ldots, \mathbf{P}^0\mathbf{a}_m]\\ &= [\mathbf{u}_1^0, \ldots, \mathbf{u}_{p-1}^0, (\mathbf{L}^0)^{-1}\mathbf{P}^0\mathbf{a}_s, \mathbf{u}_{p+1}^0, \ldots, \mathbf{u}_m^0]\\ &= [\mathbf{u}_1^0, \ldots, \mathbf{u}_{p-1}^0, \bar{\mathbf{a}}_s, \mathbf{u}_{p+1}^0, \ldots, \mathbf{u}_m^0] \equiv \mathbf{S}, \qquad (6.2\text{-}1)\end{aligned}$$

where $\bar{\mathbf{a}}_s$ can be computed by solving $\mathbf{L}^0\bar{\mathbf{a}}_s = \mathbf{P}^0\mathbf{a}_s$. **S** is an upper triangular matrix in all but the pth column, which can extend below the diagonal, as depicted in Figure 6.1.

6.3. The Bartels–Golub Update

Column p of **S** is moved to the last position and columns $p+1, \ldots, m$ are each moved forward by one position by postmultiplying **S** by a suitable permutation matrix **Q**. Thus

$$(\mathbf{L}^0)^{-1}\mathbf{P}^0\mathbf{B}\mathbf{Q} = \mathbf{S}\mathbf{Q} \equiv \mathbf{H} = [\mathbf{u}_1^0, \ldots, \mathbf{u}_{p-1}^0, \mathbf{u}_{p+1}^0, \ldots, \mathbf{u}_m^0, \mathbf{h}_m], \qquad (6.3\text{-}1)$$

where $\mathbf{h}_m \equiv \bar{\mathbf{a}}_s$. **H** is now an *upper Hessenberg matrix,* i.e., a matrix for which $h_{ij} = 0$, $i \geq j+2$. Figure 6.1 gives an example of the matrices **S** and **H** with $p = 3$.

Figure 6.1 Update of LU factors

$$\mathbf{S} = \begin{bmatrix} x & 0 & x & x & 0 & 0\\ . & x & 0 & 0 & x & x\\ . & . & x & x & 0 & x\\ . & . & 0 & x & x & 0\\ . & . & 0 & . & x & 0\\ . & . & x & . & . & x \end{bmatrix} \qquad \mathbf{H} = \begin{bmatrix} x & 0 & x & 0 & 0 & x\\ . & x & 0 & x & x & 0\\ . & . & x & 0 & x & x\\ . & . & x & x & 0 & 0\\ . & . & . & x & 0 & 0\\ . & . & . & . & x & x \end{bmatrix}.$$

Note when $\mathbf{U}^0$ is nonsingular that all its diagonal elements are nonzero. Equation (6.3-1) implies that all subdiagonal elements of **H** in columns $p, \ldots, m-1$ are nonzero.

Clearly now we can eliminate these subdiagonal elements of **H** in turn, as in the LU factorization of $\mathbf{B}^0$. The key point is that we can stabilize this process by partial pivoting (by rows) and this may be carried out as follows:

$$\mathbf{H}^{(p)} \equiv \mathbf{H}.$$

For $k = p, \ldots, m-1$,

$$\tilde{\mathbf{H}}^{(k)} = \mathbf{P}_{k,k'}\mathbf{H}^{(k)} \tag{6.3-2a}$$

$$\mathbf{H}^{(k+1)} = \tilde{\mathbf{\Gamma}}_k\tilde{\mathbf{H}}^{(k)} \tag{6.3-2b}$$

where $\mathbf{P}_{k,k'}$ denotes either the elementary permutation matrix $\mathbf{P}_{k,k'}$ or the identity matrix, if no interchange is necessary. $\tilde{\mathbf{\Gamma}}_k$ denotes an elementary lower triangular matrix, which is a special case of (5.1-6), namely,

$$\tilde{\mathbf{\Gamma}}_k = \begin{bmatrix} 1 & \cdot & \cdot \\ \cdot & \cdot & \cdot \\ \cdot & \cdot & \cdot \\ \cdot & \cdot & \cdot \\ \cdot & 1 & \cdot \\ \cdot & -\tilde{h}^{(k)}_{k+1,k}/\tilde{h}^{(k)}_{kk} & \cdot \\ \cdot & \cdot & \cdot \\ \cdot & \cdot & \cdot \\ \cdot & \cdot & 1 \end{bmatrix}. \tag{6.3-3}$$

Strategies for determining $\mathbf{P}_{k,k'}$ are analogous to those discussed in the previous chapter, but there is now much less flexibility in the choice of pivot element. Thus column interchanges are not permissible because this would destroy the Hessenberg structure. At step k the choice essentially lies between elements (k, k) and $(k+1, k)$ and one could seek a compromise between pivoting for stability and for sparsity, as discussed earlier.

The Bartels–Golub update thus takes the following form, analogous to (5.3-3),

$$\tilde{\mathbf{\Gamma}}_{m-1}\mathbf{P}_{(m-1),(m-1)'} \cdots \tilde{\mathbf{\Gamma}}_p\mathbf{P}_{p,p'}((\mathbf{L}^0)^{-1}\mathbf{P}^0\mathbf{BQ}) = \mathbf{H}^{(m)} \equiv \mathbf{U} \tag{6.3-4}$$

where $\mathbf{U}$ is an upper triangular matrix. Let us also define $\tilde{\mathbf{L}}_k \equiv \tilde{\mathbf{\Gamma}}_k^{-1}$, $k = p, \ldots, m-1$. Then

$$\mathbf{P}^0\mathbf{BQ} = \mathbf{L}^0\mathbf{P}_{p,p'}\tilde{\mathbf{L}}_p \cdots \mathbf{P}_{(m-1),(m-1)'}\tilde{\mathbf{L}}_{m-1}\mathbf{U}. \tag{6.3-5}$$

Exercise 6.3-1. Using arguments analogous to those of Sec. 5.3, show that

$$\mathbf{P}^0\mathbf{BQ} = (\mathbf{L}^0\mathbf{PL})\mathbf{U}$$

for suitably defined $\mathbf{P}$ and $\mathbf{L}$, where $\mathbf{P}$ is a permutation matrix and $\mathbf{L}$ is a unit lower triangular matrix. (Note: $(\mathbf{L}^0\mathbf{PL})$ is *not* a lower triangular matrix.)

In practice, $\mathbf{B}$ would be maintained in the *product form* (6.3-4) or (6.3-5) and used to solve the associated systems of linear equations of the form $\mathbf{Bx} = \mathbf{d}$ and $\mathbf{B}^T\mathbf{x} = \mathbf{d}$ by successive back substitution.

Bartels–Golub updating with partial pivoting (by rows) is very satisfactory from a numerical standpoint. A result formalizing this would be analogous to Theorem 5.4-1. As just mentioned, analogous rules to those of Sec. 5.6, for making a compromise between pivoting for stability and pivoting for sparsity, can be applied here. The main drawback of the update stems from the fact that one cannot predict in advance where the fill-in will occur in the submatrix of $\mathbf{H}$ corresponding to rows p through m and columns p through m, during the elimination process used to obtain $\mathbf{U}$. For example, in Figure 6.1, when the element in position $(4,3)$ of $\mathbf{H}$ is eliminated, a fill-in of elements $(4,5)$ and $(4,6)$ occurs. When $\mathbf{U}^0$ (and hence $\mathbf{H}$ and $\mathbf{U}$ that are derived from $\mathbf{U}^0$) is stored as a packed data structure, as would typically be the case, there may be a substantial data processing overhead associated with such unpredictable fill-in. We shall have more to say on this subsequently.

Exercise 6.3-2. Suppose that the matrix $\mathbf{H}$ of (6.3-1) was transformed to $\bar{\mathbf{H}} = \mathbf{Q}^T\mathbf{H}$. Characterize $\bar{\mathbf{H}}$ and give an alternative formulation of Bartels–Golub updating that transforms $\bar{\mathbf{H}}$ to an upper triangular matrix $\mathbf{U}$.

Exercise 6.3-3. Work out the simple modifications to (6.3-5) when $\mathbf{B}^0$ is initially factored in the form (5.3-1), namely, $\mathbf{P}^0\mathbf{B}^0\mathbf{Q}^0 = \mathbf{L}^0\mathbf{U}^0$.

6.4. The Forrest–Tomlin Update

As noted in the previous section, when $\mathbf{B}^0$ and consequently $\mathbf{U}^0$ are nonsingular, all relevant subdiagonal elements of $\mathbf{H}$ are nonzero. Let us successively eliminate *elements in row p* in positions (p,p), $(p,p+1),\ldots,(p,m-1)$ by elementary row operations, namely, by adding suitable multiples, in turn, of rows $p+1, p+2, \ldots, m$ to the pth row of $\mathbf{H}$. This would then define the following sequence of matrices:

$$\mathbf{H}^{(p)} \equiv \mathbf{H}.$$

For $k = p, \ldots, m-1$

$$\mathbf{H}^{(k+1)} = \mathbf{\Pi}_k\mathbf{H}^{(k)} \tag{6.4-1a}$$

Figure 6.2 Forrest–Tomlin update

$$\mathbf{H}^{(m)} = \begin{bmatrix} x & 0 & x & 0 & 0 & x \\ . & x & 0 & x & x & 0 \\ . & . & 0 & 0 & 0 & x \\ . & . & x & x & 0 & 0 \\ . & . & . & x & 0 & 0 \\ . & . & . & . & x & x \end{bmatrix}$$

where $\mathbf{\Pi}_k$ denotes an elementary upper triangular matrix, of the form

$$\mathbf{\Pi}_k = \begin{bmatrix} 1 & \cdot & \cdot & \cdot \\ \cdot & \cdot & \cdot & \cdot \\ \cdot & \cdot & \cdot & \cdot \\ \cdot & \cdot & \cdot & \cdot \\ \cdot & 1 & -h_{pk}^{(k)}/h_{k+1,k}^{(k)} & \cdot \\ \cdot & \cdot & \cdot & \cdot \\ \cdot & \cdot & \cdot & \cdot \\ \cdot & \cdot & 1 & \cdot \\ \cdot & \cdot & \cdot & \cdot \\ \cdot & \cdot & \cdot & \cdot \\ \cdot & \cdot & \cdot & 1 \end{bmatrix} \qquad (6.4\text{-}1b)$$

where the distinguished element is in row p and column $k+1$. (Because no permutations are involved, we use the notation $\mathbf{\Pi}_k$, not $\tilde{\mathbf{\Pi}}_k$.) At the end of this process, $\mathbf{H}^{(m)}$ has a pth row consisting of zero elements in all columns save the last one. For the example of Figure 6.1, it would take the form given in Figure 6.2.
Thus

$$\mathbf{H}^{(m)} = (\mathbf{\Pi}_{m-1} \cdots \mathbf{\Pi}_p)\mathbf{H} = [\mathbf{u}_1^0, \ldots, \mathbf{u}_{p-1}^0, \bar{\mathbf{u}}_{p+1}^0, \ldots, \bar{\mathbf{u}}_m^0, (\mathbf{\Pi}_{m-1} \ldots \mathbf{\Pi}_p)\mathbf{h}_m] \qquad (6.4\text{-}2)$$

where $\bar{\mathbf{u}}_j^0$ denotes the jth column of $\mathbf{U}^0$ with the element in row p set to zero and $\mathbf{h}_m$ is defined just after (6.3-1). Let us make the following definitions.

$$\mathbf{R}_j \equiv \mathbf{\Pi}_j^{-1} \qquad j = p, \ldots, m-1, \qquad (6.4\text{-}3a)$$

$$\mathbf{\Pi} \equiv (\mathbf{\Pi}_{m-1} \cdots \mathbf{\Pi}_p) \qquad \mathbf{R} \equiv (\mathbf{R}_p \cdots \mathbf{R}_{m-1}) = \mathbf{\Pi}^{-1}. \qquad (6.4\text{-}3b)$$

Thus

$$\mathbf{H}^{(m)} = \mathbf{\Pi}\mathbf{H} = [\mathbf{u}_1^0, \ldots, \mathbf{u}_{p-1}^0, \bar{\mathbf{u}}_{p+1}^0, \ldots, \bar{\mathbf{u}}_m^0, \mathbf{\Pi}\mathbf{h}_m]. \qquad (6.4\text{-}4)$$

Observe now when row p of $\mathbf{H}^{(m)}$ is permuted to the last position and rows $p+1, \ldots, m$ are all moved up one position, i.e., to rows p through $m-1$, that we obtain an upper triangular matrix, say $\mathbf{U}$. Furthermore, given $\mathbf{Q}$ defined by (6.3-1), the permutation that effects this row interchange is precisely $\mathbf{Q}^{-1}$ (or equivalently, $\mathbf{Q}^T$); see Sec. 5.1-2 and also Exercise 6.3-2. Thus

$$\mathbf{Q}^T\mathbf{H}^{(m)} = \mathbf{U}. \qquad (6.4\text{-}5)$$

Combining this with (6.4-4) and (6.3-1) gives

$$\mathbf{Q}^T\mathbf{\Pi}(\mathbf{L}^0)^{-1}\mathbf{P}^0\mathbf{B}\mathbf{Q} = \mathbf{U}$$

$$\mathbf{P}^0\mathbf{B} = \mathbf{L}^0\mathbf{R}(\mathbf{Q}\mathbf{U}\mathbf{Q}^T) = \mathbf{L}^0\mathbf{R}_p \cdots \mathbf{R}_{m-1}(\mathbf{Q}\mathbf{U}\mathbf{Q}^T). \tag{6.4-6}$$

From the form of $\mathbf{\Pi}_k$ in (6.4-1b), it is evident that

$$\mathbf{\Pi} = (\mathbf{\Pi}_{m-1} \cdots \mathbf{\Pi}_p) = (\mathbf{I} - \mathbf{e}_p\mathbf{r}^T) \tag{6.4-7}$$

where

$$\mathbf{r}^T = [0, \ldots, 0, (h_{pp}^{(p)}/h_{p+1,p}^{(p)}), \ldots, (h_{pk}^{(k)}/h_{k+1,k}^{(k)}), \ldots, (h_{p,m-1}^{(m-1)}/h_{m,m-1}^{(m-1)})]^T, \tag{6.4-8}$$

where the zero elements are in the first p positions of the foregoing vector. Also

$$\mathbf{R} = \mathbf{\Pi}^{-1} = (\mathbf{I} + \mathbf{e}_p\mathbf{r}^T). \tag{6.4-9}$$

From (6.4-6), $\mathbf{P}^0\mathbf{B}\mathbf{Q} = \mathbf{L}^0(\mathbf{R}\mathbf{Q}\mathbf{U})$, but note that $(\mathbf{RQU})$ is not upper triangular. As in Bartels–Golub updating, we would normally maintain the Forrest–Tomlin update in the product form (6.4-6) and use it to solve the associated systems of linear equations.

Forrest and Tomlin [1972] also noted that $\mathbf{\Pi}$ could be computed directly by means of the following lemma.

Lemma 6.4-1. Let $\boldsymbol{u}^T = [0, \ldots, 0, u_{p,p+1}^0, \ldots, u_{pm}^0]^T$, where u_{pj}^0 is the element of $\mathbf{U}^0$ in position (p, j). Let $\boldsymbol{r}^T$ be the solution of

$$\boldsymbol{r}^T\mathbf{U}^0 = \boldsymbol{u}^T, \tag{6.4-10}$$

and let

$$\boldsymbol{\Pi} \equiv (\mathbf{I} - \mathbf{e}_p\boldsymbol{r}^T). \tag{6.4-11}$$

Then

$$\boldsymbol{\Pi}\mathbf{H} = [\mathbf{u}_1^0, \ldots, \mathbf{u}_{p-1}^0, \bar{\mathbf{u}}_{p+1}^0, \ldots, \bar{\mathbf{u}}_m^0, \boldsymbol{\Pi}\mathbf{h}_m], \tag{6.4-12}$$

where $\mathbf{u}_j^0$, $\bar{\mathbf{u}}_j^0$, and $\mathbf{h}_m$ are defined as in (6.4-2).

Proof. We shall verify this result directly. From (6.4-10) and (6.4-11),

$$\boldsymbol{\Pi}\mathbf{U}^0 = (\mathbf{I} - \mathbf{e}_p\boldsymbol{r}^T)\mathbf{U}^0 = \mathbf{U}^0 - \mathbf{e}_p\boldsymbol{u}^T.$$

Thus

$$\boldsymbol{\Pi}\mathbf{U}^0 = [\mathbf{u}_1^0, \ldots, \mathbf{u}_{p-1}^0, \bar{\mathbf{u}}_p^0, \ldots, \bar{\mathbf{u}}_m^0]. \tag{6.4-13}$$

This matrix is identical to $\mathbf{U}^0$ in all elements save those in row p and columns $p, \ldots, m$, which are now zero. From (6.3-1),

$$\mathbf{H} = [\mathbf{u}_1^0, \ldots, \mathbf{u}_{p-1}^0, \mathbf{u}_{p+1}^0, \ldots, \mathbf{u}_m^0, \mathbf{h}_m].$$

$$\boldsymbol{\Pi}\mathbf{H} = [\boldsymbol{\Pi}\mathbf{u}_1^0, \ldots, \boldsymbol{\Pi}\mathbf{u}_{p-1}^0, \boldsymbol{\Pi}\mathbf{u}_{p+1}^0, \ldots, \boldsymbol{\Pi}\mathbf{u}_m^0, \boldsymbol{\Pi}\mathbf{h}_m].$$

Now using (6.4-13) we have

$$\boldsymbol{\Pi}\mathbf{H} = [\mathbf{u}_1^0, \ldots, \mathbf{u}_{p-1}^0, \bar{\mathbf{u}}_{p+1}^0, \ldots, \bar{\mathbf{u}}_m^0, \boldsymbol{\Pi}\mathbf{h}_m]$$

and this is the required result. ■

It remains to be shown that $\mathbf{\Pi}$ defined by (6.4-7) and $\boldsymbol{\Pi}$ defined by (6.4-11) are identical. We leave this to the following exercise.

Exercise 6.4-1. Observe that if $\mathbf{\Pi H}$ and $\boldsymbol{\Pi}\mathbf{H}$ do differ from one another, it can only be in position (p, m). Show that these elements are identical and deduce that $\mathbf{r}$ and $\boldsymbol{r}$ in (6.4-8) and (6.4-10), respectively, are the same.

The Forrest–Tomlin update has *completely predictable fill-in.* This makes for very convenient data management. For example, when $\mathbf{U}^0$ is internally represented as a packed data structure (e.g., a column list/row index structure), all that is necessary when updating $\mathbf{U}^0$ is to flag elements of row p in each column as zero, and modify the last column. Flagging elements can be done, for example, by setting their row indices to a nonpositive number that can later be used in a garbage-collection operation. There is, however, a price to be paid for this advantage. The Forrest–Tomlin update replaces partial pivoting of the Bartels–Golub update by a preassigned pivot order. It is not then possible to bound, a priori, the size of the elements of $\mathbf{r}$ in (6.4-8) (or (6.4-10)) and the Forrest–Tomlin update is therefore potentially unstable. The saving grace is that one can detect this a posteriori and take remedial action, for example, by refactorizing $\mathbf{B}$ in a stable manner.

Finally, we leave as exercises the following variants on the Forrest–Tomlin update.

Exercise 6.4-2. Consider the following.

(a) Permute row p of $\mathbf{H}$ in (6.3-1) to the bottom and move rows $p+1, \ldots, m$ up one position, i.e., to positions $p, \ldots, m-1$. (This is equivalent to premultiplying $\mathbf{H}$ by $\mathbf{Q}^T$.) Then eliminate elements of the *last row* by a sequence of elementary lower triangular matrices.

(b) Perform a Bartels–Golub update of $\mathbf{H}$ with an interchange forced at *every* step, i.e., $\mathbf{P}_{k,k'}$ is always $\mathbf{P}_{k,k+1}$, $k = p, \ldots, m-1$.

Show that both the foregoing variants are mathematically equivalent to Forrest–Tomlin updating and compare and contrast the three approaches from the point of view of data management.

Variant (b) above, coupled with Exercise 6.3-2, has the following implication. When Bartels–Golub updating is carried out with a pivot strategy that balances considerations of stability and sparsity as discussed in Sec. 6.3, it may produce a final factorization (6.3-5) that is *more sparse* than the factorization (6.4-6). The reason for this is as follows: In Forrest–Tomlin updating, fill-in may occur in row p *during the course of the update.* Of course these new nonzero elements will be subsequently eliminated, but each will contribute an elementary matrix $\mathbf{R}_k$ to the final product form. The end result may well be that the number of nonzeros in the final product form (6.4-6), coming from the matrices $\mathbf{R}_k$, $\mathbf{L}^0$, and $\mathbf{U}$, may be greater than the number of nonzeros in the product form (6.3-5), coming from $\mathbf{L}_k$, $\mathbf{L}^0$, and $\mathbf{U}$. (Note that the updates lead to different matrices $\mathbf{U}$, though, for convenience, we use the same notation.) Because the cost of solving the associated systems of linear equations is proportional to the number of nonzeros in the factorization, some of the advantages of the Forrest–Tomlin update, from the point of view of data management, may be offset. There will be further discussion of such considerations in Sec. 6.6 and in Chapter 10.

6.5. The Fletcher–Matthews Update

The techniques studied so far update the LU factorization of $\mathbf{B}^0$ in a product form. As noted earlier (see, in particular, Exercise 6.3-1), the Bartels–Golub update does not give, even implicitly, a unit lower triangular matrix $\boldsymbol{L}$ and an upper triangular matrix $\boldsymbol{U}$ such that

$$\boldsymbol{P}\mathbf{B}\boldsymbol{Q} = \boldsymbol{L}\boldsymbol{U}, \tag{6.5-1}$$

where $\boldsymbol{P}$ and $\boldsymbol{Q}$ are suitable permutation matrices. This is what the Fletcher–Matthews update sets out to do, i.e., it is, in a sense, *the* LU update. (To highlight this fact, we use $\boldsymbol{P}$, $\boldsymbol{Q}$, $\boldsymbol{L}$, and $\boldsymbol{U}$ for the updated factors in this section.)

Because the notation can become unwieldy and obscure the main point, we carry out the development in terms of a 4×4 matrix. The reader will then have no difficulty in extending results to the more general case of an $m \times m$ update. We pick up the discussion from (6.3-1), namely,

$$(\mathbf{L}^0)^{-1}(\mathbf{P}^0\mathbf{B})\mathbf{Q} = \mathbf{H} \equiv \mathbf{H}^{(1)},$$

where $\mathbf{H}^{(1)}$ is now depicted in Figure 6.3. For convenience, and without loss of generality, we both ignore sparsity and assume that the first column of $\mathbf{B}^0$ is replaced, i.e., $p = 1$. Also, as noted earlier, when $\mathbf{B}^0$ is nonsingular, $h_{k+1,k} \neq 0$ for all k.

Suppose that we carried out Bartels–Golub updating with the option of row interchange (partial pivoting by rows) but *no interchanges were actually needed* to preserve stability, i.e., $\mathbf{P}_{k,k'} \equiv \mathbf{I}$ for all k. The update would then take the form

$$\tilde{\boldsymbol{\Gamma}}_3\tilde{\boldsymbol{\Gamma}}_2\tilde{\boldsymbol{\Gamma}}_1(\mathbf{L}^0)^{-1}(\mathbf{P}^0\mathbf{B}\mathbf{Q}) = \mathbf{U}$$

where $\mathbf{U}$ is upper triangular. Thus

$$\mathbf{P}^0\mathbf{B}\mathbf{Q} = (\mathbf{L}^0\tilde{\mathbf{L}}_1\tilde{\mathbf{L}}_2\tilde{\mathbf{L}}_3)\mathbf{U}$$

where $\tilde{\boldsymbol{\Gamma}}_k$ is defined by (6.3-2) and (6.3-3) and, as usual, $\tilde{\mathbf{L}}_k \equiv \tilde{\boldsymbol{\Gamma}}_k^{-1}$. (Because no permutations are actually used, the symbol "tilde" is not strictly necessary here.) With the definitions $\boldsymbol{L} \equiv \mathbf{L}^0\tilde{\mathbf{L}}_1\tilde{\mathbf{L}}_2\tilde{\mathbf{L}}_3$, $\boldsymbol{P} \equiv \mathbf{P}^0$, $\boldsymbol{Q} \equiv \mathbf{Q}$, and $\boldsymbol{U} \equiv \mathbf{U}$, we have the update of the triangular factorization

(6.5-1). Clearly therefore the novelty in the Fletcher–Matthews update must arise because of the need for interchanges, i.e., when $\mathbf{P}_{k,k'} \neq \mathbf{I}$ for some k.

Let us therefore return to the Bartels–Golub update and now assume that an interchange was performed at the first step, so $\mathbf{P}_{1,1'} = \mathbf{P}_{1,2}$.

$$\tilde{\mathbf{H}}^{(1)} = \mathbf{P}_{1,2}\mathbf{H}^{(1)},$$
$$\bar{\mathbf{H}}^{(1)} = \tilde{\mathbf{\Gamma}}_1\tilde{\mathbf{H}}^{(1)} = \tilde{\mathbf{\Gamma}}_1\mathbf{P}_{1,2}\mathbf{H}^{(1)}. \tag{6.5-2}$$

In Bartels–Golub updating we proceed directly to the processing of the second column with $\mathbf{H}^{(2)}$ identified with $\bar{\mathbf{H}}^{(1)}$, in order to eliminate the element in position (3, 2); in Fletcher–Matthews updating, (6.5-2) represents only an intermediate stage in the processing of the first column, which is continued in a manner now to be described. For convenience, in the iterative procedure that follows, let us define $\mathbf{L}^{(1)} \equiv \mathbf{L}^0$ and denote the elements of $\mathbf{L}^{(1)}$ by $l_{ij}^{(1)}$. Then from (6.5-2) and the definition of $\mathbf{H}^{(1)}$,

$$\tilde{\mathbf{\Gamma}}_1\mathbf{P}_{1,2}(\mathbf{L}^{(1)})^{-1}\mathbf{P}^0\mathbf{B}\mathbf{Q} = \bar{\mathbf{H}}^{(1)},$$
$$\mathbf{P}^0\mathbf{B}\mathbf{Q} = (\mathbf{L}^{(1)}\mathbf{P}_{1,2}\tilde{\mathbf{L}}_1)\bar{\mathbf{H}}^{(1)}. \tag{6.5-3}$$

So far there has been no significant departure from Bartels–Golub updating. We now describe how the true triangular factors are obtained through a very simple modification.

Consider the expression in parenthesis in (6.5-3), namely, the matrix

$$\mathbf{M}^{(1)} \equiv \mathbf{L}^{(1)}\mathbf{P}_{1,2}\tilde{\mathbf{L}}_1.$$

This is obviously lower triangular in all but the first two columns, as depicted in Figure 6.3.

Figure 6.3 Fletcher–Matthews update

$$\mathbf{H}^{(1)} = \begin{bmatrix} x & x & x & x \\ x & x & x & x \\ 0 & x & x & x \\ 0 & 0 & x & x \end{bmatrix} \qquad \mathbf{M}^{(1)} = \begin{bmatrix} x & x & 0 & 0 \\ x & x & 0 & 0 \\ x & x & 1 & 0 \\ x & x & x & 1 \end{bmatrix}$$

$$\mathbf{U}^{(2)} = \begin{bmatrix} x & x & 0 & 0 \\ 0 & x & 0 & 0 \\ 0 & 0 & 1 & 0 \\ 0 & 0 & 0 & 1 \end{bmatrix} \qquad \mathbf{H}^{(2)} = \begin{bmatrix} x & x & x & x \\ 0 & x & x & x \\ 0 & x & x & x \\ 0 & 0 & x & x \end{bmatrix}.$$

The elements $m_{11}^{(1)}$ and $m_{21}^{(1)}$ of the matrix $\mathbf{M}^{(1)}$ are given by

$$m_{11}^{(1)} = l_{12}^{(1)} + l_{11}^{(1)}(\tilde{h}_{21}^{(1)}/\tilde{h}_{11}^{(1)}) = (\tilde{h}_{21}^{(1)}/\tilde{h}_{11}^{(1)}), \tag{6.5-4a}$$
$$m_{21}^{(1)} = l_{22}^{(1)} + l_{21}^{(1)}(\tilde{h}_{21}^{(1)}/\tilde{h}_{11}^{(1)}) = 1 + l_{21}^{(1)}(\tilde{h}_{21}^{(1)}/\tilde{h}_{11}^{(1)}). \tag{6.5-4b}$$

Both of the foregoing expressions use the fact that $\mathbf{L}^{(1)}$ is unit lower triangular. Observe that $m_{11}^{(1)}=0$ implies that $m_{21}^{(1)}\neq 0$, so clearly both elements cannot be simultaneously zero. We can therefore perform the triangular factorization of the nonsingular matrix $\mathbf{M}^{(1)}$ with partial pivoting restricted to the first *two* rows, *without the process breaking down* because of a *zero* divisor. This factorization of $\mathbf{M}^{(1)}$ is given by

$$\mathbf{P}_{1,\bar{1}}\mathbf{M}^{(1)} = \mathbf{L}^{(2)}\mathbf{U}^{(2)}, \tag{6.5-5}$$

where $\mathbf{P}_{1,\bar{1}}$ is either the identity matrix (when no row interchange is performed) or the elementary permutation matrix $\mathbf{P}_{1,2}$ (when the first two rows are interchanged). $\mathbf{L}^{(2)}$ is unit lower triangular; $\mathbf{U}^{(2)}$ is upper triangular and of the form depicted in Figure 6.3. Note that only the first two columns of $\mathbf{M}^{(1)}$ change during the foregoing step, which can therefore be efficiently performed in $O(m)$ operations; see also Exercise 6.5-2.

Therefore from (6.5-3),

$$\mathbf{P}^0\mathbf{B}\mathbf{Q} = \mathbf{M}^{(1)}\bar{\mathbf{H}}^{(1)} = \mathbf{P}_{1,\bar{1}}\mathbf{L}^{(2)}(\mathbf{U}^{(2)}\bar{\mathbf{H}}^{(1)}).$$

Let $\mathbf{H}^{(2)} \equiv \mathbf{U}^{(2)}\bar{\mathbf{H}}^{(1)}$ and observe that $\mathbf{H}^{(2)}$ is also upper Hessenberg with $h_{21}^{(2)}=0$. This too is depicted in Figure 6.3. Thus

$$\mathbf{P}_{1,\bar{1}}\mathbf{P}^0\mathbf{B}\mathbf{Q} = \mathbf{L}^{(2)}\mathbf{H}^{(2)}. \tag{6.5-6}$$

This completes the first cycle of the iterative procedure, which is summarized by the example of Figure 6.3. The objective has been to obtain a factorization (6.5-6) that is of the required form (6.5-1), insofar as the first column is concerned.

We can repeat the above process for the second column of $\mathbf{H}^{(2)}$, in a completely analogous manner, as follows:

$$(\mathbf{L}^{(2)})^{-1}\mathbf{P}_{1,\bar{1}}\mathbf{P}^0\mathbf{B}\mathbf{Q} = \mathbf{H}^{(2)},$$
$$\tilde{\mathbf{H}}^{(2)} = \mathbf{P}_{2,2'}\mathbf{H}^{(2)}$$

where $\mathbf{P}_{2,2'}$ is either the identity matrix or the elementary permutation matrix $\mathbf{P}_{2,3}$, which interchanges the second and third rows of $\mathbf{H}^{(2)}$.

$$\bar{\mathbf{H}}^{(2)} = \tilde{\mathbf{\Gamma}}_2\tilde{\mathbf{H}}^{(2)}$$

where $\tilde{\mathbf{\Gamma}}_2$ is defined analogously to $\tilde{\mathbf{\Gamma}}_1$.

$$\mathbf{P}_{1,\bar{1}}\mathbf{P}^0\mathbf{B}\mathbf{Q} = (\mathbf{L}^{(2)}\mathbf{P}_{2,2'}\tilde{\mathbf{L}}_2)\bar{\mathbf{H}}^{(2)} = \mathbf{M}^{(2)}\bar{\mathbf{H}}^{(2)},$$

with the definitions $\tilde{\mathbf{L}}_2 \equiv \tilde{\mathbf{\Gamma}}_2^{-1}$ and $\mathbf{M}^{(2)} \equiv \mathbf{L}^{(2)}\mathbf{P}_{2,2'}\tilde{\mathbf{L}}_2$.

$$\mathbf{P}_{1,\bar{1}}\mathbf{P}^0\mathbf{B}\mathbf{Q} = \mathbf{P}_{2,\bar{2}}\mathbf{L}^{(3)}(\mathbf{U}^{(3)}\bar{\mathbf{H}}^{(2)}),$$

where $\mathbf{P}_{2,\bar{2}}\mathbf{M}^{(2)} = \mathbf{L}^{(3)}\mathbf{U}^{(3)}$ with $\mathbf{L}^{(3)}$ unit lower triangular, $\mathbf{U}^{(3)}$ upper triangular, and $\mathbf{P}_{2,\bar{2}}$ either the identity matrix or the permutation matrix $\mathbf{P}_{2,3}$. Therefore

$$\mathbf{P}_{2,\bar{2}}\mathbf{P}_{1,\bar{1}}\mathbf{P}^0\mathbf{B}\mathbf{Q} = \mathbf{L}^{(3)}\mathbf{H}^{(3)},$$

where

$$\mathbf{H}^{(3)} \equiv \mathbf{U}^{(3)}\bar{\mathbf{H}}^{(2)}.$$

Finally, after one more such iteration,

$$\mathbf{P}_{3,\bar{3}}\mathbf{P}_{2,\bar{2}}\mathbf{P}_{1,\bar{1}}\mathbf{P}^0\mathbf{B}\mathbf{Q} = \mathbf{L}^{(4)}\mathbf{H}^{(4)} \tag{6.5-7}$$

where $\mathbf{L}^{(4)}$ is unit lower triangular and $\mathbf{H}^{(4)}$ is now upper triangular. With the definitions $\boldsymbol{P} \equiv \mathbf{P}_{3,\bar{3}}\mathbf{P}_{2,\bar{2}}\mathbf{P}_{1,\bar{1}}\mathbf{P}^0$, $\boldsymbol{Q} \equiv \mathbf{Q}$, $\boldsymbol{L} \equiv \mathbf{L}^{(4)}$, and $\boldsymbol{U} \equiv \mathbf{H}^{(4)}$, we have the required factorization,

$$\boldsymbol{P}\mathbf{B}\boldsymbol{Q} = \boldsymbol{L}\boldsymbol{U}. \tag{6.5-8}$$

As just noted, the generalization to the $m \times m$ case is completely straightforward and we leave it to the following exercise.

Exercise 6.5-1. Develop the Fletcher–Matthews update of an $m \times m$ matrix factored as $\mathbf{P}^0\mathbf{B}^0\mathbf{Q}^0 = \mathbf{L}^0\mathbf{U}^0$.

The following two exercises address important practical issues of *efficiency* and *numerical stability* that are studied, in detail, in Fletcher and Matthews [1984].

Exercise 6.5-2. Verify that the first two columns of $\mathbf{L}^{(2)}$ in (6.5-5) are linear combinations of the first two columns of $\mathbf{L}^{(1)}$ and that other columns of these two matrices are identical. Use this observation to reformulate the foregoing procedure in order to enhance its efficiency.

Exercise 6.5-3. Investigate the choice of pivot strategies (separately or in conjunction) for determining $\mathbf{P}_{1,1'}$ and $\mathbf{P}_{1,\bar{1}}$ and, more generally, $\mathbf{P}_{k,k'}$ and $\mathbf{P}_{k,\bar{k}}$, in order to enhance numerical stability of the update by constraining growth in the elements of $\mathbf{H}^{(k+1)}$.

A judicious reformulation of the Bartels–Golub extension just described and choice of pivot strategy in the factorizations

$$\bar{\mathbf{H}}^{(k)} = \tilde{\mathbf{\Gamma}}_k\mathbf{P}_{k,k'}\mathbf{H}^{(k)} \quad \text{and} \quad \mathbf{P}_{k,\bar{k}}\mathbf{M}^{(k)} = \mathbf{L}^{(k+1)}\mathbf{U}^{(k+1)} \tag{6.5-9}$$

enable Fletcher and Matthews [1984] to carry out the update so that it is efficient and numerically well behaved. Note, however, when partial pivoting is restricted at each step to rows k and $k+1$, one *cannot* place an a priori bound on the size of elements of $\mathbf{L}^{(k+1)}$, in contrast to Bartels–Golub updating. To see this, let us return to the first iteration. The pivot choice is restricted to $m_{11}^{(1)}$ and $m_{21}^{(1)}$ in (6.5-4). Thus the elements of $\mathbf{L}^{(2)}$ in positions (3, 1) and (4, 1) are of the form $m_{i1}^{(1)}/m_{11}^{(1)}$, $i=3, 4$ or $m_{i1}^{(1)}/m_{21}^{(1)}$, $i=3, 4$ and they could be large. One can, however, bound elements of $\mathbf{L}^{(k+1)}$ by permitting partial pivoting in other rows, when factorizing $\mathbf{M}^{(k)}$ in (6.5-9). The price one pays for this is that $\mathbf{U}^{(k+1)}$ is no longer necessarily a simple matrix, with just one off-diagonal element and there is an associated increase in the cost of factorizing $\mathbf{M}^{(k)}$ and forming $\mathbf{H}^{(k+1)} = \mathbf{U}^{(k+1)}\bar{\mathbf{H}}^{(k)}$. For example, at the first iteration, suppose $\mathbf{P}_{1,\bar{1}}$ were taken to be $\mathbf{P}_{1,4}$. Then $\mathbf{U}^{(2)}$ would be a full upper triangular matrix. It will, however, only be necessary to invoke this more general pivoting strategy under extreme circumstances and it is clear that many improved strategies are possible, along lines analogous to those discussed in Secs. 5.4 through 5.6, and in Sec. 5.10. (See Exercises 6.5-2 and 6.5-3.) These strategies permit some limited growth in elements, as used in Bartels–Golub updating and as discussed in Fletcher and Matthews [1984].

6.6. Sequences of Updates and Solving

Let us now briefly consider the updating of earlier factorizations over a sequence of basis changes, the solution of the associated systems of linear equations, and the relative merits of the different updating techniques introduced earlier. We revert to our more general notation used in the description of the primal simplex algorithm (Sec. 2.4) along with the use of subscripts or superscripts to indicate iteration count. Denote the starting basis by

$$\mathbf{B}^0 = [\mathbf{a}_{\beta_1}, \ldots, \mathbf{a}_{\beta_m}]. \tag{6.6-1}$$

Let us also assume that $\mathbf{B}^0$ is factorized in the more general form (5.3-1), namely, $\mathbf{P}^0\mathbf{B}^0\mathbf{Q}^0 = \mathbf{L}^0\mathbf{U}^0$.

We consider first a sequence of product form updates. When column $\mathbf{a}_{s_1}$ replaces column p_1 of the basis matrix $\mathbf{B}^0, 1 \le p_1 \le m,$ then the product form update is given by

$$\mathbf{B}^1 = [\mathbf{a}_{\beta_1}, \ldots, \mathbf{a}_{\beta_{p_1}-1}, \mathbf{a}_{s_1}, \mathbf{a}_{\beta_{p_1}+1}, \ldots, \mathbf{a}_{\beta_m}] = \mathbf{B}^0\mathbf{E}^1,$$

where $\mathbf{E}^1 = \mathbf{I} + (\tilde{\mathbf{a}}_{s_1} - \mathbf{e}_{p_1})\mathbf{e}_{p_1}^T$ and $\tilde{\mathbf{a}}_{s_1}$ is the solution of $\mathbf{B}^0\tilde{\mathbf{a}}_{s_1} = \mathbf{a}_{s_1}$ obtained using the factorization of $\mathbf{B}^0$.

Similarly, at the kth iteration, let $\mathbf{a}_{s_k}$ denote the entering column and p_k denote the index of the column of the basis marix $\mathbf{B}^{k-1}$ which is replaced. Then,

$$\mathbf{B}^k = \mathbf{B}^{k-1}\mathbf{E}^k = \mathbf{B}^0\mathbf{E}^1\mathbf{E}^2 \cdots \mathbf{E}^k, \tag{6.6-2}$$

where

$$\mathbf{E}^k = \mathbf{I} + (\tilde{\mathbf{a}}_{s_k} - \mathbf{e}_{p_k})\mathbf{e}_{p_k}^T, \tag{6.6-3}$$

and $\tilde{\mathbf{a}}_{s_k}$ is the solution of

$$\mathbf{B}^{k-1}\tilde{\mathbf{a}}_{s_k} = \mathbf{a}_{s_k}. \tag{6.6-4}$$

In order to solve a system of linear equations of the form $\mathbf{B}^k\mathbf{x} = \mathbf{d}$, we can carry out a sequence of back substitutions in the usual way, namely,

1. Solve $\mathbf{B}^0\mathbf{y}_0 = \mathbf{d}$ using the factorization of $\mathbf{B}^0$.
2. For $i = 1, \ldots, k,$ solve $\mathbf{E}^i\mathbf{y}_i = \mathbf{y}_{i-1}$ by back substitution, noting that $\mathbf{E}^i$ is a permuted lower triangular matrix (cf. (5.9-7)).
3. Set $\mathbf{x} = \mathbf{y}_k$.

With each iteration of the primal simplex algorithm, a new elementary matrix is added to the product form. The representation (6.6-2) thus grows continuously and at some iteration, say k, the cost of an update operation and the successive back substitution involved in solving the associated systems of linear equations will exceed the cost of developing a new factorization of $\mathbf{B}^k$ (*refactorizing* the basis) and using the new factors to solve the associated equations. The reason for this is that the new factored representation will contain fewer terms and will be more sparse.

Next, let us consider a sequence of Bartels–Golub updates. Again we use the foregoing notation. At the first iteration,

$$\tilde{\boldsymbol{\Gamma}}^1_{m-1}\mathbf{P}^1_{m-1,(m-1)'}\cdots\tilde{\boldsymbol{\Gamma}}^1_{p_1}\mathbf{P}^1_{p_1,p_1'}(\mathbf{L}^0)^{-1}(\mathbf{P}^0\mathbf{B}^1\mathbf{Q}^0)\mathbf{Q}^1=\mathbf{U}^1 \qquad (6.6\text{-}5a)$$

or

$$\mathbf{B}^1=(\mathbf{P}^0)^T\mathbf{L}^0[\mathbf{P}^1_{p_1,p_1'}\tilde{\mathbf{L}}^1_{p_1}\cdots\mathbf{P}^1_{m-1,(m-1)'}\tilde{\mathbf{L}}^1_{m-1}]\mathbf{U}^1(\mathbf{Q}^1)^T(\mathbf{Q}^0)^T. \qquad (6.6\text{-}5b)$$

Similarly, after k iterations,

$$\mathbf{B}^k=(\mathbf{P}^0)^T\mathbf{L}^0[\mathbf{P}^1_{p_1,p_1'}\tilde{\mathbf{L}}^1_{p_1}\cdots]\cdots[\mathbf{P}^k_{p_k,p_k'}\tilde{\mathbf{L}}^k_{p_k}\cdots$$
$$\cdots\mathbf{P}^k_{m-1,(m-1)'}\tilde{\mathbf{L}}^k_{m-1}]\mathbf{U}^k(\mathbf{Q}^k)^T\cdots(\mathbf{Q}^0)^T. \qquad (6.6\text{-}6)$$

Then, to solve a system of equations $\mathbf{B}^k\mathbf{x}=\mathbf{d}$, we proceed as follows.

1. Solve $\mathbf{L}^0\mathbf{y}=\mathbf{P}^0\mathbf{d}$ for $\mathbf{y}$.
2. Form

$$\bar{\mathbf{d}}=[\tilde{\boldsymbol{\Gamma}}^k_{m-1}\mathbf{P}^k_{m-1,(m-1)'}\cdots\tilde{\boldsymbol{\Gamma}}^1_{p_1}\mathbf{P}^1_{p_1,p_1'}]\mathbf{y},$$

   or equivalently, solve the following system by successive back substitution,

$$[\mathbf{P}^1_{p_1,p_1'}\tilde{\mathbf{L}}^1_{p_1}\cdots\mathbf{P}^k_{m-1,(m-1)'}\tilde{\mathbf{L}}^k_{m-1}]\bar{\mathbf{d}}=\mathbf{y}.$$

3. Solve $\mathbf{U}^k\bar{\mathbf{x}}=\bar{\mathbf{d}}$.
4. Form $\mathbf{x}=\mathbf{Q}^0\cdots\mathbf{Q}^k\bar{\mathbf{x}}$.

Again, the number of factors in the updated representation will eventually grow so large that it would be more efficient and also better from a numerical standpoint, to refactorize $\mathbf{B}^k$ and begin afresh.

The Forrest-Tomlin update over a sequence of basis changes can be carried out in an analogous way. The Fletcher–Matthews update is the most straightforward of all, because it produces the true LU factors at each iteration. Its distinguishing feature is that *there is no expanding file of product terms*. Thus to solve the associated systems of linear equations, one would employ successive back substitution, as in Sec. 5.7.

Let us now consider the relative merits of the different updates. The product form *update* is so simple and elegant that, in contrast to Gauss–Jordan *factorization,* there will always be room for it in the updating repertoire, despite the drawbacks noted at the end of Sec. 6.1;

namely, if some basis matrix, say $\mathbf{B}^{\bar{k}}$, is ill-conditioned, the associated ill-conditioned elementary matrix $\mathbf{E}^{\bar{k}}$ will persist in the product form representation of subsequent basis matrices, even when they are well conditioned, all the way up to the next refactorization. This is rather unsatisfactory from a numerical standpoint. In contrast, the Bartels–Golub update would produce an ill-conditioned matrix $\mathbf{U}^{\bar{k}}$, associated with the ill-conditioned matrix $\mathbf{B}^{\bar{k}}$, but on subsequent iterations, it would "forget" about this situation, i.e., subsequent well-conditioned basis matrices $\mathbf{B}^k$, $k > \bar{k}$, would have associated well-conditioned factors $\mathbf{U}^k$. As we have noted earlier, the Forrest–Tomlin update is potentially unstable on well-conditioned matrices. There is an obvious recourse, namely, refactorize when ill-conditioning is detected, a posteriori, by the appearance of large elements in the vector $\mathbf{r}$ defined by (6.4-8) or (6.4-10). Note however that this is a sufficient but not a necessary condition for ill-conditioning, as discussed in Sec. 5.8.

In terms of sparsity of representation, we have seen that updating techniques based on the LU factorization are preferable to techniques based on a Gauss–Jordan approach, notably the product form update. This means that the former will require fewer refactorizations. As between the Bartels–Golub and the Forrest–Tomlin updates, we have noted that the Bartels–Golub, with a sensible pivoting strategy, which balances considerations of stability against considerations of fill-in, may even be more effective in preserving sparsity of the factored representation. The Fletcher–Matthews update has an obvious advantage. There is no expanding file of operators and therefore no implied refactorization. However, for numerical reasons, refactorization may still be desirable. The Fletcher–Matthews update has obvious advantages for relatively small linear programming problems, where sparsity is not an issue. It is a recent development and its use for sparse updating requires further research. It would be premature, as of this writing, to assess its value for large sparse applications.

From the point of view of data management, the Forrest–Tomlin update, with its completely predictable fill-in characteristics, has some very clear advantages. Indeed this provided the initial motivation for its development. In contrast, the Bartels–Golub update requires a carefully designed data structure to handle unpredictable fill-in. We shall give further details of Forrest–Tomlin updating in Chapter 10. The implementation of Bartels–Golub updating for sparse applications is the subject of the next section.

6.7. Practical Details of Implementation

We return to the implementation of Reid [1976] (see Sec. 5.10) which was designed with the needs of Bartels–Golub updating of the LU factors very much in mind. Reall that the factorization, with pivoting in place, takes the form

$$(\tilde{\mathbf{L}}^0)^{-1}\mathbf{B}^0 = \tilde{\mathbf{U}}^0,$$

where $\tilde{\mathbf{L}}^0$ is defined by a sequence of elementary operators stored in the $\mathbf{\Gamma}$ file, and $\tilde{\mathbf{U}}^0$ is a permuted upper triangular matrix, represented as a row list/column index packed data structure and a partial column list/row index packed data structure. We now attach the superscript 0 to these quantities, as discussed at the beginning of this chapter, because we are concerned with a sequence of updates; and we attach the symbol "tilde" because we are now employing pivoting in place and are dealing with *permuted* lower and upper triangular matrices.

When column p of the basis matrix $\mathbf{B}^0$ is replaced to give $\mathbf{B}$, the above factorization becomes

$$\mathbf{B} = \tilde{\mathbf{L}}^0\tilde{\mathbf{V}}$$

where $\tilde{\mathbf{V}}$ differs from $\tilde{\mathbf{U}}^0$ only in column p, and this is $(\tilde{\mathbf{L}}^0)^{-1}$ times the entering column. Indeed, this vector is available from the previous iteration of the primal simplex algorithm and will not need to be recomputed. It can be directly read into the data structure to replace column p of $\tilde{\mathbf{U}}^0$. Then we also have

$$\mathbf{S} \equiv \mathbf{P}^0\tilde{\mathbf{V}}\mathbf{Q}^0,$$

where $\mathbf{P}^0$ and $\mathbf{Q}^0$ are the permutation matrices such that $\mathbf{U}^0 = \mathbf{P}^0\tilde{\mathbf{U}}^0\mathbf{Q}^0$ is upper triangular and defined by the arrays PMAP and QMAP of Sec. 5.10. $\mathbf{S}$ is upper triangular in all but one column, called the *spike* column. This occurs, say, in column $\bar{p}$ of $\mathbf{S}$, with QMAP$(\bar{p}) = p$. Since the implementation is row oriented, the matrix $\mathbf{S}$ is converted to one with a *spike row* as discussed in Exercise 6.3-2, namely,

$$\bar{\mathbf{S}} = \mathbf{Q}^T\mathbf{S}\mathbf{Q} = (\mathbf{Q}^T\mathbf{P}^0)\tilde{\mathbf{V}}(\mathbf{Q}^0\mathbf{Q}) \equiv \bar{\mathbf{P}}\tilde{\mathbf{V}}\bar{\mathbf{Q}},$$

where $\mathbf{Q}$ moves column $\bar{p}$ of $\mathbf{S}$ to the last position and advances each column after $\bar{p}$ by one position; $\mathbf{Q}^T$ does the same for rows. The arrays PMAP and QMAP can be revised to hold $\bar{\mathbf{P}}$ and $\bar{\mathbf{Q}}$.

Because the spike row of $\bar{\mathbf{S}}$ is sparse, it will often be possible to carry out further permutations of the rows and the columns of $\bar{\mathbf{S}}$ to advance the spike row from the last position and simultaneously to reduce its length. Details of this refinement may be found in Reid [1976]. Again, PMAP and QMAP can be revised appropriately. The rest of the implementation closely parallels the description of Bartels–Golub updating of Sec. 6.3, with indirect addressing through the PMAP array when using row indices, i.e., row index i of $\bar{\mathbf{S}}$ is converted to PMAP(i) to access the corresponding row of $\tilde{\mathbf{V}}$ in the packed data structure. At each step, as we have noted, the choice of pivot element is much simpler than the Markowitz strategy used in the LU factorization. In particular, the linked-list structure of Sec. 5.10-1 is not needed. There is no flexibility in the choice of pivot column because column interchanges would destroy the upper triangular structure in $\bar{\mathbf{S}}$ and at each step of the elimination there is a choice between two pivot rows, namely, the current spike row and the row with the index of the pivot column. The row with fewer nonzeros is chosen, subject to the pivot element meeting a suitable stability requirement. The actual elimination is carried out as described in Sec. 5.10-1 and little further detail is needed here. As the eliminations are performed, elementary operators are added to the $\mathbf{\Gamma}$ file. At the end of the process we have

$$\mathbf{B} = \tilde{\mathbf{L}}\tilde{\mathbf{U}},$$

and permutation matrices $\mathbf{P}$ and $\mathbf{Q}$ such that $\mathbf{U} = \mathbf{P}\tilde{\mathbf{U}}\mathbf{Q}$ is an upper triangular matrix. Note that $\tilde{\mathbf{L}}$ as defined by the $\mathbf{\Gamma}$ file is *not* a permuted lower triangular matrix, which is a characteristic of Bartels–Golub updating, as noted in Sec. 6.3. The $\mathbf{\Gamma}$ file only holds a permuted lower triangular matrix immediately after the initial factorization or after a refactorization.

Finally, we may observe that the solution of the associated systems of linear equations in the *FTRAN* and *BTRAN* operations is much the same as the processes described in Sec. 5.10–2.

Notes

Secs. 6.1–6.4. The main references here are Bartels [1971], Bartels and Golub [1969], Forrest and Tomlin [1972], Tomlin [1972b].

Sec. 6.5. The description here is somewhat different from that of Fletcher and Matthews [1984], and builds more directly upon the Bartels–Golub approach. In particular, it permits more direct use of the error analysis techniques of Wilkinson [1963, 1965].

Sec. 6.6. A good discussion of the relative merits of the different updates is given in Saunders [1980].

Sec. 6.7. For a column-oriented version of this procedure see Gill et al. [1986].

7
Selection Strategies: Choosing the Entering and Exiting Variables

The primal simplex algorithm computes a price vector $\boldsymbol{\pi}^0$ and a vector of reduced costs or Lagrange multipliers, $\boldsymbol{\sigma}_N^0$, associated with the nonbasic variables whose current values are $\mathbf{x}_N^0$. An improving nonbasic variable, for example the one whose reduced cost is largest in magnitude and of the appropriate sign, is then chosen to enter the basis. This is known as *pricing*.

In this chapter we study pricing in more detail. We shall first consider, in turn: (1) selection criteria, namely, the quantities on which a choice is based, (2) selection strategies for actually making the choice of entering variable. Next, we consider strategies for choosing the exiting variable. Finally, we discuss practical aspects of implementation.

7.1. Choosing the Entering Variable

Suppose $\mathbf{x}^0$ denotes the current basic feasible solution at some iteration of the simplex algorithm. The variable introduced next into the basis should bring about as large an improvement or *gain* in the objective value as possible. One might be willing to accept a smaller gain at the current iteration, if this holds the promise of more substantial gain at future iterations, but relatively little progress has been made, to date, on pricing strategies that are global in nature. Thus we concentrate on local strategies, which seek to maximize improvement during the current iteration.

Recall from Sec. 2.1 or Sec. 2.2, when the nonbasic variable x_j, $j \in N$ is introduced into the basis by changing its value x_j^0 by an amount τ, the gain in objective value is

$$\Delta z_j = \sigma_j^0 \tau,$$

where σ_j^0 is the reduced cost or Lagrange multiplier associated with the variable x_j.

It is immediately clear that no matter how large σ_j^0, this being a measure of the rate of change of objective value, there will be no actual gain if x_j happens to be a fixed variable, i.e., if $x_j^0 = l_j = u_j$, so that $\tau \equiv 0$. Again, suppose that the basis is degenerate or nearly degenerate, i.e., suppose there is a basic variable at or very close to one of its bounds. Then, in order to avoid very small values of τ, it would be advisable to select an incoming nonbasic variable that causes the degenerate or nearly degenerate variable to move away from its tight (or nearly tight) bound, even though there may be other candidates that give a larger rate of improvement. It is therefore useful to employ more than one selection criterion when making a choice. The standard approach is to incur the expense of defining additional (implicit) objective functions and, for each one, to compute an associated price vector and a vector of reduced costs. These quantities, when used in conjunction with the usual vectors $\boldsymbol{\pi}^0$ and $\boldsymbol{\sigma}_N^0$, facilitate the selection of an entering variable with a better promise of gain.

7.1-1. Selection Criteria in the Presence of Degenerate or Nearly Degenerate Basic Variables

We now discuss in more detail the point just raised, namely, how to try to avoid selecting an entering variable that results in very small values of τ.

We use the simplifying assumptions and notation of Sec. 2.2, namely, that the LP matrix $\mathbf{A}$ is partitioned into $\mathbf{A} = [\mathbf{B}^0 \mid \mathbf{N}^0]$ corresponding to basic and nonbasic variables $\mathbf{x}_B$ and $\mathbf{x}_N$ with $B = \{1, \ldots, m\}$ and $N = \{m+1, \ldots, n\}$. $\mathbf{B}^0$ is nonsingular and $\tilde{\mathbf{a}}_j$ is the solution of $\mathbf{B}^0 \tilde{\mathbf{a}}_j = \mathbf{a}_j$. $\mathbf{x}^0 = \begin{bmatrix} \mathbf{x}_B^0 \\ \mathbf{x}_N^0 \end{bmatrix}$ denotes the current basic feasible solution.

Let us define an auxiliary objective function or cost vector $\mathbf{c}^d \equiv \begin{bmatrix} \mathbf{c}_B^d \\ \mathbf{c}_N^d \end{bmatrix}$ as follows.

For $i \in B$,

$$c_i^d = \begin{cases} +1 \text{ if } |x_i^0 - l_i| \le e \max(1, |l_i|) \\ -1 \text{ if } |x_i^0 - u_i| \le e \max(1, |u_i|) \\ 0 \text{ otherwise,} \end{cases} \tag{7.1-1}$$

where e is a small constant.

For $i \in N$,

$$c_i^d = 0. \tag{7.1-2}$$

The price vector associated with $\mathbf{c}_B^d$ is

$$(\boldsymbol{\pi}^d)^T = (\mathbf{c}_B^d)^T (\mathbf{B}^0)^{-1} \tag{7.1-3}$$

and the reduced costs σ_j^d, $j \in N$, are computed as follows.

$$\sigma_j^d = c_j^d - (\boldsymbol{\pi}^d)^T \mathbf{a}_j. \tag{7.1-4a}$$

Thus σ_j^d is given by

$$\sigma_j^d = -(\mathbf{c}_B^d)^T\tilde{\mathbf{a}}_j = -\sum_{i\in\bar{L}} \tilde{a}_{ij} + \sum_{i\in\bar{U}} \tilde{a}_{ij} \tag{7.1-4b}$$

where $\bar{L}$ denotes the subset of basic varibles that are at or almost at their lower bound, as given in (7.1-1); similarly, $\bar{U}$ denotes the subset of basic variables that are at or almost at their upper bounds. Note that (7.1-4a) is used when actually computing σ_j^d, i.e., $\tilde{\mathbf{a}}_j$ is *not* computed for each $j \in N$.

As before, when a particular nonbasic variable, say x_j, is introduced into the basis by changing its value by an amount τ, the associated change in value of the auxiliary objective is

$$\Delta z_j^d = \sigma_j^d \tau \qquad j \in N, \tag{7.1-5}$$

and, for $i \in B$,

$$x_i = x_i^0 - \tilde{a}_{ij}\tau. \tag{7.1-6}$$

Suppose x_j is at its lower bound, $x_j^0 = l_j$, so that τ must be increased from zero (denoted by $\tau \uparrow 0$) and suppose the corresponding reduced cost σ_j^d in (7.1-4a) satisfies $\sigma_j^d < 0$. Then from (7.1-4b) and (7.1-6), one or both of the following two cases must be true.

1. There exists a basic variable, with value x_i^0 at or close to its lower bound, for which the corresponding $\tilde{a}_{ij} > 0$. Then x_i will *decrease* as $\tau \uparrow 0$.
2. There exists a basic variable, with value x_i^0 at or close to its upper bound, for which $\tilde{a}_{ij} < 0$. Then x_i will *increase* as $\tau \uparrow 0$.

In either case there will be an immediate loss of feasibility as $\tau \uparrow 0$. Thus x_j, $j \in N$ is only a promising candidate when $\sigma_j^d > 0$.

Exercise 7.1-1. Make analogous arguments to show when x_j is at its upper bound, and τ must be decreased from zero ($\tau \downarrow 0$), that x_j will only be a promising candidate if $\sigma_j^d < 0$.

Thus a suitable pricing strategy to handle degenerate or nearly degenerate basic variables might be to compute *two* price vectors, $\boldsymbol{\pi}^0$ and $\boldsymbol{\pi}^d$. When a nonbasic varible x_j, $j \in N$ seems to be a promising candidate on the basis of its reduced cost σ_j^0, then σ_j^d is computed from the price vector $\boldsymbol{\pi}^d$. x_j is only selected as a possible candidate when σ_j^d indicates a promise of gain, as described above.

Exercise 7.1-2. Show that the converse of the above conditions, namely, $\sigma_j^d > 0$ when $\tau \uparrow 0$ or $\sigma_j^d < 0$ when $\tau \downarrow 0$ may but do *not* necessarily imply that degenerate or near degenerate basic variables move away from their tight (or nearly tight) bounds.

7.1-2. Selection Criteria Based on Steepest-Edge

With the partition $\mathbf{A} = [\mathbf{B}^0 \mid \mathbf{N}^0]$ of the previous section, it will now be convenient to use our alternative notation,

$$(\boldsymbol{\sigma}_N^0)_k = (\mathbf{c}_N)_k - (\boldsymbol{\pi}^0)^T \mathbf{n}_k^0 \qquad k = 1, \ldots, n-m. \tag{7.1-7a}$$

Recall from Sec. 2.2, in particular, (2.2-8) that

$$(\boldsymbol{\sigma}_N^0)_k = \mathbf{c}^T\mathbf{z}_k = \mathbf{c}^T\begin{bmatrix} -(\mathbf{B}^0)^{-1}\mathbf{n}_k^0 \\ \mathbf{e}_k \end{bmatrix} \tag{7.1-7b}$$

where $\mathbf{z}_k$ normally defines an *edge* of the polytope, leading from the current vertex or basic feasible solution to an adjacent vertex. Let us recall the steepest-edge extension in Sec. 2.2, in particular (2.2-21). Consider a step from $\mathbf{x}^0$ along $\mathbf{z}_k$ to a new point $\mathbf{x}^k \equiv \mathbf{x}^0 + \mathbf{z}_k$. Thus

$$\mathbf{c}^T\mathbf{x}^k - \mathbf{c}^T\mathbf{x}^0 = \mathbf{c}^T\mathbf{z}_k = (\boldsymbol{\sigma}_N^0)_k.$$

$(\boldsymbol{\sigma}_N^0)_k$ gives the change in objective value for the step $\mathbf{z}_k$. In comparing $(\boldsymbol{\sigma}_N^0)_{k'}$ and $(\boldsymbol{\sigma}_N^0)_{k''}$, say, it would be more reasonable to normalize the corresponding steps $\mathbf{z}_{k'}$ and $\mathbf{z}_{k''}$. Let us therefore define

$$(\boldsymbol{\sigma}_N^s)_k \equiv \frac{(\boldsymbol{\sigma}_N^0)_k}{\|\mathbf{z}_k\|_2} \qquad k = 1, \ldots, n - m \tag{7.1-8}$$

where $\|.\|_2$ denotes the Euclidean norm. Since $(\boldsymbol{\sigma}_N^s)_k = \mathbf{c}^T(\mathbf{z}_k/\|\mathbf{z}_k\|_2)$ and $\mathbf{c}$ is the *gradient* of the objective function $\mathbf{c}^T\mathbf{x}$, we see that $(\boldsymbol{\sigma}_N^s)_k$ is the *directional derivative* along $\mathbf{z}_k$. Selecting the entering variable for which $(\boldsymbol{\sigma}_N^s)_k$ is largest in magnitude and of the appropriate sign (negative when $(\mathbf{x}_N^0)_k$ is at a lower bound and positive when $(\mathbf{x}_N^0)_k$ is at an upper bound) corresponds to choosing the *steepest-edge.*

At first glance, the need for $\|\mathbf{z}_k\|_2$, $k = 1, \ldots, n - m$ would seem to imply that each vector $\tilde{\mathbf{a}}_j \equiv (\mathbf{B}^0)^{-1}\mathbf{a}_j$, $j \in N$ must be computed. However, Greenberg and Kalan [1975] and, independently, Goldfarb and Reid [1977] showed how $\|\tilde{\mathbf{a}}_j\|_2$ can be *updated* from one iteration of the simplex algorithm to the next, instead of having to be computed afresh. This may be demonstrated as follows.

Suppose, as in Sec. 6.1, we replace column p of the basis matrix by the column $\mathbf{a}_s$. As usual, let $\mathbf{B}^0 = [\mathbf{a}_1, \ldots, \mathbf{a}_m]$ and let $\mathbf{B} = [\mathbf{a}_1, \ldots, \mathbf{a}_{p-1}, \mathbf{a}_s, \mathbf{a}_{p+1}, \ldots, \mathbf{a}_m]$ be the updated basis matrix. Then from (6.1-1),

$$\mathbf{B} = \mathbf{B}^0[\mathbf{I} + (\tilde{\mathbf{a}}_s - \mathbf{e}_p)\mathbf{e}_p^T],$$

where $\tilde{\mathbf{a}}_s = (\mathbf{B}^0)^{-1}\mathbf{a}_s$.

For $j \in N$ and $j \neq s$,

$$\bar{\mathbf{a}}_j \equiv \mathbf{B}^{-1}\mathbf{a}_j = [\mathbf{I} + (\tilde{\mathbf{a}}_s - \mathbf{e}_p)\mathbf{e}_p^T]^{-1}(\mathbf{B}^0)^{-1}\mathbf{a}_j = [\mathbf{I} - (\tilde{\mathbf{a}}_s - \mathbf{e}_p)\mathbf{e}_p^T/(\tilde{\mathbf{a}}_s^T\mathbf{e}_p)]\tilde{\mathbf{a}}_j.$$

Thus

$$\|\bar{\mathbf{a}}_j\|_2^2 = \tilde{\mathbf{a}}_j^T[\mathbf{I} - \mathbf{e}_p(\tilde{\mathbf{a}}_s - \mathbf{e}_p)^T/(\tilde{\mathbf{a}}_s^T\mathbf{e}_p)][\mathbf{I} - (\tilde{\mathbf{a}}_s - \mathbf{e}_p)\mathbf{e}_p^T/(\tilde{\mathbf{a}}_s^T\mathbf{e}_p)]\tilde{\mathbf{a}}_j.$$

Then it is a matter of simple algebra to expand the product of the two expressions in parentheses and collect terms, leading to

$$\|\bar{\mathbf{a}}_j\|_2^2 = \|\tilde{\mathbf{a}}_j\|_2^2 + (\gamma + 2/(\tilde{\mathbf{a}}_s)_p)(\tilde{\mathbf{a}}_j^T\mathbf{e}_p)^2 - (2/(\tilde{\mathbf{a}}_s)_p)(\tilde{\mathbf{a}}_j^T\mathbf{e}_p)(\tilde{\mathbf{a}}_j^T\tilde{\mathbf{a}}_s), \tag{7.1-9}$$

where

$$\gamma \equiv (\tilde{\mathbf{a}}_s - \mathbf{e}_p)^T(\tilde{\mathbf{a}}_s - \mathbf{e}_p)/(\tilde{\mathbf{a}}_s)_p^2.$$

Note that $(\tilde{\mathbf{a}}_s)_p$ and $\tilde{\mathbf{a}}_s^T\tilde{\mathbf{a}}_s$ and hence γ can be immediately computed. Computing $\|\bar{\mathbf{a}}_j\|_2$ efficiently, given $\|\tilde{\mathbf{a}}_j\|_2$, therefore hinges on our ability to efficiently compute $\tilde{\mathbf{a}}_j^T\mathbf{e}_p$ and $\tilde{\mathbf{a}}_j^T\tilde{\mathbf{a}}_s$ for each $j \in N$. But now observe that

$$\tilde{\mathbf{a}}_j^T\mathbf{e}_p = \mathbf{a}_j^T((\mathbf{B}^0)^{-T}\mathbf{e}_p) \tag{7.1-10a}$$

and

$$\tilde{\mathbf{a}}_j^T\tilde{\mathbf{a}}_s = \mathbf{a}_j^T((\mathbf{B}^0)^{-T}\tilde{\mathbf{a}}_s), \tag{7.1-10b}$$

where $(\mathbf{B}^0)^{-T} \equiv ((\mathbf{B}^0)^T)^{-1}$. Let us, therefore, define two new price vectors $\boldsymbol{\pi}_+$ and $\boldsymbol{\pi}_-$, as

$$\boldsymbol{\pi}_+^T = \mathbf{e}_p^T(\mathbf{B}^0)^{-1} \tag{7.1-11a}$$

and

$$\boldsymbol{\pi}_-^T \equiv \tilde{\mathbf{a}}_s^T(\mathbf{B}^0)^{-1}. \tag{7.1-11b}$$

These would, of course, be computed by solving the systems of linear equations $\boldsymbol{\pi}_+^T\mathbf{B}^0 = \mathbf{e}_p^T$ and $\boldsymbol{\pi}_-^T\mathbf{B}^0 = \tilde{\mathbf{a}}_s^T$ in a numerically stable way. Thus

$$\tilde{\mathbf{a}}_j^T\mathbf{e}_p = \boldsymbol{\pi}_+^T\mathbf{a}_j \quad \text{and} \quad \tilde{\mathbf{a}}_j^T\tilde{\mathbf{a}}_s = \boldsymbol{\pi}_-^T\mathbf{a}_j. \tag{7.1-12}$$

Therefore, for $j \in N$, $j \neq s$,

$$\|\bar{\mathbf{a}}_j\|_2^2 = \|\tilde{\mathbf{a}}_j\|_2^2 + (\gamma + 2/(\tilde{\mathbf{a}}_s)_p)(\boldsymbol{\pi}_+^T\mathbf{a}_j)^2 - (2/(\tilde{\mathbf{a}}_s)_p)(\boldsymbol{\pi}_+^T\mathbf{a}_j)(\boldsymbol{\pi}_-^T\mathbf{a}_j). \tag{7.1-13}$$

The update of $\|\mathbf{z}_k\|_2^2$, $k = 1, \ldots, n - m$, $(m + k) \neq s$ follows immediately.

The update of $\|\bar{\mathbf{a}}_p\|_2$, corresponding to the remaining nonbasic variable x_p, is left to the following exercise.

Exercise 7.1-3. Observe that $\tilde{\mathbf{a}}_p = (\mathbf{B}^0)^{-1}\mathbf{a}_p = \mathbf{e}_p$. Use this and (7.1-9) to develop an expression analogous to (7.1-13) to update $\|\bar{\mathbf{a}}_p\|_2$.

7.1-3. Selection Criteria Based on Considerations of Scale Invariance

Given a linear program (1.2-2), consider a rescaling of variables (*column scaling*) of the form $\mathbf{x} = \mathbf{S}\mathbf{x}'$, where $\mathbf{S} = \text{diag}[s_1, \ldots, s_n] > 0$ is a diagonal matrix with positive elements. Let $\mathbf{A}$, $\mathbf{b}$, $\mathbf{c}$, $\mathbf{B}^0$, $\mathbf{N}^0$, $\mathbf{x}_B^0$, $\boldsymbol{\pi}^0$, σ_j^0, $\mathbf{a}_j$, $\mathbf{z}_k$, $\mathbf{l}$, $\mathbf{u}$ denote the usual quantities as in Sec. 7.1-2 and Chapter 2, and let us attach the symbol "prime" to the corresponding transformed quantities under the change of scale. Then

1. $\mathbf{A}' = \mathbf{A}\mathbf{S}$; $\mathbf{b}' = \mathbf{b}$; $\mathbf{c}' = \mathbf{S}\mathbf{c}$; $\mathbf{c}_B' = \mathbf{S}_B\mathbf{c}_B$, where $\mathbf{S}_B = \text{diag}[s_1, \ldots, s_m] > 0$; $\mathbf{l}' = \mathbf{S}^{-1}\mathbf{l}$; $\mathbf{u}' = \mathbf{S}^{-1}\mathbf{u}$.
2. $\mathbf{B}'^0 = \mathbf{B}^0\mathbf{S}_B$; $\mathbf{N}'^0 = \mathbf{N}^0\mathbf{S}_N$ where $\mathbf{S}_N = \text{diag}[s_{m+1}, \ldots, s_n] > 0$; $\mathbf{x}_B'^0 = \mathbf{S}_B^{-1}\mathbf{x}_B^0$.
3. $(\boldsymbol{\pi}'^0)^T = (\mathbf{c}_B')^T(\mathbf{B}'^0)^{-1} = \mathbf{c}_B^T(\mathbf{B}^0)^{-1} = (\boldsymbol{\pi}^0)^T$.
4. $\sigma_j'^0 = c_j' - (\boldsymbol{\pi}'^0)^T\mathbf{a}_j' = (c_j - (\boldsymbol{\pi}^0)^T\mathbf{a}_j)s_j = s_j\sigma_j^0$, $j \in N$.
5. $\tilde{\mathbf{a}}_j' = \mathbf{S}_B^{-1}\tilde{\mathbf{a}}_j s_j$, $j \in N$.

The price vector $\boldsymbol{\pi}^0$ is *invariant* under the column scaling, i.e., it does not depend on the choice of units for the variables. If $s_j = 1, j \in N,$ then the reduced costs $\sigma_j^0, j \in N$ are invariant regardless of $\mathbf{S}_B$. We shall say $\boldsymbol{\sigma}_N^0$ is B-column scale invariant. However, $\boldsymbol{\sigma}_N^0$ is *not* invariant when $\mathbf{S}_N \neq \mathbf{I}$. We shall say that it is not N-column scale invariant. The relative worth of introducing a particular variable $x_j, j \in N$ into the basis, as measured by its reduced cost, can thus be *arbitrarily* changed by its column scale (choice of units).

Consider now a *row scaling* of the constraints $\mathbf{Ax} = \mathbf{b}$ given by $\mathbf{RAx} = \mathbf{Rb}$ where $\mathbf{R} = \text{diag}[r_1, \ldots, r_m] > 0$. The objective function will remain unscaled. If we again attach the symbol "prime" to transformed quantities, we have

6. $\mathbf{A}' = \mathbf{RA}$; $\mathbf{b}' = \mathbf{Rb}$; $\mathbf{c}' = \mathbf{c}$.
7. $\mathbf{B}'^0 = \mathbf{RB}^0$; $\mathbf{N}'^0 = \mathbf{RN}^0$; $\mathbf{x}_B'^0 = \mathbf{x}_B^0$.
8. $(\boldsymbol{\pi}'^0)^T = (\mathbf{c}_B')^T(\mathbf{B}'^0)^{-1} = \mathbf{c}_B^T(\mathbf{B}^0)^{-1}\mathbf{R}^{-1} = (\boldsymbol{\pi}^0)^T\mathbf{R}^{-1}$.
9. $\sigma_j'^0 = c_j' - (\boldsymbol{\pi}'^0)^T\mathbf{a}_j' = c_j - (\boldsymbol{\pi}^0)^T\mathbf{a}_j = \sigma_j^0, j \in N$.
10. $\tilde{\mathbf{a}}_j' = (\mathbf{B}'^0)^{-1}\mathbf{a}_j' = (\mathbf{B}^0)^{-1}\mathbf{a}_j = \tilde{\mathbf{a}}_j, j \in N; \mathbf{z}_k' = \mathbf{z}_k, k = 1, \ldots, n - m$.

The reduced costs are invariant under row scaling, i.e., they do not depend on the choice of units for the right-hand side. However, the price vector is not invariant. For a linear program with constraints in the alternative form $\mathbf{Ax} \leq \mathbf{b}$, note also that the relative worth of introducing *logical* variables or slacks into the basis could be arbitrarily changed by a row scaling $\mathbf{RAx} \leq \mathbf{Rb}$.

In this section we study criteria that seek to lessen the dependence on choice of units for the variables or for the right-hand side. Let us first consider the invariance properties of steepest-edge pricing (7.1-8), namely,

$$(\boldsymbol{\sigma}_N^s)_k = (\boldsymbol{\sigma}_N^0)_k/(\|\tilde{\mathbf{a}}_{m+k}\|_2^2 + 1)^{1/2}. \tag{7.1-14}$$

Note immediately from items (9) and (10) above that $(\boldsymbol{\sigma}_N^s)_k$ is row scale invariant. In order to obtain a selection criterion that is N-column scale invariant, we modify (7.1-14). Define

$$(\bar{\boldsymbol{\sigma}}_N^s)_k \equiv (\boldsymbol{\sigma}_N^0)_k/\|\tilde{\mathbf{a}}_{m+k}\|_2. \tag{7.1-15}$$

Using (2.2-14) and (2.2-16) we see that $(\bar{\boldsymbol{\sigma}}_N^s)_k$ is a measure of the change in objective value per unit (Euclidean) distance traversed by *the basic variables*. (Note that if $\mathbf{a}_j = \mathbf{0}$ for some j in the linear program (1.2-2), then the sign of c_j immediately determines whether the corresponding variable x_j is at lower or upper bound in the optimal solution, or whether the problem has an unbounded optimum. Therefore we may assume, in this section, that such variables are excluded. Since the basis matrix is nonsingular, we then have $\tilde{\mathbf{a}}_{m+k} \neq \mathbf{0}$.) From items (4) and (5) above, it is easily verified that $(\bar{\boldsymbol{\sigma}}_N^s)_k$ in (7.1-15) is N-column scale invariant.

However, it is still not B-column scale invariant. Let us now discuss selection criteria that seek to achieve this property, as well.

Under the usual nondegeneracy assumption, $\mathbf{x}_B^0 > 0$, define a cost vector $\mathbf{c}^g = \begin{bmatrix} \mathbf{c}_B^g \\ \mathbf{c}_N^g \end{bmatrix}$ as follows.

$$(\mathbf{c}_B^g)_i = 1/(\mathbf{x}_B^0)_i \qquad i = 1, \ldots, m \tag{7.1-16a}$$

$$(\mathbf{c}_N^g)_j = 0 \qquad j \in N. \tag{7.1-16b)$$

The associated reduced costs, say σ_j^g, $j \in N$ are given by

$$\sigma_j^g = -\sum_{i=1}^{m} (\tilde{a}_{ij})/(\mathbf{x}_B^0)_i \qquad j \in N. \tag{7.1-17}$$

We can again easily verify that σ_j^g is row scale invariant and B-column scale invariant. However it is not N-column scale invariant because $\sigma_j'^g = s_j \sigma_j^g$, where we again use a "prime" to distinguish a transformed quantity.

Now let

$$\rho_j^g \equiv \sigma_j^0 / \sigma_j^g \qquad j \in N. \tag{7.1-18}$$

σ_j^0 denotes the usual reduced cost, which as we indicated above, is row scale invariant and B-column scale invariant. Therefore these invariance relations hold true for ρ_j^g. In addition, we may now verify that ρ_j^g is N-column scale invariant, i.e., $\sigma_j'^0/\sigma_j'^g = (s_j\sigma_j^0)/(s_j\sigma_j^g) = \sigma_j^0/\sigma_j^g$. Thus ρ_j^g is both row scale and column (i.e., B- and N-) scale invariant. The criterion (7.1-18) was proposed by Greenberg [1978b].

Given $\mathbf{x}_B^0 > \mathbf{0}$ and $\mathbf{D}_0 \equiv \text{diag}[x_1^0, \ldots, x_m^0]$, a related idea for obtaining row and column scale invariance, motivated by the method of Karmarkar [1984], is to define a *normalized linear program* by means of the transformation

$$\mathbf{x}_B = \mathbf{D}_0 \mathbf{x}_B^n \qquad \mathbf{x}_N = \mathbf{x}_N^n. \tag{7.1-19}$$

Then $(\mathbf{x}_B^n)^0 = \mathbf{D}_0^{-1}\mathbf{x}_B^0 = \mathbf{1}$, where $\mathbf{1}$ denotes the vector whose components are all unity. Also $(\mathbf{c}_B^n)^T = \mathbf{c}_B^T \mathbf{D}_0$ and $\tilde{\mathbf{a}}_j^n = \mathbf{D}_0^{-1}(\mathbf{B}^0)^{-1}\mathbf{a}_j$, $j \in N$. How does $\tilde{\mathbf{a}}_j^n$ depend on the scale of the original linear program prior to the normalization? The answer, in effect, has just been discussed. Like σ_j^g in (7.1-17), $\tilde{\mathbf{a}}_j^n$ is both row and B-column scale invariant.

Now define steepest-edge reduced costs, analogous to (7.1-14), but this time on the normalized linear program.

$$(\boldsymbol{\sigma}_N^n)_k = (\boldsymbol{\sigma}_N^0)_k / \|\mathbf{z}_k^n\|_2 = (\boldsymbol{\sigma}_N^0)_k / \|\tilde{\mathbf{a}}_{m+k}^n\|_2^2 + 1)^{1/2}, \tag{7.1-20}$$

where $(\boldsymbol{\sigma}_N^0)_k$ are the usual reduced costs, these being unaltered by the normalization. In order to obtain both row and column scale invariance, we proceed as we did earlier in going from (7.1-14) to (7.1-15), namely, define

$$(\bar{\boldsymbol{\sigma}}_N^n)_k \equiv (\boldsymbol{\sigma}_N^0)_k / \|\tilde{\mathbf{a}}_{m+k}^n\|_2. \tag{7.1-21}$$

Note that (7.1-20) and (7.1-21) may also be obtained by replacing quantities defined in the Euclidean norm in (7.1-14) and (7.1-15), by quantities defined in the following norm.

$$\|\tilde{\mathbf{a}}_{m+k}\|_{\mathbf{D}_0^{-2}} \equiv (\tilde{\mathbf{a}}_{m+k}^T \mathbf{D}_0^{-2} \tilde{\mathbf{a}}_{m+k})^{1/2} = \|\tilde{\mathbf{a}}_{m+k}^n\|_2. \qquad (7.1\text{-}22)$$

Exercise 7.1-4. Can one develop a method for updating the quantities in (7.1-22) analogous to the method given in Sec. 7.1-2 for updating $\|\tilde{\mathbf{a}}_{m+k}\|_2$?

Exercise 7.1-5. Study the extension of (7.1-22) when some basic variables are degenerate.

7.2. Selection Strategies

Given the price vector $(\boldsymbol{\pi}^0)^T = \mathbf{c}_B^T(\mathbf{B}^0)^{-1}$, the simplest procedure for picking the entering variable at Step P3 of the primal simplex algorithm of Sec. 2.4 is to compute reduced costs $\sigma_j^0, j \in N$, and to choose the variable whose reduced cost is largest in magnitude and of the appropriate sign. This implicitly assumes the following:

1. A *single* selection criterion, namely $\boldsymbol{\sigma}_N^0$, is used to pick a *single* candidate.
2. *Every* nonbasic variable is priced.

More sophisticated strategies depart from both these assumptions.

7.2-1. Multiple Pricing

Here we depart from assumption (1) above. As we saw in Sec. 7.1, there are good reasons for employing more then one selection criterion, each with an associated price vector, and using them in combination, to decide whether a particular nonbasic variable x_j is a suitable candidate for insertion into the basis. In addition, instead of picking just one candidate, a number of suitable candidates may be chosen, forming a list of, say, K members. The tactic of using several criteria and/or picking several candidates is called *multiple pricing*. In order to avoid confusion, we shall refer to the former as *multiple pricing criteria* and the latter as *multiple pricing candidates.*

7.2-2. Partial and Sectional Pricing

Here we depart from assumption (2) above. Instead of scanning the entire set of nonbasic variables to select a list of candidates, one prices out only a subset at a time, beginning typically with the nonbasic variable where one previously left off the scan. One tactic is to stop after finding K suitable candidates. Another is to keep replacing candidates in the list and stop, for example, when the ratio of the number of replacements to the number of variables priced out becomes suitably small. The tactic of employing a suitable *stopping rule* to scan only a portion of the LP matrix each time is called *partial pricing*. In addition, the LP matrix **A** can be partitioned into *sections,* and the multiple and partial pricing tactics confined to a small number of sections each time, possibly just a single one if enough candidates are found within it.

7.2-3. Major and Minor Iterations

Applying the above tactics in order to obtain a list of suitable candidates is termed a *major iteration*. Then a set of *minor iterations* of the simplex algorithm can be conducted over just this restricted set of variables, i.e., nonbasic variables not in the list are temporarily ignored. This defines a *subproblem*. Further variations are possible depending on whether the subproblem is totally or partially optimized during the minor iterations, before proceeding to the next major iteration.

It is clear that many variations are possible within the above three themes and the following pricing template, due to Greenberg [1978b], summarizes the general approach.

Price Template

T1. Price out variable x_j by the specified multiple pricing criteria and accept or reject it. If x_j is rejected, go to Step T3.

T2. Apply the multiple pricing candidate mechanism to see if the acceptable variable should be inserted into the candidate list. This might be done, for example, when fewer than K candidates have been generated so far, or when x_j is "better" (as judged by the selection criteria) than one of the K members of the list.

T3. Apply the stopping rule described by the partial pricing mechanism to decide whether to terminate pricing, and either proceed to the next step of the simplex algorithm or go to step T4 below.

T4. Select the next activity to be priced by the sectional pricing rule and return to Step T1.

At the optimal solution, the entire matrix will have been scanned and the resulting candidate list found to be empty.

7.3. Choosing the Exiting Variable When the Basis Is Feasible

The basic procedure for choosing the exiting or *blocking* variable was illustrated by Example 2.1-1 and given in Step P5 of the primal simplex algorithm of Sec. 2.4. We restate it here as follows.

$$P \leftarrow \underset{i=1,\ldots,m \ \& \ |(\tilde{\mathbf{a}}_s)_i| \geq tolpiv}{\operatorname{argmin}} \begin{cases} (x^0_{\beta_i} - l_{\beta_i})/|(\tilde{\mathbf{a}}_s)_i| \text{ when } (\sigma^0_s)(\tilde{\mathbf{a}}_s)_i < 0. \\ (u_{\beta_i} - x^0_{\beta_i})/|(\tilde{\mathbf{a}}_s)_i| \text{ when } (\sigma^0_s)(\tilde{\mathbf{a}}_s)_i > 0. \\ (u_s - l_s) \end{cases} \quad (7.3\text{-}1)$$

Similarly θ_s is defined by replacing "argmin" by "min" in (7.3-1). Note that it makes practical sense to exclude potential pivots that are below a certain tolerance, say *tolpiv,* so as to avoid instability in the basis update.

We seek now to be more specific about how to define ties, i.e., the set P, and how to resolve them and thus choose p. As we have noted, ties come about in the presence of degeneracy, i.e., when two or more basic variables simultaneously meet their bounds. Since only one is removed from the basis, at the next iteration there will be a basic variable at its bound, i.e., the basis will become degenerate. In practice, we are of course much more likely to encounter *near* ties and near degeneracy, rather than *exact* ties and exact degeneracy.

We discuss a degeneracy resolution technique due to Harris [1975] for using to advantage the fact that near ties are quite likely to occur in practice.

Example 7.3-1. Let us return to Example 2.1-1. Figure 7.1 below can be superimposed over the last column of Figure 2.1, and depicts a particular set of values t_1,t_2, and t_3.

We see that the blocking threshold is given, for this case, by

$$\theta_s = \min(t_1, t_2, t_3, u_s - l_s) = t_1$$

and $(\mathbf{x}_B)_1$ is blocking variable. Now, in practice, there is a feasibility tolerance, say $tolx > 0$, which one is willing to associate with each bound u_i and l_i and to consider a solution to be feasible when it satisfies

$$\mathbf{l} - tolx\,\mathbf{1} \leq \mathbf{x}^0 \leq \mathbf{u} + tolx\,\mathbf{1},$$

where **1** denotes the vector whose components are all unity. (We consider a fixed absolute tolerance here and a relative tolerance in an exercise following.) This is simply an acknowledgement of the fact that there is always some uncertainty in problem data and that we compute in finite precision arithmetic, so bounds have a little bit of fuzziness or slack

Figure 7.1 Degeneracy resolution

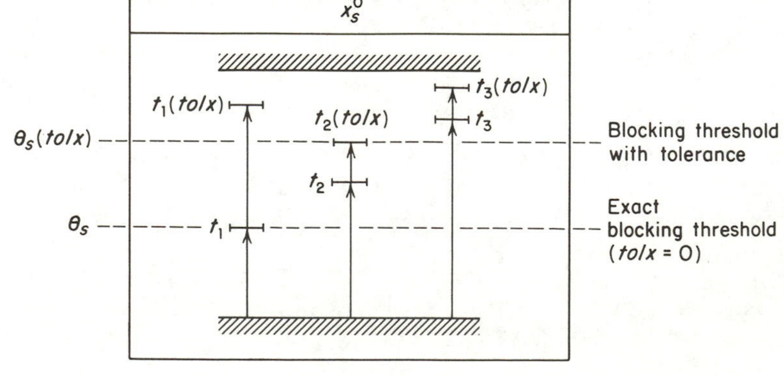

inherent in them. When this slight relaxation of the bounds occurs, the corresponding threshold levels are given by

$$t_1(tolx) = ((\mathbf{x}_B^0)_1 - (\mathbf{l}_B)_1 + tolx)/|(\tilde{\mathbf{a}}_s)_1| \tag{7.3-2a}$$

$$t_2(tolx) = ((\mathbf{u}_B)_2 + tolx - (\mathbf{x}_B^0)_2)/|(\tilde{\mathbf{a}}_s)_2| \tag{7.3-2b}$$

$$t_3(tolx) = ((\mathbf{x}_B^0)_3 - (\mathbf{l}_B)_3 + tolx)/|(\tilde{\mathbf{a}}_s)_3|, \tag{7.3-2c}$$

and $t_i = t_i(0)$, $i = 1, 2, 3$. In each case there is a shift in the threshold level (always an increase) which is inversely proportional to the size of the potential pivot element $|(\tilde{\mathbf{a}}_s)_i|$. Clearly the relative ordering of $t_i(tolx)$ can be different from that of t_i and, for the example, we see that the blocking threshold with tolerance *tolx* is

$$\theta_s(tolx) = \min(t_1(tolx), t_2(tolx), t_3(tolx), u_s - l_s + tolx) = t_2(tolx).$$

Any basic variable x_i whose exact threshold level t_i does not exceed $\theta_s(tolx)$ is considered to be a member of the set P, i.e.,

$$P = \{i \in B \mid t_i \leq \theta_s(tolx)\}, \tag{7.3-3}$$

and for the example in Figure 7.1 we have $P = \{1, 2\}$. P defines a set of candidates for exiting variable and we may, for example, choose the one whose pivot element is largest in magnitude. As we have seen in previous chapters, this will improve the stability of the update when the basis is revised. Thus,

$$p = \underset{i \in P}{\operatorname{argmax}} |(\tilde{\mathbf{a}}_s)_i|. \tag{7.3-4}$$

The general situation is now clear. It is summarized by the following two-pass procedure.

Pass 1.

Let $I \equiv \{i \mid 1 \leq i \leq m, |(\tilde{\mathbf{a}}_s)_i| \geq tolpiv\}$.
Compute $t_i(tolx)$, $i \in I$ as follows.

$$t_i(tolx) = \begin{cases} (x_{\beta_i}^0 - l_{\beta_i} + tolx)/|(\tilde{\mathbf{a}}_s)_i| \text{ when } (\sigma_s^0)(\tilde{\mathbf{a}}_s)_i < 0 \\ (u_{\beta_i} + tolx - x_{\beta_i}^0)/|(\tilde{\mathbf{a}}_s)_i| \text{ when } (\sigma_s^0)(\tilde{\mathbf{a}}_s)_i > 0 \end{cases} \tag{7.3-5}$$

Set $\theta_s(tolx) = \min_{i \in I}(t_i(tolx), u_s - l_s + tolx)$.

Pass 2.

Compute $t_i = t_i(0)$ given by (7.3-5) with $tolx = 0$, and form

$$P = \{i \mid i \in I \text{ and } t_i \leq \theta_s(tolx)\}.$$

Choose $p \in P$ for which $|(\tilde{\mathbf{a}}_s)_p| \geq |(\tilde{\mathbf{a}}_s)_i|$, $i \in P$.

It should be clear that the Harris [1975] strategy described above ensures a step t_p and a pivot $(\tilde{\mathbf{a}}_s)_p$ that are both at least as large, in magnitude, as those given by the strategy without thresholds, namely (7.3-1).

Exercise 7.3-1. Devise an analogous strategy to the above two-pass procedure, when a relative tolerance of the form $(u_i + tolx[\max(1, |u_i|)]$ is placed on an upper bound and $(l_i - tolx[\max(1, |l_i|)])$ on a lower bound.

7.4. Practical Details of Implementation

We conclude by mentioning a few further details concerning implementation. We assume that the linear program is in computational canonical form (3.2-3) and represented by the internal data structure of Figure 3.3.

7.4-1. PRICE Module

A particular implementation, using the pricing template and techniques discussed in Secs. 7.1 and Sec. 7.2, can be kept quite simple. On the other hand, subtle and complex strategies could be devised, if required. A few qualifying observations are all that we make here.

1. In the computational canonical form, the free logical variable corresponding to the original objective is always basic. It defines the new cost row in which all other variables have a zero coefficient. Thus for this representation of the problem, reduced costs are given by $-(\boldsymbol{\pi}^0)\mathbf{a}_j$, $j \in N$, where $\mathbf{a}_j$ is the jth column of the LP matrix (3.2-3).
2. Fixed variables should not be priced.
3. A variable that is nonbasic and pegged between its bounds, or a free variable that has not yet entered the basis, can move in either direction. Such a variable is thus always an acceptable candidate when its reduced cost is nonzero.
4. The test on reduced costs, used to determine whether a variable is an acceptable candidate, should be implemented using a small tolerance. For example, $(\sigma_s^0) \geq 0$ could be implemented as $(\sigma_s^0) \geq -tol\sigma$ and $(\sigma_s^0) \leq 0$ as $(\sigma_s^0) \leq tol\sigma$. See also Chapter 9, Sec. 9.2.

7.4-2. CHUZR Module

This is an acronym for "Choose Row" and is a standard name for the module that determines the exiting variable. The implementation can be carried out in a convenient and elegant manner by appending one more elements to each of the arrays, say Y and X, holding the vectors $\tilde{\mathbf{a}}_s$ and $\mathbf{x}_B^0$, respectively, and to the array JH holding the indices of the basic variables, as shown in Figure 7.2. Following the Fortran naming conventions of the Communication Data Structure of Chapter 3, in particular Figure 3.3, we shall denote the number of rows by NROWS and identify the entering variable x_s by the pointer JXIN.

JH[NROWS + 1] $\leftarrow$ JXIN and points to the entering variable, in

Figure 7.2 *CHUZR*

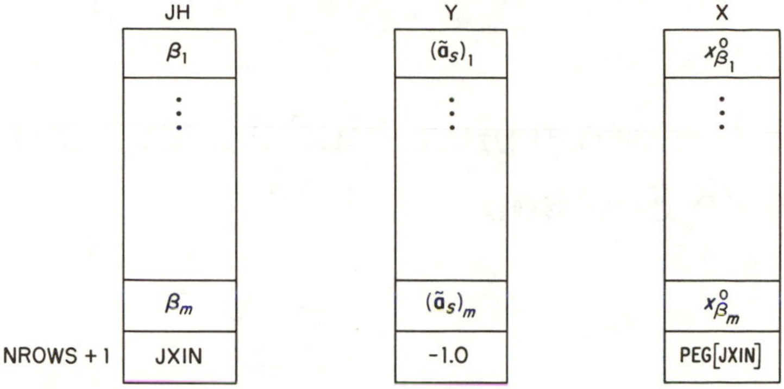

particular, to the bounds u_s and l_s in BU[JXIN] and BL[JXIN]. The last element of the array Y is set to -1.0, and the last element of the array X is set to the current value of the variable x_s, which will be at a bound, or defined by PEG[JXIN] when x_s is pegged between bounds. Machine representations of infinity in the bounds could be chosen to be some large number, say INF, that will not overflow when one computes 2(INF)/*tolpiv*. The two-pass algorithm of Harris [1975] can then be implemented in a straightforward manner with all bounds treated as though they are finite, i.e., no special tests are needed. Upon exit, if JP = NROWS + 1 then the entering variable is blocking. If JP $\leq$ NROWS and the corresponding variable x_{β_p} is blocking at an INF or $-$INF bound, then the linear program has an unbounded optimal solution.

Note

Sec. 7.1. The selection criterion in the presence of degeneracy is due to Greenberg [1978b]. The scale invariance argument can be extended, leading to an affine variant of the method of Karmarkar [1984]. For a discussion, see Nazareth [1987].

8

Selection Strategies: Finding an Initial Feasible Solution

When the nonbasic variables are fixed at one of their bounds, or pegged at some suitable values off their bounds, thereby defining $\mathbf{x}_N^0$, the corresponding basic variables $\mathbf{x}_B^0$ are given, in the usual way, by solving $\mathbf{B}^0\mathbf{x}_B^0 = \mathbf{b} - \mathbf{N}^0\mathbf{x}_N^0$. In order to initiate the simplex method, we have assumed, so far, that $\mathbf{x}_B^0$ is feasible. Let us now suppose that it need *not* satisfy the bounds $\mathbf{l}_B \leq \mathbf{x}_B^0 \leq \mathbf{u}_B$. Such a basic solution is much easier to find, because the only restriction placed upon it is that the corresponding basis matrix $\mathbf{B}^0$ must have linearly independent columns.

In order to obtain a basic feasible solution from this basic solution, we extend the simplex method, in particular the primal simplex algorithm of Sec. 2.4, by replacing the original cost vector $\mathbf{c}_B$ by another, say $\mathbf{c}_B^f$, chosen so that the associated reduced costs, σ_j^f, $j \in N$ are *a measure of the rate of change of the sum of infeasibility*. This overall approach is already quite familiar in another context, see Sec. 7.1. However, before we proceed with the details, let us give some background material. This will serve both to motivate our development and to give the reader some idea of the historical backgound that led up to the techniques now in use. These are then described in the main body of this chapter.

8.1. Background

For simplicity, let us restrict attention in this section to the linear program with nonnegative bounds, namely,

$$\begin{aligned} \text{minimize } & \mathbf{c}^T\mathbf{x} \\ \text{s.t. } & \mathbf{A}\mathbf{x} = \mathbf{b} \\ & \mathbf{x} \geq \mathbf{0}. \end{aligned} \tag{8.1-1}$$

We also assume, without loss of generality, that the sign of each constraint is chosen to ensure $b_i \geq 0$, $i = 1, \ldots, m$ in (8.1-1).

The "traditional" approach for initiating the simplex method was to add a variable, termed an "artificial" variable, say y_i, $i = 1, \ldots, m$, to each constraint of (8.1-1) and to replace the objective function by an "artificial" objective function, $\sum_{i=1}^{m} y_i = \mathbf{1}^T\mathbf{y}$, where $\mathbf{1}$ denotes the vector whose components are all unity. This results in the Phase 1 problem,

$$\begin{aligned} \text{minimize} \quad & w = \mathbf{1}^T\mathbf{y} \\ \text{s.t.} \quad & \mathbf{I}\mathbf{y} + \mathbf{A}\mathbf{x} = \mathbf{b} \\ & \mathbf{y} \geq \mathbf{0} \qquad \mathbf{x} \geq \mathbf{0}. \end{aligned} \tag{8.1-2}$$

For this problem, the initial basic solution, $[\mathbf{b}, \mathbf{0}]$, is obtained by setting the nonbasic variables, in this case $\mathbf{x}$, to their zero bound values, and it is feasible because $\mathbf{b} \geq \mathbf{0}$, by assumption. The primal simplex algorithm of Sec. 2.4 can thus be immediately applied to solve (8.1-2). Any feasible solution of (8.1-1), say $\mathbf{x}$, has a corresponding feasible solution, namely, $(\mathbf{0}, \mathbf{x})$ of (8.1-2), and the associated value of the objective function of (8.1-2) is $w = 0$. Thus, if the simplex algorithm applied to (8.1-2) terminates with a *minimum* value $w^0 > 0$, (8.1-1) cannot have a feasible solution. If, on the other hand, $w^0 = 0$ and the associated optimal solution of (8.1-2), say $(\mathbf{y}^0, \mathbf{x}^0)$ is nondegenerate, then the optimal basis of (8.1-2) cannot contain any $\mathbf{y}$ variables. We therefore have a basic feasible solution $\mathbf{x}^0$ for (8.1-1). The original objective and constraints can then be restored, and the primal simplex algorithm initiated from $\mathbf{x}^0$. This constitutes the start of Phase 2 and at the end of it an optimal solution $\mathbf{x}^*$ will be found, or the optimum will be discovered to be unbounded. Some technicalities arise when $w^0 = 0$ but $(\mathbf{y}^0, \mathbf{x}^0)$ is degenerate, because there could be some "artificial" variables in the basis at zero level. We shall not concern ourselves with this further, because our aim in mentioning the "artificial" variable approach is primarily to motivate techniques that have since supplanted it. These have, in effect, pinpointed the essence of the approach and generalized it, thereby rendering obsolete the term "artificial."

Two variants on the foregoing idea have often been suggested. The first, called the big-M method, seeks to simultaneously reduce the original and the "artificial" objective, by minimizing a weighted sum $\mathbf{c}^T\mathbf{x} + M(\mathbf{1}^T\mathbf{y})$, where M is a large positive constant. However, when M is too large, it will dominate the calculations involving the objective function, i.e., the original objective will only appear as a minor perturbation of $M(\mathbf{1}^T\mathbf{y})$ and, from a numerical standpoint, this is extremely unsatisfactory. Too small a value of M, on the other hand, could give an optimal solution that is infeasible, because some $\mathbf{y}$ variables remain in the basis at a nonzero level. Because of difficulties in making an appropriate choice for M, the big-M method has little to recommend it outside of specialized usage by an experienced practitioner.

The second variant seeks to introduce only a single "artificial" variable, say y_0, into the problem (8.1-1), instead of a set of m variables

y_i, $i = 1, \ldots, m$ as in (8.1-2). Suppose, with the usual partition $\mathbf{A} = [\mathbf{B}^0 \mid \mathbf{N}^0]$ and $\mathbf{B}^0$ nonsingular, we express the constraints of (8.1-1) as

$$\mathbf{B}^0\mathbf{x}_B + \mathbf{N}^0\mathbf{x}_N = \mathbf{b},$$

and suppose that the corresponding basic solution $\mathbf{x}_B^0 = (\mathbf{B}^0)^{-1}\mathbf{b}$ has some negative components. (Recall that we are confining attention in this section to nonnegative bounds.) Let us introduce an "artificial" variable y_0, with corresponding column $-\mathbf{B}^0\mathbf{1}$ into (8.1-1), thus defining the problem,

$$\begin{aligned} &\text{minimize } y_0 \\ &\quad \text{s.t. } \mathbf{B}^0\mathbf{x}_B + \mathbf{N}^0\mathbf{x}_N - \mathbf{B}^0\mathbf{1}y_0 = \mathbf{b} \\ &\qquad \mathbf{x}_B \geq \mathbf{0} \qquad \mathbf{x}_N \geq \mathbf{0} \qquad y_0 \geq 0. \end{aligned} \tag{8.1-3}$$

Now introduce the variable y_0 into the basis in place of the variable x_{β_p}, $\beta_p \in B$, with p chosen so that $((\mathbf{B}^0)^{-1}\mathbf{b})_p \leq ((\mathbf{B}^0)^{-1}\mathbf{b})_i$, $i = 1, \ldots, m$. By assumption $((\mathbf{B}^0)^{-1}\mathbf{b})_p < 0$. Denote the resulting basis by $\mathbf{B}$. Then analogously to (6.1-1),

$$\mathbf{B} = \mathbf{B}^0 - \mathbf{a}_{\beta_p}\mathbf{e}_p^T - (\mathbf{B}^0\mathbf{1})\mathbf{e}_p^T, \tag{8.1-4}$$

where $\mathbf{a}_{\beta_p}$ is the replaced pth column of the basis matrix. Thus

$$\mathbf{B} = \mathbf{B}^0(\mathbf{I} + (-\mathbf{1} - \mathbf{e}_p)\mathbf{e}_p^T) \tag{8.1-5}$$

$$\begin{aligned} \mathbf{B}^{-1}\mathbf{b} &= (\mathbf{I} - (\mathbf{1} + \mathbf{e}_p)\mathbf{e}_p^T)^{-1}(\mathbf{B}^0)^{-1}\mathbf{b} \\ &= (\mathbf{I} - (\mathbf{1} + \mathbf{e}_p)\mathbf{e}_p^T)(\mathbf{B}^0)^{-1}\mathbf{b} \\ &= (\mathbf{B}^0)^{-1}\mathbf{b} - (\mathbf{1} + \mathbf{e}_p)((\mathbf{B}^0)^{-1}\mathbf{b})_p, \end{aligned} \tag{8.1-6}$$

From the choice of p, we now observe that $\mathbf{B}^{-1}\mathbf{b} \geq \mathbf{0}$, and we can therefore use the corresponding basis matrix $\mathbf{B}$ to initiate Phase 1 of the primal simplex algorithm applied to (8.1-3). Using identical arguments to those given for (8.1-2), Phase 1 will either discover that the original linear program (8.1-1) is infeasible or will terminate with a basis with which to initiate Phase 2.

The obvious advantage of the second variant (8.1-3) over (8.1-2), namely, that it introduces only a single additional variable rather than m such variables, is not as great as would appear at first sight. First, large matrices are stored in packed form, so it is the number of nonzeros introduced into the matrix that determines the additional storage needed; as we can see, both methods introduce m nonzeros. Second, a sizeable linear program often has $n \gg m$; in this case, the numbers of columns in the two formulations do not differ significantly ($n + m$ versus $n + 1$). Third, there is no particular reason why the linear program (8.1-3) should take *fewer* iterations to discover a feasible solution of (8.1-1) than the linear program (8.1-2). Fourth, we saw in Chapter 3 that by associating a different logical variable with each general constraint, one can transform a linear program into a very convenient computational canonical form; various options on types of row, bounds on variables, and ranges on constraints can then be handled in a uniform and flexible manner. For these reasons, (8.1-2) is generally preferable to (8.1-3).

We seek therefore to extract the germ of the idea involved in (8.1-2) and employ it in a method for finding a basic feasible solution *that utilizes the computational canonical form and fits in a natural way into the cycle of operations of the primal simplex algorithm* outlined in Sec. 2.4. We would also like to avoid any artificial initialization procedures so as to be able to reinvoke Phase 1 whenever there is a subsequent loss of feasibility. For example, suppose rounding error arising from an ill-conditioned basis matrix $\mathbf{B}^0$ leads to an erroneous solution $\tilde{\mathbf{a}}_s$ of $\mathbf{B}^0\tilde{\mathbf{a}}_s = \mathbf{a}_s$. This could, in turn, lead to an incorrect choice of exiting variable and a loss of feasibility.

To motivate the technique sought, let us make the following *interpretation* of the "artificial" variable technique used in (8.1-2). Suppose at some intermediate stage in Phase 1 we have a basis given by the index set $B = \{\beta_1, \ldots, \beta_m\}$ composed of some artificial and some regular (structural) variables. Let us assume that the corresponding basis matrix $\mathbf{B}^0$ is nondegenerate. Let us also define the index set $\bar{V}$ as follows:

$$\bar{V} \equiv \{i \mid 1 \leq i \leq m \text{ and basic variable } i \text{ is an artificial } y_{\beta_i}\}. \quad (8.1\text{-}7)$$

The cost vector, say $\mathbf{c}_B^f$, at the current iteration of Phase 1, comes from the artificials in the basis and is given by

$$(\mathbf{c}_B^f)_i \equiv \begin{cases} 1 & \text{if } i\text{th basic variable is an artificial, i.e., } i \in \bar{V}, \\ 0 & \text{otherwise,} \end{cases} \quad (8.1\text{-}8)$$
$$(\mathbf{c}_N^f)_j \equiv 0 \qquad \text{for all } j.$$

Now we may observe that the constraints of (8.1-2) are *not* equivalent to those of (8.1-1). In order to obtain an equivalent set of constraints, we must replace those of (8.1-2) by

$$\begin{aligned} \mathbf{Iy} + \mathbf{Ax} &= \mathbf{b} \\ \mathbf{0} \geq \mathbf{y} \geq \mathbf{0} \qquad \mathbf{x} &\geq \mathbf{0} \end{aligned} \quad (8.1\text{-}9)$$

i.e., we must introduce both *upper* and lower bounds on the artificial variables. We see that artificial variables are logical variables (Chapter 3) whose upper bounds $\mathbf{y} \leq \mathbf{0}$ are not explicitly stated in the program. Let us therefore rewrite (8.1-7) and (8.1-8) as

$$\bar{V} \equiv \{i \mid 1 \leq i \leq m \quad \text{and } i\text{th basic variable violates upper bound}\} \quad (8.1\text{-}10)$$

$$(\mathbf{c}_B^f)_i \equiv \begin{cases} 1 & i\text{th basic variable violates upper bound,} \\ 0 & \text{otherwise.} \end{cases} \quad (8.1\text{-}11)$$

The price vector and the reduced costs associated with $\mathbf{c}_B^f$ and a column of $\mathbf{A}$, say $\mathbf{a}_j$, are given in the usual way as follows.

$$(\boldsymbol{\pi}^f)^T = (\mathbf{c}_B^f)^T(\mathbf{B}^0)^{-1} \tag{8.1-12a}$$

$$\sigma_j^f = 0 - (\boldsymbol{\pi}^f)^T\mathbf{a}_j = -(\mathbf{c}_B^f)^T\tilde{\mathbf{a}}_j = -\sum_{i\in\bar{V}} \tilde{a}_{ij}, \tag{8.1-12b}$$

where $\tilde{\mathbf{a}}_j \equiv (\mathbf{B}^0)^{-1}\mathbf{a}_j$ and $\tilde{a}_{ij}$ denotes its ith element. σ_j^f is the rate of change of the artificial objective row $\mathbf{1}^T\mathbf{y}$ in (8.1-2) when x_j is altered in value. Within the setting of (8.1-9), σ_j^f is also the *rate of change of the sum of infeasibility*. This can be easily seen as follows.

Let $y_{\beta_i}^0$, $i \in \bar{V}$ denote the current values of the (logical) basic variables that violate their (upper) bound. When $\mathbf{a}_j$ is introduced into the basis, we have in the usual way (see, for example, (2.2-19)),

$$y_{\beta_i} = y_{\beta_i}^0 - \tilde{a}_{ij}\tau \qquad i \in \bar{V}. \tag{8.1-13}$$

The basis is assumed to be nondegenerate. Therefore $y_{\beta_i}^0 > 0$ implies that $y_{\beta_i} > 0$, for $|\tau|$ sufficiently small. The contribution to infeasibility made by the logical variable y_{β_i} is $(y_{\beta_i}^0 - \tilde{a}_{ij}\tau - 0)$, for $|\tau|$ sufficiently small. Thus

$$\textit{total sum of infeasibility} = \sum_{i\in\bar{V}} (y_{\beta_i}^0 - \tilde{a}_{ij}\tau),$$

and differentiating with respect to τ,

$$\textit{rate of change of sum of infeasibility} = -\sum_{i\in\bar{V}} \tilde{a}_{ij}. \tag{8.1-14}$$

From (8.1-12b), it follows that σ_j^f is the rate of change of the sum of infeasibility, when x_j, $j \in N$ is introduced.

We therefore see that the artificial variable approach (8.1-2) may be *interpreted* as a technique for setting up a suitable cost vector (8.1-11), which is defined in terms of logical variables that violate an upper bound of the constraints of (8.1-9) with associated reduced costs that measure the rate of change of sum of infeasibility. This interpretation, given within the restricted setting of (8.1-1) and (8.1-9), in fact, applies to any basic variable, be it logical or structural, that violates a bound. The foregoing interpretation therefore provides the gist of the technique used in practice, to which we now turn. As we shall see, there is no need whatsoever to distinguish certain variables as being "artificial" and henceforthe we drop usage of this outmoded term.

8.2. Choosing the Entering Variable When the Basis Is Infeasible

Let us revert to our canonical form (1.2-2), namely,

$$\begin{aligned} &\text{minimize } \mathbf{c}^T\mathbf{x} \\ &\text{s.t. } \mathbf{A}\mathbf{x} = \mathbf{b} \\ &\mathbf{l} \le \mathbf{x} \le \mathbf{u} \end{aligned}$$

with the usual partition of **A** and our standard notation for basic and nonbasic variables and other associated quantities. With the nonbasic variables at levels $\mathbf{x}_N^0$, we have as usual,

$$\mathbf{B}^0\mathbf{x}_B^0 = \mathbf{b} - \mathbf{N}^0\mathbf{x}_N^0.$$

The difference now is that $\mathbf{x}_B^0$ *may be infeasible.* Suppose some nonbasic variable x_j, $j \in N$ is changed from its current level by an amount τ. Then

$$x_{\beta_i} = x_{\beta_i}^0 - \tilde{a}_{ij}\tau \quad \text{where} \quad \tilde{\mathbf{a}}_j = (\mathbf{B}^0)^{-1}\mathbf{a}_j.$$

We shall assume that the basis is nondegenerate, i.e., that no basic variable $x_{\beta_i}^0$ is at a bound. We have three cases to consider.

1. $x_{\beta_i}^0 > u_{\beta_i}$. For $|\tau|$ sufficiently small, $x_{\beta_i} > u_{\beta_i}$ and x_{β_i} also violates its upper bound. The contribution to infeasibility by the variable x_{β_i} is
$$(x_{\beta_i}^0 - \tilde{a}_{ij}\tau) - u_{\beta_i}. \tag{8.2-1}$$
2. $x_{\beta_i}^0 < l_{\beta_i}$. For $|\tau|$ sufficiently small, $x_{\beta_i} < l_{\beta_i}$. The contribution to infeasibility by the variable x_{β_i} is
$$l_{\beta_i} - (x_{\beta_i}^0 - \tilde{a}_{ij}\tau). \tag{8.2-2}$$
3. $l_{\beta_i} < x_{\beta_i}^0 < u_{\beta_i}$. Then for $|\tau|$ sufficiently small, x_{β_i} remains feasible.

Let us define $\bar{V}$ and $\underline{V}$ as follows.

$$\bar{V} \equiv \{i \mid \beta_i \in B \text{ and } x_{\beta_i}^0 \text{ violates its upper bound}\} \tag{8.2-3a}$$

$$\underline{V} \equiv \{i \mid \beta_i \in B \text{ and } x_{\beta_i}^0 \text{ violates its lower bound}\}. \tag{8.2-3b}$$

It then follows from (8.2-1) and (8.2-2) that

$$\textit{sum of infeasibility} = \sum_{i \in \bar{V}} (x_{\beta_i}^0 - \tilde{a}_{ij}\tau - u_{\beta_i}) + \sum_{i \in \underline{V}} (l_{\beta_i} - x_{\beta_i}^0 + \tilde{a}_{ij}\tau).$$

$$\textit{rate of change of sum of infeasibility} = -\sum_{i \in \bar{V}} \tilde{a}_{ij} + \sum_{i \in \underline{V}} \tilde{a}_{ij}. \tag{8.2-4}$$

More specifically, (8.2-4) measures the rate of *increase* of the sum of infeasibility as τ *increases.*

Now, in an analogous manner to (8.1-11), let us define a cost vector $\mathbf{c}_B^f$ as follows

$$(\mathbf{c}_B^f)_i \equiv \begin{cases} 1 & \text{if } i \in \bar{V}, \\ -1 & \text{if } i \in \underline{V}, \\ 0 & \text{otherwise,} \end{cases} \tag{8.2-5}$$

$$(\mathbf{c}_N^f)_j \equiv 0 \quad \text{for all } j.$$

Then the price vector and reduced costs associated with $\mathbf{c}_B^f$ are

$$(\boldsymbol{\pi}^f)^T = (\mathbf{c}_B^f)^T(\mathbf{B}^0)^{-1}$$

and

$$\sigma_j^f = -(\boldsymbol{\pi}^f)^T\mathbf{a}_j = -(\mathbf{c}_B^f)^T\tilde{\mathbf{a}}_j = -\sum_{i \in \bar{V}} \tilde{a}_{ij} + \sum_{i \in \underline{V}} \tilde{a}_{ij}. \tag{8.2-6}$$

We see that (8.2-4) and (8.2-6) are identical. We conclude that the cost vector (8.2-5) has associated reduced costs that measure the rate of change of sum of infeasibility when a nonbasic variable is introduced into an infeasible basis $\mathbf{B}^0$. Thus whenever a basis is found to be infeasible, we simply replace the cost vector by the cost vector (8.2-5) and proceed in the usual way. Eventually, a feasible basis will be found at which point the original objective function can be restored. Alternatively, the linear program will be found to be infeasible because the pricing mechanism discovers that there is no improving variable that reduces the sum of infeasibility.

Exercise 8.2-1. Devise a cost vector analogous to (8.2-5), say $\mathbf{c}_B^m$, whose associated reduced cost σ_j^m measures the rate of change of the basic variable with *maximum* infeasibility when the variable x_j corresponding to σ_j^m is altered in value.

Exercise 8.2-2. Devise a strategy that uses both σ_j^f and the usual reduced costs σ_j^0, and seeks to simultaneously reduce infeasibility and the objective value of the original problem, whenever possible.

Exercise 8.2-3. Suppose that $x_j, j \in N$ is at its lower bound, i.e., $\tau \uparrow 0$. Show in the presence of degeneracy that σ_j^f given by (8.2-5) does not exceed the rate of change of sum of infeasibility.

8.3. Choosing the Exiting Variable When the Basis Is Infeasible

Let us return to Example 2.1-1, which we now modify so that some basic variables are infeasible, as shown in Figure 8.1. Now when basic variables become blocking, the threshold levels are as follows.

1. $(\mathbf{x}_B)_1$ cannot be blocking, so $t_1 = \infty$.

Figure 8.1 Choosing the exiting variable for infeasible basis

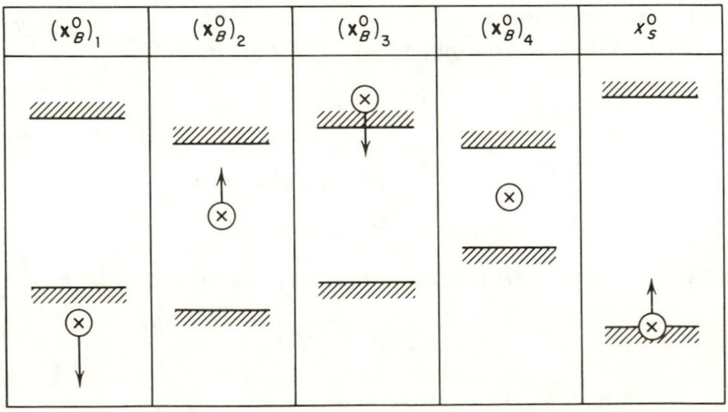

2. $(\mathbf{x}_B)_2$ meets its upper bound when t attains the value $t_2 = ((\mathbf{x}_B^0)_2 - (\mathbf{u}_B)_2)/(\tilde{\mathbf{a}}_s)_2$.
3. $(\mathbf{x}_B)_3$ becomes blocking at its *lower* bound when t attains the value $t_3 = ((\mathbf{x}_B^0)_3 - (\mathbf{l}_B)_3)/(\tilde{\mathbf{a}}_s)_3$.
4. $(\mathbf{x}_B)_4$ does not change and cannot be blocking.

More generally, we observe the following:

1. A basic variable, say $(\mathbf{x}_B)_i$, which is infeasible and moves even further away from feasibility as t increases from zero, cannot be blocking. Because of the way the entering variable is chosen as described in Sec. 8.2, other basic variables are moving *toward* feasibility at a faster rate than $(\mathbf{x}_B)_i$ is moving away from feasibility.
2. A basic variable that moves toward feasibility is permitted to move through its nearer bound and only becomes blocking at its farther bound.

The following exercise gives a simple procedure for taking these considerations into account.

Exercise 8.3-1. Consider a linear program with finite upper and lower bounds, for which the current basis defined by $B = \{1, \ldots, m\}$ is *feasible.* The following procedure, termed CHUZR, discovers the blocking variable.

Given $\mathbf{x}_B^0$, the entering variable x_s, σ_s^0, and $\tilde{\mathbf{a}}_s = (\mathbf{B}^0)^{-1}\mathbf{a}_s$.

1. Set $t \leftarrow (u_s - l_s)$; $p \leftarrow s$; $i \leftarrow 1$
2. If $(\tilde{\mathbf{a}}_s)_i = 0$ go to Step 3, otherwise do the following.
   2.1. Set $\tau \leftarrow t(-\text{sign}(\sigma_s^0))$ and $z \leftarrow x_i^0 - \tau(\tilde{\mathbf{a}}_s)_i$.
   2.2. If $z < l_i$ then set $t \leftarrow (x_i^0 - l_i)/|(\tilde{\mathbf{a}}_s)_i|$; $p \leftarrow i$.
   2.3. If $z > u_i$ then set $t \leftarrow (u_i - x_i^0)/|(\tilde{\mathbf{a}}_s)_i|$; $p \leftarrow i$.
3. Set $i \leftarrow i + 1$. If $i \leq m$ return to Step 2, otherwise exit. Upon termination, p identifies the blocking variable.

Show when $\mathbf{x}_B^0$ is infeasible (with σ_s^f defined by (8.2-6) being used in place of σ_s^0 to identify the entering variable) that the previous procedure requires the following two simple modifications in order to handle considerations (1) and (2) discussed just before this exercise.

(a) Change the test in Step 2.2 from $z < l_i$ to ($z < l_i$ and $x_i^0 \geq l_i$).
(b) Change the test in Step 2.3 from $z > u_i$ to ($z > u_i$ and $x_i^0 \leq u_i$).

The procedure given in the previous exercise does not permit basic variables that are feasible to enter the infeasible domain, i.e., the *number* of infeasibilities is not permitted to increase. It may however be advantageous to allow τ to increase even though a feasible variable goes through a bound and becomes infeasible, because it is counteracted by other infeasible variables that continue to move toward feasibility at a greater rate. We should seek, therefore, to minimize the *sum of infeasibility,* rather than reduce the number of infeasibilities. We illustrate the approach through an example.

Example 8.3-1. This example is again similar to Example 2.1-1, but specific numerical values will now be attached to bounds and to the basic variables. To keep the example more manageable, we shall assume that all bounds have the same upper and the same lower values, as shown in Figure 8.2. The values of the basic variables are also shown in the figure and the entering variable x_s is at its lower bound. As usual, $(\mathbf{x}_B)_i = (\mathbf{x}_B^0)_i - \tilde{a}_{is}\tau$ with $\tilde{\mathbf{a}}_s = (\mathbf{B}^0)^{-1}\mathbf{a}_s$. Since x_s is at its lower bound, $t = |\tau| = \tau > 0$, so we can equivalently write $(\mathbf{x}_B)_i = (\mathbf{x}_B^0)_i - \tilde{a}_{is}t,\ t \uparrow 0$.

$|\tilde{a}_{is}|$ determines the rate of change of $(\mathbf{x}_B)_i$ as t increases, and the direction of change depends on the sign of $\tilde{a}_{is}$. For our example, the directions of change are indicated by the arrows in Figure 8.2. From (8.2-4), the initial rate of change (increase) of infeasibility is

$$-\tilde{a}_{1s} + \tilde{a}_{2s} - \tilde{a}_{4s} = -4. \qquad (8.3\text{-}1)$$

If we define $\mathbf{c}_B^f$ as in (8.2-5), we have $(\mathbf{c}_B^f)^T = (1, -1, 0, 1)$, and the rate of change (increase) of infeasibility is

$$\sigma_s^f = -(\mathbf{c}_B^f)^T\tilde{\mathbf{a}}_s = -4. \qquad (8.3\text{-}2)$$

Figure 8.2 Status of basic variables

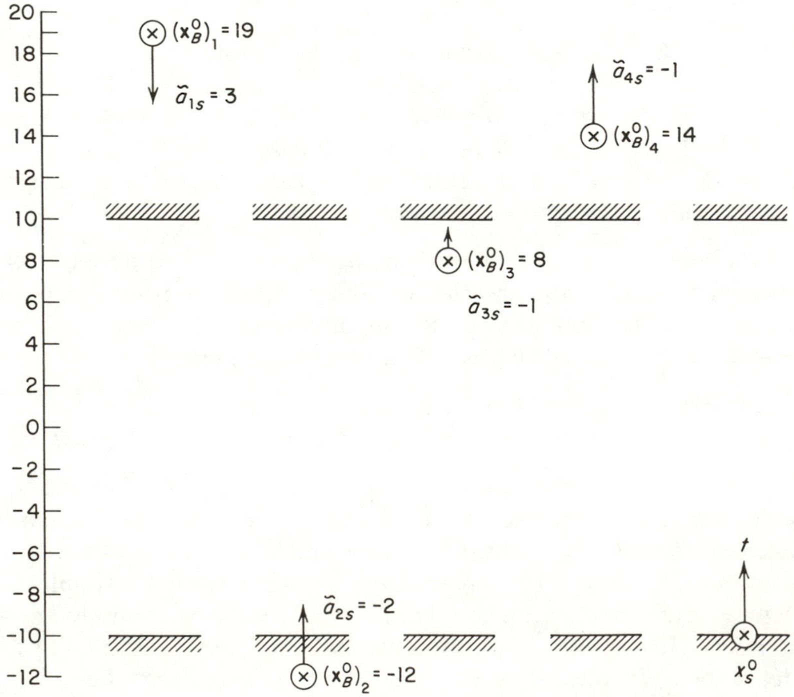

This, of course, matches (8.3-1) and provides the mechanism for choosing the entering variable x_s, as discussed in Sec. 8.2. σ_s^f must obviously be *negative* because $t = \tau \geq 0$ and we want a decrease of infeasibility as $t \uparrow 0$. Note also that the infeasibility of $(\mathbf{x}_B)_4$ increases as $t \uparrow 0$, but that this is counteracted by the variables $(\mathbf{x}_B)_1$ and $(\mathbf{x}_B)_2$ that move toward feasibility at a faster rate.

When t has increased to the value 1, $(\mathbf{x}_B)_2$ will have reached its lower bound and will no longer contribute to the sum of infeasibility. The other basic variables $(\mathbf{x}_B)_1$, $(\mathbf{x}_B)_3$, and $(\mathbf{x}_B)_4$ will take on values 16, 9, and 15 respectively. Only $(\mathbf{x}_B)_1$ and $(\mathbf{x}_B)_4$ now contribute to infeasibility and the rate of increase of infeasibility is now

$$-\tilde{a}_{1s} - \tilde{a}_{4s} = -2. \tag{8.3-3}$$

Since it is negative, there is still a decrease in infeasibility as t increases from 1.

When t increases to the value 2, the basic variable $(\mathbf{x}_B)_3$ will have reached its upper bound and the basic variables will have assumed values 13, −8, 10, and 16 respectively. Any subsequent increase of t will cause $(\mathbf{x}_B)_3$ to become infeasible, contributing to the rate of increase of infeasibility, which now becomes

$$-\tilde{a}_{1s} - \tilde{a}_{3s} - \tilde{a}_{4s} = -1. \tag{8.3-4}$$

Finally, when t reaches the value 3, $(\mathbf{x}_B)_1$ becomes feasible at its upper bound and the rate of increase of infeasibility becomes

$$-\tilde{a}_{3s} - \tilde{a}_{4s} = +2. \tag{8.3-5}$$

Since this is now positive, any further increase in t beyond the value 3, will cause a net *increase* in the sum of infeasibility of the basic variables.

The foregoing example illustrates the main approach to minimizing the sum of infeasibility. When an improving variable x_s is selected as the entering variable as described in Sec. 8.2, the reduced cost σ_s^f gives the initial rate of change (increase) of the sum of infeasibility (as τ increases) and it must, of course, be negative (initial rate of *decrease* is positive) when x_s is at its lower bound, and positive when x_s is at its upper bound. This net decrease in the sum of infeasibility continues at a linear rate as $t = |\tau|$ increases from zero, until either an *infeasible* basic variable, say $(\mathbf{x}_B)_i$, encounters a bound and enters the feasible domain, or a *feasible* basic variable encounters a bound and enters the infeasible domain, for any further increase in t. In either case, the total rate of *decrease* in the sum of infeasibility (as t increases) must *diminish* by an amount $|\tilde{a}_{is}|$, but it may still remain positive. A series of such thresholds can be crossed until either all the basic variables have become feasible or the rate of decrease of infeasibility (as a function of t) becomes negative (rate of increase becomes positive). We see that the sum of infeasibility, as a

Figure 8.3 $PL(t)$

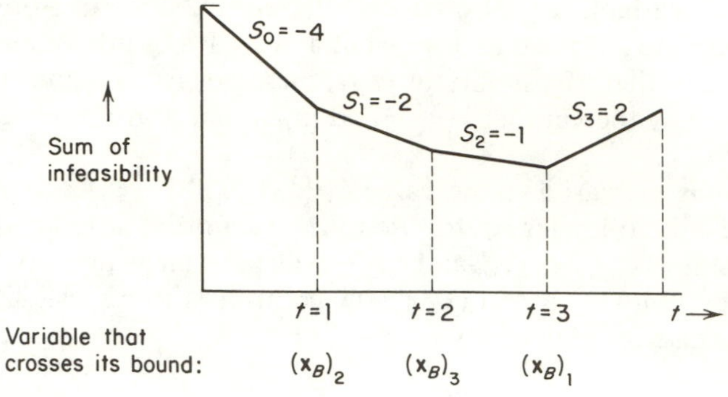

function of t, is piecewise linear and convex. We shall denote it by $PL(t)$. For the above Example 8.3-1, $PL(t)$ is shown in Figure 8.3. Clearly if one sought to minimize the sum of infeasibility, $(\mathbf{x}_B)_1$ would be chosen as the exiting variable.

The foregoing example illustrates two further points.

1. The basic variables can encounter their thresholds *in any order* as t increases from zero. Thus the first to cross its bound is $(\mathbf{x}_B)_2$, then $(\mathbf{x}_B)_3$ and finally $(\mathbf{x}_B)_1$.
2. Each basic variable has either zero, one or two thresholds associated with it. Thus $(\mathbf{x}_B)_1$ and $(\mathbf{x}_B)_2$ are infeasible, move toward feasibility, and have two finite bounds. Hence each has two thresholds. $(\mathbf{x}_B)_3$ is feasible and moves toward its finite upper bound. It, therefore, has one threshold associated with it. $(\mathbf{x}_B)_4$ is infeasible and moves away from feasibility. It, therefore, has no thresholds associated with it.

The general procedure can now be summarized by the following, fairly straightforward, two-pass procedure.

Two-Pass CHUZR

Pass 1. For each basic variable $(\mathbf{x}_B)_i$, $i = 1, \ldots, m$ determine a set of upto two threshold levels from the associated bounds and define corresponding triples of the form $[t, i, \alpha]$ where $t = |\tau| > 0$ denotes the threshold level, i denotes the row index of the basic variable $(\mathbf{x}_B)_i$ and $\alpha \equiv |\bar{a}_{is}|$. As previously noted, there may be no threshold levels (infeasible and moves away from feasibility), one threshold level (feasible and moves toward finite bound or infeasible with only one finite bound and moves toward feasibility), or two threshold levels (infeasible with two finite bounds and moves toward feasibility). Each threshold has a corresponding triple. If a basic variable has more than one triple, they differ only in the first member t. (We have discussed repeatedly how the thresholds are actually defined in terms of the bounds, $(\mathbf{x}_B^0)_i$, $\tilde{a}_{is}$ and, in this case, σ_s^f in place of σ_s^0, and it does not bear repetition.)

Any triple for which $t > (u_s - l_s)$ is discarded.

Pass 2. Sort the entire list of triples in order of increasing values of t. Let us denote the elements of this sorted list by $[t_k, i_k, \alpha_k]$, $k = 1, \ldots, \bar{k}$.

Form the partial sums that represent successive slopes of $PL(t)$, namely,

$$S_k = -|\sigma_s^f| + \sum_{i=1}^{k} \alpha_i \qquad k = 0, \ldots, \bar{k},$$

where we make the convention that $\sum_{i=1}^{0} \alpha_i = 0$.

Find the index K such that $S_{K-1} < 0$ and $S_K \geq 0$. The exiting variable then has index i_K.

Exercise 8.3-2. Show that $S_{\bar{k}} \geq 0$. Deduce that the above procedure is well defined, since $S_0 < 0$. Suppose when t reaches some positive value in the above procedure that all basic variables have entered the feasible domain. Show that there must be an index, say K, for which $S_K = 0$. Does the converse hold, namely, if an index K exists for which $S_K = 0$, is the problem necessarily feasible?

8.4. Practical Details of Implementation

We can replace the two-pass scheme at the end of the previous section, by a *one-pass* scheme, which builds up the partial sums S_k, as each threshold level is generated. This can be conveniently implemented by means of a doubly linked list, in which each node is of the form

t	i	α	Σ	*fp*	*bp*

t, i, and α are the triples mentioned in the earlier procedure. *fp* and *bp* are the forward and back pointers, respectively, of the list. As each triple is generated, it is inserted into the list in the correct position, in order of nondecreasing values of t. In this sorted list, the Σ field of each node holds the sum of its α field plus the α fields of all nodes preceding it in the list. (This is the same as the sum of its α field and the Σ field of the node immediately preceding it in the list.) At the end of the procedure, the Σ field of each node in the list will hold the slopes of $PL(t)$, namely, $S_0, S_1, S_2, \ldots$.

The complete procedure is summarized in the flowchart given in Figure 8.4. We distinguish certain nodes, using an identifying capital letter C, T, etc. We also attach the letter as a subscript to the quantities, in the fields of the corresponding node. The nodes are as follows.

Figure 8.4 *CHUZR* module

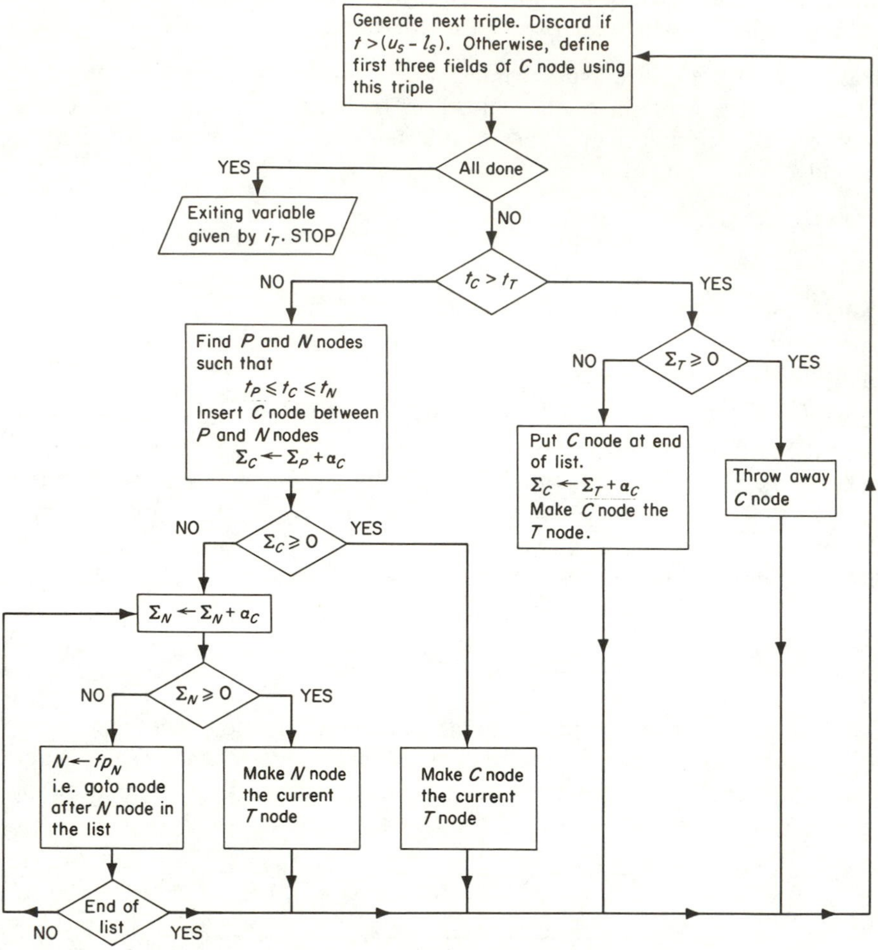

1. T or tail node. This is the last *relevant* member of the list and is defined to be the node with *least* threshold level for which the Σ field is nonnegative. If none as yet exists, then it is the node with the highest threshold value generated so far.
2. C or current node. This is the node that holds the triple generated at the current iteration of the procedure.
3. P or previous node and the N or next node. These are the two nodes between which the C node is inserted.
4. H or head node. This is used to initiate the process and is defined as follows.

$-\infty$	0	$-\lvert\sigma_s^f\rvert$	$-\lvert\sigma_s^f\rvert$	*fp*	*nil*

The first box of the flowchart generates successively triples as discussed in Pass 1 of the two-pass procedure of the previous section. The procedure given in the flowchart of Figure 8.4 then carries out a "depth-first" search, to discover whether the new triple is relevant. If it is, the procedure forms and inserts the new C node in its correct position. It then updates the partial sums in the Σ field of each node after the one where the C node was inserted.

When the procedure terminates, the i field of the T node defines the index of the exiting basic variable.

Notes

Secs. 8.2–8.3. These sections draw substantially on Greenberg [1978b].

Sec. 8.4. The one-pass scheme is essentially due to D. Rarick (see Tomlin [1975a]). We give a more efficient version based on depth-first search, see Tarjan [1972].

9

Practical Implementation

We have developed all the necessary techniques required to implement the primal simplex algorithm of Sec. 2.4, and we now summarize details of a practical implementation that may be composed from them. We first give the main procedural steps, closely paralleling the algorithm of Sec. 2.4, and then discuss a number of further issues concerning effective implementation.

9.1. An Implementation of the Primal Simplex Algorithm

B and N denote the indices of the current basic and nonbasic variables, $\mathbf{B}^0$ denotes the current basis matrix, $\mathbf{N}^0$ the matrix of nonbasic columns, $\mathbf{x}_N^0$ the current values of the nonbasic variables, $\mathbf{x}_B^0$ the corresponding values of the basic variables, s the index of the entering variable chosen during the current iteration with the aid of the price vector $\boldsymbol{\pi}^0$ and reduced costs σ_j^0, $j \in N$, p the index of the basic column that exits from the basis at the end of the current iteration.

The initialization and main cycle of the implementation may be summarized as follows.

1. SETUP. Convert the initial data defining the linear program, usually represented externally in MPS format, into a column list/row index data structure and other arrays holding the bounds. These internally represent the problem in computational canonical form, as described in Sec. 3.3, in particular, Figure 3.3.

2. INITIALIZE. Select a starting basis composed of all logical variables or some other set of linearly independent columns (see Sec. 9.2-2 below). This defines the index set B that is held in the array JH.

2.1. FACTOR. Set up the data structure for the basis matrix $\mathbf{B}^0 = (\mathbf{a}_{\beta_1}, \ldots, \mathbf{a}_{\beta_m})$ corresponding to B, as described in Sec. 5.10, and factorize $\mathbf{B}^0$.

2.2. MODRHS. Although it is usually the case that nonbasic variables are at one of their finite bounds, we have noted in Secs. 2.2 and 3.3

that such variables may be initially "pegged" at any convenient value between their bounds. The current values of the nonbasic variables, $\mathbf{x}_N^0$, are held in the arrays KINBAS and PEG. (The latter normally takes its values from BL and BU.) Form the modified right-hand side

$$\tilde{\mathbf{b}} = \mathbf{b} - \sum_{k=1}^{n-m} \mathbf{a}_{\eta_k} x_{\eta_k}^0. \tag{9.1-1}$$

2.3. SOLVE. Using the factorization of $\mathbf{B}^0$, as described in Sec. 5.10-2, solve

$$\mathbf{B}^0 \mathbf{x}_B^0 = \tilde{\mathbf{b}} \tag{9.1-2}$$

to obtain the starting solution $\mathbf{x}_B^0$.

3. FORMC. Check whether $\mathbf{x}_B^0$ is feasible, i.e., whether it satisfies $l_{\beta_i} \leq x_{\beta_i}^0 \leq u_{\beta_i}$, $i = 1, \ldots, m$.

If $\mathbf{x}_B^0$ is feasible, form the cost vector $(\mathbf{c}_B)_i$, $i = 1, \ldots, m$ as

$$(\mathbf{c}_B)_i = \begin{cases} -1 & \text{if } i = \text{IOBJ} \\ 0 & \text{if } i \neq \text{IOBJ}, \end{cases}$$

where IOBJ is the index of the objective row in the computational canonical form.

Otherwise, if $\mathbf{x}_B^0$ is infeasible, form the cost vector, which we now again denote by $(\mathbf{c}_B)_i$, $i = 1, \ldots, m$, as described in Sec. 8.2, namely,

$$(\mathbf{c}_B)_i = \begin{cases} 1 & \text{if } (\mathbf{x}_B^0)_i \text{ violates its upper bound,} \\ -1 & \text{if } (\mathbf{x}_B^0)_i \text{ violates its lower bound,} \\ 0 & \text{if } (\mathbf{x}_B^0)_i \text{ does not violate a bound.} \end{cases}$$

4. BTRAN. From the price vector $\boldsymbol{\pi}^0$ by solving the system

$$(\boldsymbol{\pi}^0)^T \mathbf{B}^0 = \mathbf{c}_B^T, \tag{9.1-3}$$

as described in Secs. 5.10-2 and 6.7. Monitor the accuracy of the solution as discussed in Sec. 9.2-3, and determine whether the basis should be refactorized on grounds of numerical stability.

The name *BTRAN* comes from the way this step was originally implemented, as a transformation of $\mathbf{c}_B^T$ by the elementary matrices in the product form representation of $(\mathbf{B}^0)^{-1}$, held in a file traversed from back to front.

5. PRICE. Compute reduced costs σ_j^0, $j \in N$, price out nonbasic columns using one of the strategies discussed in Secs. 7.1, 7.2, and 7.4-1, and select the entering variable x_s. (See also Sec. 9.2-4 on major/minor iterations.) If there is no improving variable, stop. At this point if $\mathbf{x}_B^0$ is infeasible then the linear program is infeasible, otherwise $\mathbf{x}_B^0$ and $\mathbf{x}_N^0$ define the optimal solution.

6. *FTRAN.* Solve the system

$$\mathbf{B}^0\tilde{\mathbf{a}}_s = \mathbf{a}_s \tag{9.1-4}$$

for $\tilde{\mathbf{a}}_s$, using the factorization of $\mathbf{B}^0$ as described in Secs. 5.10-2 and 6.7. Also save the intermediate vector $(\tilde{\mathbf{L}}^0)^{-1}\mathbf{a}_s$ corresponding to (5.10-14) for use in the *UPDATE* step. Monitor the accuracy of the solution as described in Sec. 9.2-3, and decide whether the basis should be refactorized on grounds of numerical stability.

The name *FTRAN* comes from the way this step was originally implemented, as a transformation of $\mathbf{a}_s$ by elementary matrices in a file holding the product form representation of $(\mathbf{B}^0)^{-1}$, traversed in a forward direction.

7. *CHUZR.* Find the index p of the variable x_{β_p}, $1 \le p \le m$, which exits from the current basis. If $\mathbf{x}_B^0$ is feasible then use techniques discussed in Secs. 7.3 and 7.4-2, otherwise use techniques discussed in Secs. 8.3 and 8.4. For the former case, if there is no blocking variable, stop. The solution is unbounded from below. If the entering variable x_s is the first to become blocking, then proceed directly to Step 9 because no basis change occurs.

The name *CHUZR* is an acronym for Choose Row.

8. *UPDATE.* Determine whether the basis matrix, obtained by replacing column $\mathbf{a}_{\beta_p}$ by $\mathbf{a}_s$, should be refactorized on grounds of sparsity of representation (efficiency); see Refactorization Frequency, Sec. 9.2-5. If the decision is to refactorize, update the status arrays JH and KINBAS identifying B and N and go to Step 2.1. Otherwise, update the factorization as described in Secs. 6.3 and 6.7.

9. *UPSOLN.* Update the status arrays JH and KINBAS identifying the new basis and the solution $\mathbf{x}^0$ and go to Step 3.

9.2. Further Implementation Issues

9.2-1. Organizing Storage and Access to Data

A common technique is to allocate all storage within a work array that is partitioned by a suitable storage map, namely, a set of pointers to starting locations of individual arrays needed by the procedure, each with an appropriate number of bytes that depend on whether the array is integer, single, or double precision floating-point (see, for example, Murtagh and Saunders [1983]). Another technique is to arrange all access to data by routines that implement the above individual steps of the simplex algorithm through interface subroutines, which are the only ones to actually touch the data. This makes the implementation largely independent of the data structures and it is then relatively easy to unplug one set of data structures and substitute another (see Marsten [1981]).

9.2-2. *Crashing a Starting Basis*

This is the task of choosing a starting basis that is well conditioned and as close to feasibility and triangularity as possible. A common strategy is to choose as many *free* logicals and structurals as possible and, from the remaining variables, to give preference to structurals over logicals, subject to the above goals.

9.2-3. *Tolerances*

These provide the means for establishing the margin of error in finite precision arithmetic, when making comparisons between a number and zero, or between two numbers, as required at various steps of the simplex algorithm. The right choice of tolerance often has a considerable impact on efficiency of the implementation, as it applies to a given problem. Key tolerances and particular ways of utilizing them are as follows.

Primal Feasibility Tolerance (tolx). A structural or logical variable x_j is considered to be feasible provided

$$l_j - tolx \leq x_j \leq u_j + tolx.$$

tolx is also used in the *CHUZR* step to resolve degeneracy, see Sec. 7.3.

Optimality Tolerance (tolσ). A nonbasic variable $x_j, j \in N$ is not a candidate, i.e., is considered nonimproving, when its reduced cost σ_j^0 satisfies

$$\sigma_j^0 \geq -tol\sigma(\|\boldsymbol{\pi}^0\|_2) \text{ and } x_j \text{ is at its lower bound}$$

or

$$\sigma_j^0 \leq tol\sigma(\|\boldsymbol{\pi}^0\|_2) \text{ and } x_j \text{ is at its upper bound.}$$

Pivot Tolerance (tolpiv). This is the smallest pivot (in magnitude) permitted in *CHUZR*.

Accuracy of Computation (tolrow). Given a computed basic solution $\mathbf{x}_B^0$ associated with the basis $\mathbf{B}^0 = [\mathbf{a}_{\beta_1}, \ldots, \mathbf{a}_{\beta_m}]$, and associated settings for the nonbasic variables $\mathbf{x}_N^0$, the residual for the system (9.1-2) is given by

$$\mathbf{r} \equiv \mathbf{B}^0\mathbf{x}_B^0 - \tilde{\mathbf{b}},$$

where $\tilde{\mathbf{b}}$ is defined by (9.1-1).

If

$$|r_k| \leq tolrow\left[\sum_{i=1}^{m} |(\mathbf{a}_{\beta_i})_k x_{\beta_i}^0|\right] \qquad k = 1, \ldots, m, \qquad (9.1\text{-}5)$$

then the current basis factorization is judged to be satisfactory for solving the associated systems of equations. Analogous statements apply to the solution of (9.1-3) or (9.1-4). If (9.1-5) does not hold, then the current factorization is judged to be inaccurate. If obtained by a sequence of updates, the basis matrix can be refactorized.

9.2-4. Major/Minor Iterations

As discussed in Sec. 7.2-3, it is common practice to select several candidates with a promise of gain at step *PRICE* and to restrict minor iterations to just these candidates. At a major iteration, a new candidate list is formed.

9.2-5. Refactorization Frequency

A refactorization, instead of an update of the basis representation, can be triggered for two principal reasons. First, refactorization is warranted when the current LU representation is judged to be inaccurate, for example, when it fails the test (9.1-5), in solving the equations in the *FTRAN* or *BTRAN* steps, or when there is a sudden unexpected loss of feasibility. Second, refactorization may be necessary for reasons of computational efficiency. With each update of the LU factorization, the number of factors in the **Γ** file grows and the density of **U** increases. At some stage it will be profitable to refactorize the current basis, because this will lead to a representation with substantially fewer nonzeros and hence to more efficient *FTRAN* and *BTRAN* operations. This would more than offset the increased expense of a factorization over that of an update. A common tactic, when a basis matrix is factorized, is to keep track of the time it takes to accomplish this task along with the time for the next *FTRAN* and *BTRAN* operations (say t_F). This is compared with the time for an update and the succeeding *FTRAN* and *BTRAN* operations (say t_U). When $t_U > t_F$, the next basis is refactorized.

9.2-6. Scaling

Scaling of a linear program improves the condition number of basis matrices and makes for more meaningful comparisons in the simplex algorithm, for example, between the reduced costs of nonbasic variables. Considerations of scale invariance were discussed in Sec. 7.1-3. Tomlin [1975b] gives a number of techniques for explicit rescaling of a linear program.

9.2-7. Degeneracy and Convergence

Many linear programs that arise in practice turn out to be degenerate. Even though many of the strategies discussed earlier for choosing the entering and exiting variables cannot, in theory, be guaranteed to ensure convergence in the presence of degeneracy, for all practical purposes they always do converge, i.e., cycling is almost never a problem in practice. The reasons for this are not hard to see. Degeneracy occurs when basic variables are at a bound. However, a tiny perturbation of the right-hand side would generally eliminate degeneracy and yet leave the problem unchanged for all practical purposes. Rounding error during computations is generally sufficient to introduce small ad hoc perturbations. Finite precision arithmetic is, therefore, a degeneracy resolution technique in itself, and many implementations simply ignore the remote possibility of cycling. Degeneracy can, however, slow down convergence, and techniques such as those of Harris [1975] (Sec. 7.3) are extremely useful in a practical setting, to both reduce the number of iterations and to improve numerical stability. See also Wolfe [1963].

Note

Further discussion of implementation may be found in Murtagh and Saunders [1983], Marsten [1981], Tomlin [1975a], Nazareth [1986a].

10
Mathematical Programming Systems in Practice

We conclude Part II of this book with a very brief survey in order to provide the reader with additional references dealing with implementation at a more advanced level.

10.1. Mathematical Programming Systems

When solving a large-scale linear program it is common to think in terms of a three-stage process as follows.

1. *Specification* of the model or problem and its generation in MPS format using a high level modeling language (matrix generator).
2. *Solution* of the problem using an LP optimizer and output of the solution.
3. *Interpretation* of the solution and presentation in the terminology of the original model using a report generator.

The process is summarized in Figure 10.1.

10.2. Model Specification and Interpretation

For a linear programming model to be successful, it must be user oriented, i.e., its content and structure must be easy to understand, it should be relatively easy to modify, and the results it produces should be both verifiable and easily communicated to others. On the other hand, the model must be rigorously specified in a standard format (usually MPS format, see Chapter 3), which an LP optimizer can accept. In particular, the MPS format serves as a communication interface that makes it possible to use different optimizers interchangeably. Many systems have been proposed to help bridge the gap between user and optimizer and to help manage the large volumes of data that, in turn, determine the coefficients in the model equations. Such systems are often called

Figure 10.1 Overview of an MPS

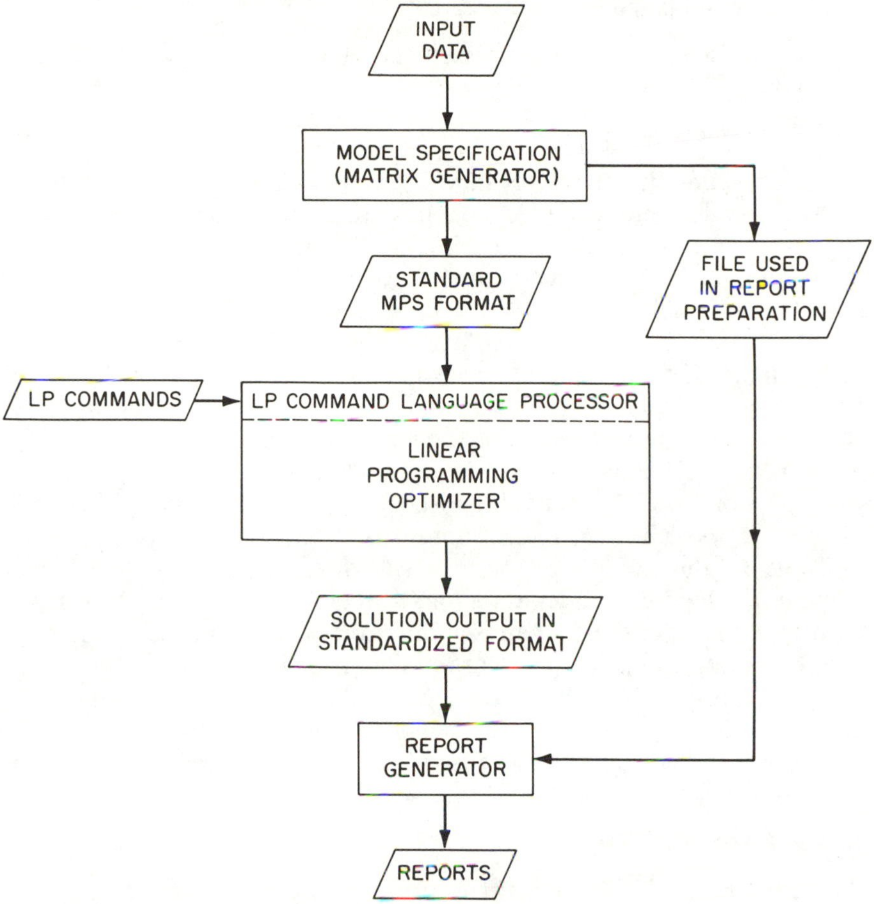

matrix/report generators, though the term is a little restrictive because it does not cover all the above needs. (For examples of matrix/report generators, see the notes at the end of this chapter.)

10.3. The Linear Programming Optimizer

Implementation at an advanced level must be able to do the following.

1. Manage a two-level hierarchy of computer storage where much of the problem resides in secondary memory (usually disk) and is brought into main memory as needed.
2. Provide adaptive solution strategies (tactics), which tailor the solution techniques used at various steps of the simplex algorithm to the characteristics of the particular linear program being solved.

3. Provide a wide range of user options through a suitable LP command language.

These issues are well discussed in Benichou et al. [1977] and Greenberg [1978b].

10.3-1. Problem Setup

When operating in a two-level storage environment where the LP matrix is stored on disk and individual columns are brought into central memory as needed, it is important to reduce the overhead associated with input/output (disk reads and writes). In order to *localize* access to data, it may be preferable to structure data as a single sequentially allocated linear list, in which each variable length node holds the nonzero elements of a column of the matrix in a packed form. For example, consider the *basic sequential columnar form,* Greenberg [1978a]. Here the jth node corresponds to the jth column of the matrix and contains a data item for each nonzero element. This, in turn, consists of two fields, containing the row index and the element value respectively. A header in the first field of the node gives the number of nonzeros in the column. For the matrix of Table 3.3 this would take the form shown in Figure 10.2.

Note that Figure 10.2 depicts a *logical* representation, i.e., it must still be mapped into a physical representation that is determined by the word length and other characteristics of the computer in use.

Other more sophisticated versions of the foregoing scheme take advantage of the fact that coefficients in the LP matrix often repeat themselves, a phenomenon that is sometimes known as *supersparsity*. For a discussion, see Greenberg [1978a].

10.3-2. Basis Handling

We noted in Sec. 5.5-2 that static (a priori) strategies for choosing a pivot order simply require a *Boolean* representation of a basis matrix, i.e., a

Figure 10.2 Basic sequential columnar form

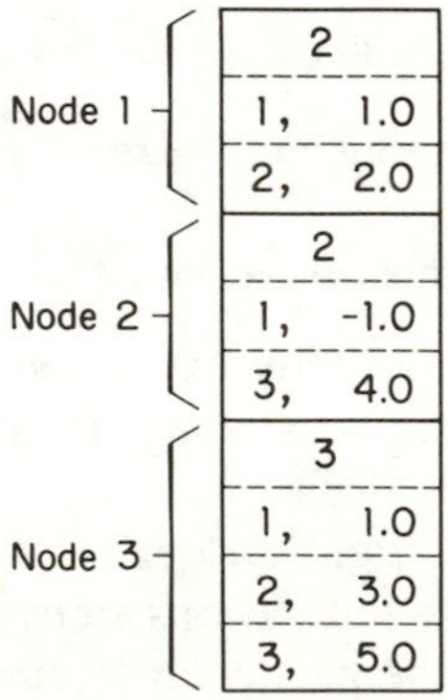

matrix of the same dimensions as the original matrix, which has a unit element in every position where the original matrix has a nonzero element, and zero everywhere else. The Boolean representation can be stored much more compactly because no floating-point numbers are involved.

When *factorizing* a very large sparse basis matrix, it is common to determine an initial preassigned pivot order using such a Boolean representation. An efficient way to proceed is as follows. Find (implicitly, of course) the permutation of rows and columns that gives a block lower triangular matrix with as many and as small blocks as possible, using the following two-step procedure.

1. Find the permutation of rows and columns (not necessarily symmetric, meaning not necessarily the same permutation of rows as is used for columns) so that the permuted matrix, say $\tilde{\mathbf{B}}^0$, has nonzeros on the diagonal. This is called finding a maximum transversal. See Duff [1981] for a discussion of suitable techniques.
2. Find the symmetric permutation of rows and columns of $\tilde{\mathbf{B}}^0$ to block lower triangular form, using the algorithm of Tarjan [1972]. See Duff and Reid [1978] for an implementation. This is a refinement of the algorithm of Sargent and Westerberg [1964] discussed in Sec. 5.5-2 and is computationally more efficient, because it avoids the potentially expensive node collapsing operations.

It is worth noting again that the widely used form of Figure 5.1 is just a special case of the foregoing block lower traingular form, one that is much easier to find.

Having found a block lower triangular form, one can process it further in one of two ways. Following Hellerman and Rarick [1971, 1972], permutations of rows and columns within each block or "bump" are sought, so that superdiagonal elements are confined to relatively few columns (called "spikes"). An alternative approach is to apply Gaussian elimination, in a Boolean manner, to the block lower triangular matrix in order to choose a pivot sequence that will limit the amount of fill-in. Details are given by Benichou et al. [1977] and Duff and Reid [1978]. An efficient data structure is described by George and Ng [1984].

Then the *numerical* phase is carried out using, whenever possible, the preassigned pivot order that was determined by the set of permutations discovered previously, subject to meeting stability requirements. Columns that hold pivots that are rejected on the grounds of stability are processed last within a bump. For a discussion of the "bump and spike" version see Saunders [1976]. Note that fill-in during the LU factorization can only occur in spike columns. For a version based on a single bump of the form depicted in Figure 5.1 and (Boolean) Gaussian elimination, see Benichou et al. [1977].

When *updating* a very large sparse basis matrix whose factors reside in secondary memory, the Forrest–Tomlin update [1972] has significant advantages. These stem from the fact that the **L** and **U** factors can both be maintained in sequentially allocated linear lists (called the **L** and **U** files) whose nodes do *not* change arbitrarily in length when fill-in occurs. These files could be designed along the lines discussed, for example, in Sec. 10.3-1. Suppose, for purposes of discussion, that $\mathbf{B}^0 = \mathbf{L}^0\mathbf{U}^0$, the **L** file is given by

$$\mathbf{L}^0 = \mathbf{L}_1^0\mathbf{L}_2^0 \cdots \mathbf{L}_m^0,$$

and the column-oriented **U** file by

$$\mathbf{U}^0 = \mathbf{U}_m^0\mathbf{U}_{m-1}^0 \cdots \mathbf{U}_{p+1}^0\mathbf{U}_p^0\mathbf{U}_{p-1}^0 \cdots \mathbf{U}_1^0.$$

(See also Sec. 5.1-3.) Suppose that column p of the basis matrix $\mathbf{B}^0$ is replaced and a Forrest–Tomlin update is performed, along the lines described in Sec. 6.4. This will involve the following changes to the **L** and **U** files. The node holding $\mathbf{U}_p^0$ is flagged as deleted. Nodes that hold $\mathbf{U}_{p+1}^0$ through $\mathbf{U}_m^0$ are modified by having the element in row p flagged as zero. Let us now denote them by $\bar{\mathbf{U}}_j^0, j = p+1, \ldots, m$. A new node, say $\bar{\mathbf{U}}_{m+1}^0$, which holds the packed column $\mathbf{\Pi h}_m$ (see (6.4-4)) is added to the **U** file. Finally, the node holding the elementary matrix **R** is added to the **L** file. (In contrast, in Bartels–Golub updating, the nodes in the **U** file would change unpredictably.) It is also possible to avoid a separate scan of the files and to organize the Forrest–Tomlin update so that the modification of the files is carried out during the succeeding *BTRAN* operation. Thus the cost of the update essentially amounts to a doubling of the arithmetic in about half of this *BTRAN* operation. For a full discussion, see Forrest and Tomlin [1972] and Tomlin [1972b].

As we have noted in Sec. 6.4, the Forrest–Tomlin update is potentially unstable and accuracy checks are essential, see Tomlin [1975c]. Saunders [1976] proposes a stable update based on the "bump and spike" form, in which most of the information defining the **L** and **U** factors resides in sequentially allocated linear lists maintained in secondary storage. Gay [1978] gives a variant that incorporates some of the techniques of Reid [1982, 1976], which we described in Sec. 5.10.

For an excellent survey of advanced basis handling techniques, see the article of Saunders [1980].

10.3-3. Strategies

Advanced pricing strategies (including sectional pricing) are discussed in Greenberg [1978b]. Computationally efficient refinements of the steepest-edge techniques of Sec. 7.1-2 (called *DEVEX* techniques, because they were originated by Harris [1975] for the DEVEX LP code) for choosing the entering variable and of the techniques of Sec. 7.3 for choosing the exiting variable are given by Benichou et al. [1977]. The latter article also discusses degeneracy resolution tactics and adaptive strategies tailored to the linear program being solved.

Notes

Sec. 10.1. Examples of Mathematical Programming Systems and the firms that distribute them are: MPSX/370 (International Business Machines) and MPS III (Ketron).

Sec. 10.2. Examples of matrix/report generators and the organizations or firms that distribute them are GAMS—General Algebraic Modelling System (World Bank), GAMMA (Bonner and Moore), MAGEN, PDS, and OMNI (Haverley Systems), DATAFORM (Ketron), DATAMAT (National Bureau of Economic Research), MGG (Scicon), ALPS (United Computing Systems).

Part III

Optimization Principle + Simplex Method = LP Algorithm

We turn now to three fundamental concepts or principles of linear programming: duality, decomposition, and homotopy. In conjunction with the simplex method, each leads to a key linear programming algorithm.

So far, we have emphasized the form of the linear programming problem most suited to computation. Henceforth, however, we shall frame the discussion in terms of whichever form of the linear program enables the ideas to be presented most directly. The reader will have grown accustomed to the simplex method within the setting of the computational canonical form and he or she should have no difficulty in making the conceptual generalization to this case. Computational details concerning bound constraints are given, when necessary, in the concluding section of each chapter.

11
The Duality Principle and the Simplex Method

11.1. The Diet Problem Revisited

Let us return to the diet problem of Chapter 1, Sec. 1.1 and formulate another linear program based on the same data. We shall see that this linear program bears a very special relationship to the linear program (1.1-1) previously formulated.

We now look at the problem through the eyes of a supplier of *nutrients,* who wishes to associate a *price* with each of the nutrients of the problem, namely, protein (π_1 cents/gram), carbohydrate (π_2 cents/gram), calcium (π_3 cents/milligram), iron (π_4 cents/milligram) and vitamin A (π_5 cents/international unit). Thus the return obtained by selling the *equivalent* of the nutrients contained in 100 grams of poultry is

$$20\pi_1 + 0\pi_2 + 8\pi_3 + 1.4\pi_4 + 80\pi_5.$$

Let us require that the prices charged by the supplier are *fair,* i.e., that the return that he or she obtains by selling the nutrient equivalent of 100 grams of poultry *does not exceed* the amount that the consumer would have to pay for 100 grams of poultry itself. Therefore

$$20\pi_1 + 0\pi_2 + 8\pi_3 + 1.4\pi_4 + 80\pi_5 \leq 40.$$

Similar constraints can be formulated for the other two foods, namely spinach and potatoes.

As the data of the problem states, the consumer needs at least 65 grams of protein, 90 grams of carbohydrate, 200 milligrams of calcium, 10 milligrams of iron, and 5000 international units of vitamin A. If the supplier provided them directly to the consumer, the supplier would be paid

$$65\pi_1 + 90\pi_2 + 200\pi_3 + 10\pi_4 + 5000\pi_5$$

and would, of course, seeks to set (fair) prices so as to maximize his or her return.

The supplier might, therefore, formulate the following linear program.

$$\begin{aligned}
\text{maximize}\quad & w = 65\pi_1 + 90\pi_2 + 200\pi_3 + 10\pi_4 + 5000\pi_5 \\
\text{s.t.}\quad & 20\pi_1 + 8\pi_3 + 1.4\pi_4 + 80\pi_5 \le 40 \\
& 3\pi_1 + 3\pi_2 + 83\pi_3 + 2\pi_4 + 7300\pi_5 \le 15 \\
& 2\pi_1 + 18\pi_2 + 7\pi_3 + 0.6\pi_4 \le 10 \\
& \pi_i \ge 0 \qquad i = 1, \ldots, 5.
\end{aligned} \tag{11.1-1}$$

We observe that (11.1-1) and (1.1-1) are both defined in terms of the same set of data and that they bear a special relationship to one another. These two linear programs are a particular instance of the *duality principle of linear programming that postulates the structural correspondence between a given linear program* (*called the primal and usually stated in terms of minimizing its objective*) *and an associated linear program* (*called the dual and usually stated in terms of maximizing its objective*). The dual program has some important properties relative to the primal. The reader will very likely have guessed, for example, that the cost to the buyer (say z^*) in the optimal solution of (1.1-1) *equals* the return to the supplier (say w^*) in the optimal solution of (11.1-1). This chapter defines and explores these duality relationships in some detail and develops associated computational procedures.

11.2. Dual Programs and Duality Theory

Consider the following linear program.

$$\text{Find } \mathbf{x} \in R^n \text{ s.t. } \mathbf{Ax} \ge \mathbf{b}, \mathbf{x} \ge \mathbf{0}, \text{ and } \mathbf{c}^T\mathbf{x} \text{ is minimized,} \tag{11.2-1}$$

where $\mathbf{A}$ is an $m \times n$ matrix with $m \ge n$, $\mathbf{b}$ is an m vector and $\mathbf{c}$ is an n vector.

Let us associate with this linear program, a new linear program that is obtained by using the right-hand side vector to define the (new) objective function and the transpose of the LP matrix and the objective (of the original linear program) to define the (new) constraints. This operation, termed *dualizing,* leads to the following linear program.

$$\text{Find } \boldsymbol{\pi} \in R^m \text{ s.t. } \mathbf{A}^T\boldsymbol{\pi} \le \mathbf{c}, \boldsymbol{\pi} \ge \mathbf{0}, \text{ and } \mathbf{b}^T\boldsymbol{\pi} \text{ is maximized.} \tag{11.2-2}$$

Let us reexpress (11.2-2), using the usual transformations, to obtain an equivalent linear program.

$$\text{Find } \boldsymbol{\pi} \in R^m \text{ s.t. } (-\mathbf{A}^T)\boldsymbol{\pi} \ge -\mathbf{c}, \boldsymbol{\pi} \ge \mathbf{0}, \text{ and } (-\mathbf{b})^T\boldsymbol{\pi} \text{ is minimized.} \tag{11.2-3}$$

This is of the same form as (11.2-1) and we can now verify, when the above dualizing operation is reapplied to (11.2-3), that we are led back to the program with which we started in the first place. Dualizing (11.2-3) we obtain

Find $\mathbf{z} \in R^n$ s.t. $(-\mathbf{A}^T)^T\mathbf{z} \leq (-\mathbf{b})$, $\mathbf{z} \geq \mathbf{0}$, and $(-\mathbf{c})^T\mathbf{z}$ is maximized,

or equivalently,

$$\text{Find } \mathbf{z} \in R^n \text{ s.t. } \mathbf{Az} \geq \mathbf{b},\ \mathbf{z} \geq \mathbf{0}, \text{ and } \mathbf{c}^T\mathbf{z} \text{ is minimized.} \qquad (11.2\text{-}4)$$

When we identify $\mathbf{z}$ and $\mathbf{x}$, we see that (11.2-1) and (11.2-4) are identical.

Consider therefore the pair of *dual programs* as follows.

$$\begin{aligned} (\boldsymbol{P})\text{: minimize } & \mathbf{c}^T\mathbf{x} \\ \text{s.t. } & \mathbf{Ax} \geq \mathbf{b} \\ & \mathbf{x} \geq \mathbf{0}. \end{aligned} \qquad (11.2\text{-}5a)$$

and

$$\begin{aligned} (\boldsymbol{D})\text{: maximize } & \mathbf{b}^T\boldsymbol{\pi} \\ \text{s.t. } & \mathbf{A}^T\boldsymbol{\pi} \leq \mathbf{c} \\ & \boldsymbol{\pi} \geq \mathbf{0}. \end{aligned} \qquad (11.2\text{-}5b)$$

The linear program with the minimizing objective is termed the *primal* and the one with the maximizing objective is termed the *dual*. Note that with each constraint $(\mathbf{a}^i)^T\mathbf{x} \geq b_i$ of $(\boldsymbol{P})$ we associate a dual variable π_i of $(\boldsymbol{D})$ and with each primal variable x_j of $(\boldsymbol{P})$ we associate a constraint $(\mathbf{a}_j)^T\boldsymbol{\pi} \leq c_j$ of $(\boldsymbol{D})$. Also, we have just demonstrated that the *dual of the dual is the primal*.

The following three theorems are fundamental to the duality theory of linear programming.

Theorem 11.2-1. Weak Duality. Let $\hat{\mathbf{x}}$ be any feasible solution of $(\boldsymbol{P})$ and $\hat{\boldsymbol{\pi}}$ be any feasible solution of $(\boldsymbol{D})$. Then

$$\mathbf{c}^T\hat{\mathbf{x}} \geq \mathbf{b}^T\hat{\boldsymbol{\pi}}. \qquad (11.2\text{-}6a)$$

Proof. $\mathbf{A}\hat{\mathbf{x}} \geq \mathbf{b}$ and $\hat{\boldsymbol{\pi}} \geq \mathbf{0}$. Thus $\hat{\boldsymbol{\pi}}^T\mathbf{A}\hat{\mathbf{x}} \geq \mathbf{b}^T\hat{\boldsymbol{\pi}}$.
$\mathbf{A}^T\hat{\boldsymbol{\pi}} \leq \mathbf{c}$ and $\hat{\mathbf{x}} \geq \mathbf{0}$. Thus $\hat{\mathbf{x}}^T\mathbf{A}^T\hat{\boldsymbol{\pi}} \leq \mathbf{c}^T\hat{\mathbf{x}}$.

Hence

$$\mathbf{c}^T\hat{\mathbf{x}} \geq \hat{\mathbf{x}}^T\mathbf{A}^T\hat{\boldsymbol{\pi}} = \hat{\boldsymbol{\pi}}^T\mathbf{A}\hat{\mathbf{x}} \geq \mathbf{b}^T\hat{\boldsymbol{\pi}}. \quad \blacksquare \qquad (11.2\text{-}6b)$$

Corollary 11.2-1. If $(\boldsymbol{P})$ has an unbounded optimal solution, then $(\boldsymbol{D})$ is infeasible. Similarly, if $(\boldsymbol{D})$ has an unbounded optimal solution then $(\boldsymbol{P})$ is infeasible.

Exercise 11.2-1. Show, by example, that $(\boldsymbol{P})$ and $(\boldsymbol{D})$ can both be infeasible, i.e., that the converse of Corollary 11.2-1 is not true.

Theorem 11.2-2. Strong Duality. Suppose $(\boldsymbol{P})$ and $(\boldsymbol{D})$ both have feasible solutions, say $\hat{\mathbf{x}}$ and $\hat{\boldsymbol{\pi}}$. Then there exists $\mathbf{x}^*$ that is optimal for (***P***) and $\boldsymbol{\pi}^*$ optimal for (***D***) such that

$$\mathbf{c}^T\mathbf{x}^* = \mathbf{b}^T\boldsymbol{\pi}^*. \tag{11.2-7}$$

Proof. Since (***P***) is feasible by assumption and, by Theorem 11.2-1, $\mathbf{b}^T\hat{\boldsymbol{\pi}}$ is a lower bound on the optimal solution of (***P***), it follows that (***P***) must have an optimal solution, say $\mathbf{x}^*$ (see also Theorem 1.3-3 or Corollary 2.5-1, if necessary).

Let us define index sets I and L associated with the active constraints of (***P***) at $\mathbf{x}^*$ as follows.

$$I \equiv \{i \mid 1 \le i \le m \quad \text{and} \quad (\mathbf{a}^i)^T\mathbf{x}^* = b_i\} \tag{11.2-8}$$

and

$$L \equiv \{j \mid 1 \le j \le n \quad \text{and} \quad (\mathbf{e}^j)^T\mathbf{x}^* = 0\}. \tag{11.2-9}$$

Then it follows from the optimality conditions for (***P***) given in Corollary 1.4-1 that

$$\mathbf{c} = \sum_{i \in I} \pi_i^* \mathbf{a}^i + \sum_{j \in L} \sigma_j^* \mathbf{e}^j \tag{11.2-10a}$$

and

$$\pi_i^* \ge 0 \qquad \sigma_j^* \ge 0 \qquad \text{for all } i, j. \tag{11.2-10b}$$

(Note that we now use σ_j^* in place of λ_j^*.)

Therefore, using (11.2-8) and (11.2-9),

$$\mathbf{c}^T\mathbf{x}^* = \sum_{i \in I} \pi_i^*(\mathbf{a}^i)^T\mathbf{x}^* + \sum_{j \in L} \sigma_j^*(\mathbf{e}^j)^T\mathbf{x}^* = \sum_{i \in I} \pi_i^* b_i.$$

Let us define

$$\pi_i^* \equiv 0, \; 1 \le i \le m, \text{ and } i \notin I. \tag{11.2-11}$$

Then

$$\mathbf{c}^T\mathbf{x}^* = \mathbf{b}^T\boldsymbol{\pi}^*. \tag{11.2-12}$$

It also follows immediately from (11.2-10) and (11.2-11) that

$$\mathbf{A}^T\boldsymbol{\pi}^* \le \mathbf{c} \quad \text{and} \quad \boldsymbol{\pi}^* \ge \mathbf{0},$$

and thus $\boldsymbol{\pi}^*$ is feasible for (***D***).

Therefore, again using Theorem 11.2-1, we see that (11.2-12) implies that $\mathbf{x}^*$ is optimal in (***P***) and $\boldsymbol{\pi}^*$ is optimal in (***D***). ■

Corollary 11.2-2. (***P***) has a finite optimal solution if and only if (***D***) has a finite optimal solution.

Theorem 11.2-3. Complementary Slackness. Suppose that $\hat{\mathbf{x}}$ is feasible for (***P***) and $\hat{\boldsymbol{\pi}}$ is feasible for (***D***). Then $\hat{\mathbf{x}}$ and $\hat{\boldsymbol{\pi}}$ are optimal for (***P***) and (***D***), respectively, if and only if,

$$\hat{\boldsymbol{\pi}}^T(\mathbf{A}\hat{\mathbf{x}}-\mathbf{b})=0 \quad \text{and} \quad \hat{\mathbf{x}}^T(\mathbf{c}-\mathbf{A}^T\hat{\boldsymbol{\pi}})=0. \tag{11.2-13}$$

Proof. Suppose (11.2-13) holds. Then $\mathbf{b}^T\hat{\boldsymbol{\pi}}=\mathbf{c}^T\hat{\mathbf{x}}$. Since $\hat{\mathbf{x}}$ and $\hat{\boldsymbol{\pi}}$ are feasible for ($\boldsymbol{P}$) and ($\boldsymbol{D}$), respectively, it follows from Theorem 11.2-1 that they are also optimal in their respective programs.

Suppose that (11.2-13) does *not* hold. From feasibility of $\hat{\mathbf{x}}$ and $\hat{\boldsymbol{\pi}}$ and (11.2-6b) we have

$$\mathbf{b}^T\hat{\boldsymbol{\pi}} \le \hat{\boldsymbol{\pi}}^T\mathbf{A}\hat{\mathbf{x}} \le \mathbf{c}^T\hat{\mathbf{x}}.$$

The assumption thus implies that $\mathbf{b}^T\hat{\boldsymbol{\pi}} < \mathbf{c}^T\hat{\mathbf{x}}$. From Theorem 11.2-2, $\hat{\mathbf{x}}$ and $\hat{\boldsymbol{\pi}}$ cannot both be optimal for ($\boldsymbol{P}$) and ($\boldsymbol{D}$), respectively. This completes the proof. ∎

An alternative statement of the complementary slackness conditions (11.2-13) is

$$\hat{x}_j > 0 \quad \text{if and only if} \quad (\mathbf{a}_j)^T\hat{\boldsymbol{\pi}} = c_j \tag{11.2-14a}$$

and

$$\hat{\pi}_i > 0 \quad \text{if and only if} \quad (\mathbf{a}^i)^T\hat{\mathbf{x}} = b_i \tag{11.2-14b}$$

for $1 \le i \le m$, $1 \le j \le n$.

11.2-1. Variants

Consider the linear program of the form

$$\begin{aligned} \text{minimize}\quad & \mathbf{c}^T\mathbf{x} \\ \text{s.t.}\quad & \mathbf{A}\mathbf{x}=\mathbf{b} \\ & \mathbf{x} \ge \mathbf{0}. \end{aligned} \tag{11.2-15}$$

This can be equivalently stated as

$$\begin{aligned} \text{minimize}\quad & \mathbf{c}^T\mathbf{x} \\ \text{s.t.}\quad & \begin{bmatrix} \mathbf{A} \\ -\mathbf{A} \end{bmatrix}\mathbf{x} \ge \begin{bmatrix} \mathbf{b} \\ -\mathbf{b} \end{bmatrix} \\ & \mathbf{x} \ge \mathbf{0}. \end{aligned}$$

This is of the form (11.2-1) and the associated dual program is

$$\begin{aligned} \text{maximize}\quad & \mathbf{b}^T\boldsymbol{\pi}_1 - \mathbf{b}^T\boldsymbol{\pi}_2 \\ \text{s.t.}\quad & \mathbf{A}^T\boldsymbol{\pi}_1 - \mathbf{A}^T\boldsymbol{\pi}_2 \le \mathbf{c} \\ & \boldsymbol{\pi}_1, \boldsymbol{\pi}_2 \ge \mathbf{0}. \end{aligned} \tag{11.2-16}$$

Define $\boldsymbol{\pi} \equiv \boldsymbol{\pi}_1 - \boldsymbol{\pi}_2$ and note that $\boldsymbol{\pi}$ is of arbitrary sign. Therefore (11.2-16) is equivalent to

$$\begin{aligned} \text{maximize}\quad & \mathbf{b}^T\boldsymbol{\pi} \\ \text{s.t.}\quad & \mathbf{A}^T\boldsymbol{\pi} \le \mathbf{c}. \end{aligned} \tag{11.2-17}$$

Table 11.1.

PRIMAL (min)	DUAL (max)
ith constraint is an inequality ($\geq$)	ith variable is nonnegative ($\geq$)
ith constraint is an inequality ($\leq$)	ith variable is nonpositive ($\leq$)
ith constraint is an equality ($=$)	ith variable is unrestricted in sign
jth variable is nonnegative ($\geq$)	jth constraint is an inequality ($\leq$)
jth variable is nonpositive ($\leq$)	jth constraint is an inequality ($\geq$)
jth variable is unrestricted in sign	jth constraint is an equality ($=$)

The programs (11.2-15) and (11.2-17) are dual programs. Note again the association between constraints (variables) of the primal and variables (constraints) of the dual. This time *equality* constraints are present and are associated with *unrestricted* variables.

Indeed, for every primal program (expressed in terms of minimizing an objective) there is a dual program (expressed in terms of maximizing an objective) and a summary of the corresponding relationships between primal and dual are shown in Table 11.1.

Exercise 11.2-2. State and prove analogues of Theorems 11.2-1 through 11.2-3 for the primal–dual pair (11.2-15) and (11.2-17).

Exercise 11.2-3. Find the dual of the linear program (1.2-2).

Exercise 11.2-4. Consider the following linear program

$$\begin{aligned} \text{minimize } & \mathbf{c}^T\mathbf{x} \\ \text{s.t. } & \mathbf{A}\mathbf{x} = \mathbf{0} \\ & \mathbf{e}^T\mathbf{x} = 1 \\ & \mathbf{x} \geq \mathbf{0}, \end{aligned} \tag{11.2-18}$$

where $\mathbf{A}$ is an $m \times n$ matrix and $\mathbf{e}$ is a vector whose components are all unity. Given a feasible basis matrix

$$\mathbf{B}^0 = \begin{bmatrix} \mathbf{a}_1 \cdots \mathbf{a}_{m+1} \\ 1 \cdots \quad 1 \end{bmatrix}$$

with corresponding solution $\mathbf{x}^0$, define the price vector $\boldsymbol{\pi}^0$ (which has $m+1$ components) in the usual way as the solution of $(\boldsymbol{\pi}^0)^T\mathbf{B}^0 = \mathbf{c}_B^T$, and define reduced costs as

$$\sigma_j^0 = c_j - (\boldsymbol{\pi}^0)^T \begin{bmatrix} \mathbf{a}_j \\ 1 \end{bmatrix} \qquad j = 1, \ldots, n.$$

Let

$$\sigma_s^0 = \min_{1 \leq j \leq n} \sigma_j^0.$$

Suppose that (11.2-18) is modified by replacing $\mathbf{c}$ by $(\mathbf{c}+\sigma_s^0\mathbf{e})$. Show that $\boldsymbol{\pi}^0$ as just defined is feasible for the dual of this modified linear program. Hence show that $(\mathbf{c}^T\mathbf{x}^0+\sigma_s^0)$ is a *lower* bound on the objective value of (11.2-18).

11.3. Ties to Optimality Conditions and Simplex Algorithms

It should be apparent from the proof of Theorem 11.2-2, that there is a very close tie between the optimality conditions of Chapter 1 and the above duality theory. Theorem 1.4-1, in particular Corollary 1.4-1, gives necessary and sufficient conditions for optimality. Theorem 11.2-3 also gives necessary and sufficient conditions for optimality, namely, that the pair $\hat{\mathbf{x}}$ and $\hat{\boldsymbol{\pi}}$ be primal feasible and dual feasible, respectively, and that they satisfy the complementary slackness conditions (11.2-14). Logic demands that these two sets of conditions be equivalent.

Consider now the primal–dual pair (11.2-15) and (11.2-17). When the primal simplex algorithm is applied to (11.2-15) and the current basic feasible solution, say $\mathbf{x}^0$, is defined by B and N in the usual way, with nonbasics at their bounds (in this case, $x_j^0=0,\ j\in N$), then the associated price vector is $(\boldsymbol{\pi}^0)^T=\mathbf{c}_B^T(\mathbf{B}^0)^{-1}$. Let us define reduced costs in the usual way as

$$\sigma_j^0=c_j-(\boldsymbol{\pi}^0)^T\mathbf{a}_j \qquad j=1,\ldots,n.$$

Clearly, $\sigma_j^0=0,\ j\in B$. Since $x_j^0=0,\ j\in N$ we have

$$x_j^0\sigma_j^0=x_j^0(c_j-(\boldsymbol{\pi}^0)^T\mathbf{a}_j)=0 \qquad j=1,\ldots,n.$$

Thus the pair of solutions $\mathbf{x}^0$ and $\boldsymbol{\pi}^0$ satisfy the complementary slackness conditions (11.2-13) or (11.2-14). Note that $\mathbf{A}\mathbf{x}^0=\mathbf{b}$ implies that $(\boldsymbol{\pi}^0)^T(\mathbf{A}\mathbf{x}^0-\mathbf{b})\equiv 0$. In addition, $\mathbf{x}^0$ is obviously primal feasible, but $\boldsymbol{\pi}^0$ is *not* necessarily feasible for the dual (11.2-17), i.e., $\sigma_j^0\geq 0$ for all $j\in N$ only when $\mathbf{x}^0$ is optimal. We see that the primal simplex algorithm maintains primal feasibility and complementary slackness and looks for dual feasibility. This is depicted by the first row of Table 11.2. The second row is what we turn to next, in the remainder of this chapter. The next two rows are dealt with in the following two chapters.

Table 11.2.

	Primal Feasible	Dual Feasible	Complementary Slackness
Primal simplex	X	?	X
Dual simplex	?	X	X
Decomposition	X	X	?
Self-dual	?	?	X

11.4. Solving the Dual Program

We have noted that the canonical form (1.2-2) is the most suited to computation. However, for the moment, let us confine ourselves to the case of nonnegative bounds $\mathbf{x} \geq \mathbf{0}$. Details for the more general case will be given in Sec. 11.6. The primal–dual pair we are concerned with is (11.2-15) and (11.2-17), namely,

$$\begin{array}{llll} (P)\text{: minimize } \mathbf{c}^T\mathbf{x} & & (D)\text{: maximize } \mathbf{b}^T\boldsymbol{\pi} & \\ \text{s.t. } \mathbf{A}\mathbf{x} = \mathbf{b} & & \text{s.t. } \mathbf{A}^T\boldsymbol{\pi} \leq \mathbf{c} & \quad (11.4\text{-}1) \\ \mathbf{x} \geq \mathbf{0} & & & \end{array}$$

To solve (D) we would naturally resort to the simplex method. As we have emphasized in Chapter 2, in particular, in the introduction to that chapter, the essence of the procedure is to find a solution defined by a set of active (tight and linearly independent) constraints that define a vertex, determine the associated multipliers, identify an active constraint whose multiplier indicates that there would be an improvement if a (feasible) step were to be taken off it, determine the (unique) direction when this constraint is relaxed and all other active constraints remain active, and finally move off the vertex along this direction until a new constraint and associated vertex is encountered. Furthermore, we may note that the dual program (D) above is precisely of the form (2.3-1) when we identify $\boldsymbol{\pi}$ with $\bar{\mathbf{x}}$, $\mathbf{A}^T$ with $\bar{\mathbf{A}}$ (hence $\mathbf{a}_k$ with $\bar{\mathbf{a}}^k$ and $\mathbf{a}^k$ with $\bar{\mathbf{a}}_k$), $\mathbf{b}$ with $\bar{\mathbf{c}}$ and $\mathbf{c}$ with $\bar{\mathbf{b}}$, m with $\bar{n}$ and n with $\bar{m}$. Let us recount the steps *in the new notation,* then state an algorithm.

11.4-1. Simplex Method Applied to the Dual

As in Sec. 2.3, let us partition the constraints of (D) into active and inactive constraints, namely, $\mathbf{A}^T = \begin{bmatrix} (\mathbf{B}^0)^T \\ (\mathbf{N}^0)^T \end{bmatrix}$, where $(\mathbf{B}^0)^T$ is assumed to be nonsingular. Let $B \equiv \{1, \ldots, m\}$ and $N \equiv \{m+1, \ldots, n\}$. Also partition the right-hand side of (D) as $\mathbf{c} = \begin{bmatrix} \mathbf{c}_B \\ \mathbf{c}_N \end{bmatrix}$ and let us assume that the associated solution $\boldsymbol{\pi}^0$ is feasible, so that

$$(\mathbf{B}^0)^T\boldsymbol{\pi}^0 = \mathbf{c}_B \quad \text{and} \quad (\mathbf{N}^0)^T\boldsymbol{\pi}^0 \leq \mathbf{c}_N. \tag{11.4-2}$$

Thus

$$\boldsymbol{\pi}^0 = \mathbf{c}_B^T(\mathbf{B}^0)^{-1} \quad \text{and} \quad (\boldsymbol{\sigma}_N^0)^T \equiv \mathbf{c}_N^T - (\boldsymbol{\pi}^0)^T\mathbf{N}^0 \geq \mathbf{0}. \tag{11.4-3}$$

Let us now denote the Lagrange multipliers $\bar{\boldsymbol{\sigma}}^0$ of Sec. 2.3 by $\mathbf{x}_B^0$ for reasons that the reader will very likely guess at and that will, in any case, become apparent very shortly. Then rewriting (2.3-4) gives

$$(\mathbf{x}_B^0)^T(\mathbf{B}^0)^T = \mathbf{b}^T \quad \text{or} \quad \mathbf{x}_B^0 = (\mathbf{B}^0)^{-1}\mathbf{b}. \tag{11.4-4}$$

Note that $\mathbf{x}_B^0$ can have components of *arbitrary* sign and that $\begin{bmatrix}\mathbf{x}_B^0\\ \mathbf{0}\end{bmatrix}$ corresponds to a basic solution of (P), but not necessarily a basic feasible solution. Also $\boldsymbol{\pi}^0$ and $\boldsymbol{\sigma}_N^0$ correspond to the associated price vector and vector of reduced costs on the primal (P). This accounts for our choice of notation.

If $\mathbf{x}_B^0 \geq \mathbf{0}$, then the solution $\boldsymbol{\pi}^0$ is optimal for (D). Otherwise, one of the m active constraints for which $(\mathbf{x}_B^0)_{\bar{s}} < 0$ is relaxed and the associated direction $\bar{\mathbf{d}}$ is obtained by solving (2.3-7), namely,

$$(\mathbf{B}^0)^T\bar{\mathbf{d}} = \operatorname{sign}(x_{\bar{s}}^0)\mathbf{e}_{\bar{s}} = -\mathbf{e}_{\bar{s}}. \tag{11.4-5}$$

Finally, the largest step $\bar{\theta}_{\bar{s}} \geq 0$ from $\boldsymbol{\pi}^0$ that can be taken along $\bar{\mathbf{d}}$ subject to not violating any constraint of (D) is

$$\begin{bmatrix}\bar{\theta}_{\bar{s}}\\ \bar{p}\end{bmatrix} = \begin{matrix}\min\limits_{j\in\bar{E}}\\ \\ \operatorname{argmin}\limits_{j\in\bar{E}}\end{matrix}\left\{(c_j - \mathbf{a}_j^T\boldsymbol{\pi}^0)/\mathbf{a}_j^T\bar{\mathbf{d}},\right.$$

where

$$\bar{E} \equiv \{j \mid \mathbf{a}_j^T\bar{\mathbf{d}} > 0 \qquad j \in N\}. \tag{11.4-6}$$

Using (11.4-3), this can be written

$$\begin{bmatrix}\bar{\theta}_{\bar{s}}\\ \bar{p}\end{bmatrix} = \begin{matrix}\min\limits_{j\in\bar{E}}\\ \\ \operatorname{argmin}\limits_{j\in\bar{E}}\end{matrix}\left\{\sigma_j^0/\mathbf{a}_j^T\bar{\mathbf{d}}.\right. \tag{11.4-7}$$

The new dual solution $\boldsymbol{\pi}$ is

$$\boldsymbol{\pi} = \boldsymbol{\pi}^0 + \bar{\theta}_{\bar{s}}\bar{\mathbf{d}} \tag{11.4-8a}$$

and

$$\sigma_j \equiv c_j - \boldsymbol{\pi}^T\mathbf{a}_j = \sigma_j^0 - \bar{\theta}_{\bar{s}}(\mathbf{a}_j^T\bar{\mathbf{d}}) \qquad j \in N. \tag{11.4-8b}$$

In particular, using $\sigma_{\bar{s}}^0 = 0$ and (11.4-5), we obtain

$$\sigma_{\bar{s}} = \bar{\theta}_{\bar{s}}.$$

Finally, note that $\bar{s}$ identifies the row of $(\mathbf{B}^0)^T$ (column of $\mathbf{B}^0$) that is replaced and $\bar{p}$ the row of $(\mathbf{N}^0)^T$ (column of $\mathbf{N}^0$) with which it is replaced. Therefore, consistent with our earlier primal notation, we shall use p in place of $\bar{s}$ and s in place of $\bar{p}$.

11.4-2. The Dual Simplex Algorithm

The dual simplex algorithm simply carries out the above computations on the primal representation (P) of the problem *using the primal terminol-*

ogy. It is the counterpart of the primal simplex algorithm of Sec. 2.4, in particular, its specialization to nonnegative bounds in Exercise 2.4-1. Again we use the same extension of notation, so that B now denotes the set of indices $\{\beta_1, \ldots, \beta_m\}$ of the m basic variables and N denotes the set of indices $\{\eta_1, \ldots, \eta_{n-m}\}$ of the $n - m$ nonbasic variables that are, in this case, at their zero bounds. We assume also that the corresponding dual solution is feasible.

Algorithm Dual Simplex

Step D1 [*Initialize*]. Solve $(\mathbf{B}^0)^T\boldsymbol{\pi}^0 = \mathbf{c}_B$ for $\boldsymbol{\pi}^0$, which is the initial solution of the dual and the price vector of the primal. Form $\sigma^0_{\eta_k} \leftarrow c_{\eta_k} - (\boldsymbol{\pi}^0)^T\mathbf{a}_{\eta_k}$, $\eta_k \in N$. These are the slacks associated with the dual constraints or the reduced costs associated with the price vector $\boldsymbol{\pi}^0$ of the primal. By assumption, $\sigma^0_{\eta_k} \geq 0$ for all $\eta_k \in N$.

Solve $\mathbf{B}^0\mathbf{x}^0_B = \mathbf{b}$ for $\mathbf{x}^0_B$, namely, the basic solution in the primal or the multipliers associated with the active constraints of the dual.

Step D2 [*Choose the exiting variable*]. Form $P \leftarrow \{i \mid x^0_{\beta_i} < 0,\ \beta_i \in B\}$. If $P = \varphi$ then stop because $\begin{bmatrix} \mathbf{x}^0_B \\ \mathbf{0} \end{bmatrix}$ is optimal. Otherwise, choose any member $p \in P$, typically by the largest coefficient rule, namely, $p \leftarrow \underset{i \in P}{\operatorname{argmax}} |x^0_{\beta_i}|$. (Note: φ denotes the empty set.)

Step D3 [*Find the search direction* $\bar{\mathbf{d}}$ *in the space of dual variables*]. Solve $(\mathbf{B}^0)^T\bar{\mathbf{d}} = -\mathbf{e}_p$ for $\bar{\mathbf{d}}$. (In the terminology of Chapter 9, this is a *BTRAN* operation.)

Step D4 [*Form* $\mathbf{w}$]. Form $w_{\eta_k} \leftarrow \bar{\mathbf{d}}^T\mathbf{a}_{\eta_k}$, $\eta_k \in N$. (In the terminology of Chapter 9, this corresponds to a *PRICE* operation.)

Step D5 [*Choose the entering variable*]. Form $\bar{E} = \{\eta_k \mid w_{\eta_k} > 0,\ \eta_k \in N\}$. If $\bar{E} = \varphi$ then stop, the dual is unbounded from above, i.e., the primal is infeasible (see Corollary 11.2-1). Otherwise, let

$$\bar{\theta}_p \leftarrow \min_{\eta_k \in \bar{E}} \sigma^0_{\eta_k}/w_{\eta_k},$$

$$s \leftarrow \eta_q = \underset{\eta_k \in \bar{E}}{\operatorname{argmin}}\, \sigma^0_{\eta_k}/w_{\eta_k},$$

where ties are resolved arbitrarily.

Step D6 [*Find the direction in the space of the primal variables*]. Solve $\mathbf{B}^0\tilde{\mathbf{a}}_s = \mathbf{a}_s$ for $\tilde{\mathbf{a}}_s$. (In the terminology of Chapter 9, this is an *FTRAN* operation.)

Step D7 [*Determine the new basis and revise associated quantities*]. The new basis matrix is (mathematically) defined by (6.1-2), so update $\mathbf{x}^0_B$ to

$(\mathbf{I}+(\tilde{\mathbf{a}}_s-\mathbf{e}_p)\mathbf{e}_p^T)^{-1}(\mathbf{B}^0)^{-1}\mathbf{b}$; namely, revise the basic solution vector as follows:

$$\mathbf{x}_B^0 \leftarrow \mathbf{x}_B^0 - \frac{\mathbf{x}_{\beta_p}^0}{(\tilde{\mathbf{a}}_s)_p}(\tilde{\mathbf{a}}_s-\mathbf{e}_p).$$

Revise the price vector and reduced costs (primal terminology) by (11.4-8), namely,

$$\boldsymbol{\pi}^0 \leftarrow \boldsymbol{\pi}^0 + \bar{\theta}_p\bar{\mathbf{d}}$$

$$\sigma_{\eta_k}^0 \leftarrow \sigma_{\eta_k}^0 - \bar{\theta}_p w_{\eta_k} \qquad \eta_k \in N \quad \text{and} \quad \eta_k \neq \eta_q \qquad \sigma_{\beta_p}^0 \leftarrow \bar{\theta}_p.$$

Update $\mathbf{B}^0$, $\mathbf{N}^0$, B, and N as follows:

$$\mathbf{B}^0 \leftarrow \{\mathbf{a}_{\beta_1}, \ldots, \mathbf{a}_{\beta_{p-1}}, \mathbf{a}_s, \mathbf{a}_{\beta_{p+1}}, \ldots, \mathbf{a}_{\beta_m}\}$$
$$\mathbf{N}^0 \leftarrow \{\mathbf{a}_{\eta_1}, \ldots, \mathbf{a}_{\eta_{q-1}}, \mathbf{a}_{\beta_p}, \mathbf{a}_{\eta_{q+1}}, \ldots, \mathbf{a}_{\eta_{n-m}}\}$$
$$B \leftarrow \{\beta_1, \ldots, \beta_{p-1}, s, \beta_{p+1}, \ldots, \beta_m\}$$
$$N \leftarrow \{\eta_1, \ldots, \eta_{q-1}, \beta_p, \eta_{q+1}, \ldots, \eta_{n-m}\}$$

Return to Step D2.

It should be clear that the dual simplex algorithm maintains dual feasibility and complementary slackness and looks for primal feasibility, because it is mathematically equivalent to the primal simplex algorithm applied to the dual program. It therefore corresponds to the second row of Table 11.2.

11.5. Sensitivity Analysis

Consider again the pair of dual programs (11.4-1) and suppose that optimal solutions $\mathbf{x}^*$ of (P) and $\boldsymbol{\pi}^*$ of (D) have been found, using either the primal or dual simplex algorithms. *Sensitivity analysis is the systematic study of variation of* $\mathbf{x}^*$ *and* $\boldsymbol{\pi}^*$ *with changes in the coefficients* $\mathbf{A}$, $\mathbf{b}$, *and* $\mathbf{c}$. In particular, it is concerned with finding bounds on the coefficient values that maintain feasibility and optimality of $\mathbf{x}^*$ and $\boldsymbol{\pi}^*$ in (P) and (D), respectively, and with systematic methods, usually based on use of the primal simplex and dual simplex algorithms, to restore feasibility and optimality when the coefficient changes exceed these bounds.

If $\mathbf{B}^0$ is an optimal basis for (P) or $(\mathbf{B}^0)^T$ for (D), then

$$\mathbf{x}^* = (\mathbf{B}^0)^{-1}\mathbf{b} \geq \mathbf{0} \qquad (\boldsymbol{\pi}^*)^T = \mathbf{c}_B^T(\mathbf{B}^0)^{-1} \tag{11.5-1}$$

and

$$\sigma_j^* = c_j - (\boldsymbol{\pi}^*)^T\mathbf{a}_j \geq 0 \qquad j \in N. \tag{11.5-2}$$

Again, for simplicity, assume that $B = \{1, \ldots, m\}$ and $N = \{m + 1, \ldots, n\}$.

1. Variation in $\mathbf{b}$ does not affect the value of $\boldsymbol{\pi}^*$ or σ_j^*. When $\mathbf{b}$ changes to $(\mathbf{b}+\Delta\mathbf{b})$, then $(\mathbf{B}^0)^{-1}(\mathbf{b}+\Delta\mathbf{b})$ will be optimal provided it is feasible, i.e., provided that $(\mathbf{B}^0)^{-1}(\mathbf{b}+\Delta\mathbf{b}) \geq \mathbf{0}$. In particular, if a single coefficient of $\mathbf{b}$, say b_k, is altered to $b_k+\delta$, then feasibility is maintained when

$$\mathbf{b}_k^{(-1)}\delta + \mathbf{x}^* \geq \mathbf{0}, \tag{11.5-3}$$

where $\mathbf{b}_k^{(-1)}$ denotes the kth column of $(\mathbf{B}^0)^{-1}$. Thus, δ must satisfy

$$\max_{b_{ik}^{(-1)}>0} (-x_i^*/b_{ik}^{(-1)}) \leq \delta \leq \min_{b_{ik}^{(-1)}<0} (x_i^*/-b_{ik}^{(-1)}). \tag{11.5-4}$$

When $\Delta\mathbf{b}$ is such that $((\mathbf{B}^0)^{-1}\Delta\mathbf{b}+\mathbf{x}^*)$ has at least one negative element, then we can use the dual simplex algorithm to restore optimality.

2. (a) When c_j, $j \in N$ is changed, then $\mathbf{x}^*$ and $\boldsymbol{\pi}^*$ remain unaltered. Thus, if c_j is changed to $c_j+\delta$, the solution $\mathbf{x}^*$ remains optimal provided that $(c_j+\delta)-(\boldsymbol{\pi}^*)^T\mathbf{a}_j \geq 0$, $j \in N$, i.e.,

$$\delta \geq -\sigma_j^* \qquad j \in N. \tag{11.5-5}$$

When this bound is exceeded, the primal simplex algorithm can be used to restore optimality.

(b) If c_j, $j \in B$ is changed, then $\mathbf{x}^*$ remains unaltered. Thus, if $\mathbf{c}_B$ is changed to $\mathbf{c}_B+\Delta\mathbf{c}_B$, $\mathbf{x}^*$ remains optimal provided that

$$c_j - (\mathbf{c}_B+\Delta\mathbf{c}_B)^T(\mathbf{B}^0)^{-1}\mathbf{a}_j \geq 0 \qquad j \in N. \tag{11.5-6}$$

Exercise 11.5-1. Suppose a single element of $\mathbf{c}$, say c_k, is changed to $c_k+\delta$. Establish ranges on δ so that $\mathbf{x}^*$ remains optimal.

3. When $\mathbf{a}_j$, $j \in N$ is changed, then $\mathbf{x}^*$ and $\boldsymbol{\pi}^*$ remain unaltered. Thus, if $\mathbf{a}_j$ is changed to $\mathbf{a}_j+\Delta\mathbf{a}_j$, $\mathbf{x}^*$ remains optimal provided that

$$c_j - (\boldsymbol{\pi}^*)^T(\mathbf{a}_j+\Delta\mathbf{a}_j) \geq 0 \qquad j \in N. \tag{11.5-7}$$

Exercise 11.5-2. When only the kth element of $\mathbf{a}_j$ changes, say to $a_{kj}+\delta$, establish bounds on δ so that $\mathbf{x}^*$ remains optimal.

For a further discussion of sensitivity analysis see, for example, Shapiro [1979] and Chvatal [1983].

11.6. Practical Details of Implementation

11.6-1. Dual Simplex Algorithm for General Bounds

Consider now the canonical form most suited to computation, namely,

$$\begin{aligned} (P)\colon \text{ minimize } & \mathbf{c}^T\mathbf{x} \\ \text{s.t. } & \mathbf{A}\mathbf{x} = \mathbf{b} \\ & \mathbf{l} \leq \mathbf{x} \leq \mathbf{u}. \end{aligned} \tag{11.6-1}$$

In the usual way, we shall assume that unbounded variables have a machine representation of infinity in the corresponding elements of $\mathbf{l}$ and $\mathbf{u}$.

The reformulation of the simplex method applied to the dual of (11.6-1), resulting in the dual simplex algorithm for solving (11.6-1), very closely parallels the development in Sec. 11.4. The details are, however, somewhat more taxing and we give them here.

By writing the bounds of (11.6-1) as $\mathbf{x} \geq \mathbf{l}$ and $-\mathbf{x} \geq -\mathbf{u}$, it is easily verified (see Table 11.1) that the dual of (11.6-1) is

$$\begin{aligned} (D)\text{: maximize } & \mathbf{b}^T\boldsymbol{\pi} + \mathbf{l}^T\boldsymbol{\lambda} - \mathbf{u}^T\boldsymbol{\mu} \\ \text{s.t. } & \mathbf{A}^T\boldsymbol{\pi} + \mathbf{I}\boldsymbol{\lambda} - \mathbf{I}\boldsymbol{\mu} = \mathbf{c} \\ & \boldsymbol{\lambda} \geq \mathbf{0} \qquad \boldsymbol{\mu} \geq \mathbf{0}. \end{aligned} \tag{11.6-2}$$

As we did previously, we have available a vertex (basic feasible solution of (D)) associated with a square system of active constraints of (11.6-2). We determine the corresponding multipliers and choose an active constraint whose multiplier indicates that there will be an increase in the objective value of (11.6-2) when the constraint is dropped from the active set. We then proceed to a new vertex. Again the formulation simplifies because of structure in the equations defining a vertex and can be organized to apply *directly* to the primal program (11.6-1).

Let us introduce suitable index sets associated with a partition of constraints and variables. The dual program (11.6-2) has $n + 2n$ constraints, namely, n equations and $2n$ bounds, and $m + 2n$ variables. Because the n equality constraints have full rank and are thus always active, a vertex of (11.6-2) is defined by $(m + 2n) - n = m + n$ additional active bounds, which together with the equality constraints define a square nonsingular system of linear equations. Thus there are $2n - (m + n) = n - m$ inactive bounds in (D). To establish correspondences later, recall when the *primal* simplex algorithm is applied to (11.6-1) that the primal variables are partitioned into m basic variables and $n - m$ nonbasic variables. The latter correspond to the $n - m$ active bound constraints in (P). Thus there are $2n - (n - m) = m + n$ inactive primal bound constraints. The nonbasic variables of the primal are further partitioned into ι variables at their lower bound and $n - m - \iota$ variables at their upper bound.

Let us define $B = \{1, \ldots, m\}$, $L = \{m + 1, \ldots, m + \iota\}$ and $U = \{m + \iota + 1, \ldots, n\}$. We assume this ordering of variables for convenience of description. Let us correspondingly partition $\boldsymbol{\lambda}$ and $\boldsymbol{\mu}$ as follows:

$$\boldsymbol{\lambda} = \begin{bmatrix} \boldsymbol{\lambda}_B \\ \boldsymbol{\lambda}_L \\ \boldsymbol{\lambda}_U \end{bmatrix} \begin{matrix} m \\ \iota \\ n - m - \iota \end{matrix} \qquad \boldsymbol{\mu} = \begin{bmatrix} \boldsymbol{\mu}_B \\ \boldsymbol{\mu}_L \\ \boldsymbol{\mu}_U \end{bmatrix} \begin{matrix} m \\ \iota \\ n - m - \iota \end{matrix}$$

The $m + n$ active bounds of (D) are taken to correspond to $\boldsymbol{\lambda}_B$, $\boldsymbol{\lambda}_U$, $\boldsymbol{\mu}_B$, and $\boldsymbol{\mu}_L$ so that

$$\begin{bmatrix} \mathbf{I}_B \\ \mathbf{I}_U \end{bmatrix} \boldsymbol{\lambda} = \mathbf{0} \quad \text{and} \quad \begin{bmatrix} \mathbf{I}_B \\ \mathbf{I}_L \end{bmatrix} \boldsymbol{\mu} = \mathbf{0} \tag{11.6-3}$$

where

$$\mathbf{I}_B = \begin{bmatrix} \mathbf{e}_1^T \\ \vdots \\ \mathbf{e}_m^T \end{bmatrix} \qquad \mathbf{I}_L = \begin{bmatrix} \mathbf{e}_{m+1}^T \\ \vdots \\ \mathbf{e}_{m+\iota}^T \end{bmatrix} \qquad \mathbf{I}_U = \begin{bmatrix} \mathbf{e}_{m+\iota+1}^T \\ \vdots \\ \mathbf{e}_n^T \end{bmatrix} \tag{11.6-4}$$

and where $\mathbf{e}_j^T$, in each of the above epressions, denotes the jth row of the $n \times n$ identity matrix. Let us also partition $\mathbf{A}^T$ as

$$\mathbf{A}^T = \begin{bmatrix} (\mathbf{B}^0)^T \\ (\mathbf{N}_L^0)^T \\ (\mathbf{N}_U^0)^T \end{bmatrix} \begin{matrix} m \\ \iota \\ n-m-\iota \end{matrix} \qquad \mathbf{c} = \begin{bmatrix} \mathbf{c}_B \\ \mathbf{c}_L \\ \mathbf{c}_U \end{bmatrix} \tag{11.6-5}$$

where $(\mathbf{B}^0)^T$ is an $m \times m$ matrix that is assumed to be nonsingular. In keeping with our earlier notation, in particular, Sec. 1.5, let us also define

$$\boldsymbol{\sigma}_N = \begin{bmatrix} \boldsymbol{\lambda}_L \\ -\boldsymbol{\mu}_U \end{bmatrix} \quad \text{and} \quad \mathbf{c}_N = \begin{bmatrix} \mathbf{c}_L \\ \mathbf{c}_U \end{bmatrix} \tag{11.6-6}$$

The vertex, say $\boldsymbol{\pi}^0$, $\boldsymbol{\lambda}^0$, and $\boldsymbol{\mu}^0$ defined by the above active constraints of the dual program (11.6-2), is found by solving the equations defined by (11.6-3) and the equality constraints of (11.6-2). Using the partition (11.6-5), we have

$$\begin{bmatrix} (\mathbf{B}^0)^T & \mathbf{I}_B & \cdot & \cdot & -\mathbf{I}_B & \cdot & \cdot \\ (\mathbf{N}_L^0)^T & \cdot & \mathbf{I}_L & \cdot & \cdot & -\mathbf{I}_L & \cdot \\ (\mathbf{N}_U^0)^T & \cdot & \cdot & \mathbf{I}_U & \cdot & \cdot & -\mathbf{I}_U \\ \cdot & \mathbf{I}_B & \cdot & \cdot & \cdot & \cdot & \cdot \\ \cdot & \cdot & \cdot & \mathbf{I}_U & \cdot & \cdot & \cdot \\ \cdot & \cdot & \cdot & \cdot & \mathbf{I}_B & \cdot & \cdot \\ \cdot & \cdot & \cdot & \cdot & \cdot & \mathbf{I}_L & \cdot \end{bmatrix} \begin{bmatrix} \boldsymbol{\pi}^0 \\ \boldsymbol{\lambda}_B^0 \\ \boldsymbol{\lambda}_L^0 \\ \boldsymbol{\lambda}_U^0 \\ \boldsymbol{\mu}_B^0 \\ \boldsymbol{\mu}_L^0 \\ \boldsymbol{\mu}_U^0 \end{bmatrix} = \begin{bmatrix} \mathbf{c}_B \\ \mathbf{c}_L \\ \mathbf{c}_U \\ \cdot \\ \cdot \\ \cdot \\ \cdot \end{bmatrix} \tag{11.6-7}$$

This can be more simply written, using (11.6-6) as

$$\begin{bmatrix} (\mathbf{B}^0)^T & \cdot \\ (\mathbf{N}^0)^T & \mathbf{I} \end{bmatrix} \begin{bmatrix} \boldsymbol{\pi}^0 \\ \boldsymbol{\sigma}_N^0 \end{bmatrix} = \begin{bmatrix} \mathbf{c}_B \\ \mathbf{c}_N \end{bmatrix} \tag{11.6-8}$$

Thus

$$(\boldsymbol{\pi}^0)^T = \mathbf{c}_B^T (\mathbf{B}^0)^{-1} \tag{11.6-9a}$$

and

$$\boldsymbol{\sigma}_N^0 = \mathbf{c}_N - (\mathbf{N}^0)^T \boldsymbol{\pi}^0. \tag{11.6-9b}$$

We see that these are the usual price vector and vector of reduced costs of (P). The assumption that $\boldsymbol{\pi}^0$, $\boldsymbol{\lambda}^0$, $\boldsymbol{\mu}^0$ (or equivalently, $\boldsymbol{\pi}^0$, $\boldsymbol{\sigma}_N^0$) are feasible for (D) implies that

$$\sigma_j^0 = c_j - (\boldsymbol{\pi}^0)^T \mathbf{a}_j \qquad j \in N \tag{11.6-10a}$$

with

$$\sigma_j^0 \geq 0 \qquad j \in L \quad \text{and} \quad \sigma_j^0 \leq 0 \qquad j \in U. \tag{11.6-10b}$$

Next, Lagrange multipliers, say $(\mathbf{x}^0)^T$, $(\mathbf{y}^0)^T$, and $(\mathbf{z}^0)^T$, corresponding to the above vertex are obtained by solving

$$[(\mathbf{x}^0)^T, (\mathbf{y}^0)^T, (\mathbf{z}^0)^T] \begin{bmatrix} (\mathbf{B}^0)^T & \mathbf{I}_B & \cdot & \cdot & -\mathbf{I}_B & \cdot & \cdot \\ (\mathbf{N}_L^0)^T & \cdot & \mathbf{I}_L & \cdot & \cdot & -\mathbf{I}_L & \cdot \\ (\mathbf{N}_U^0)^T & \cdot & \cdot & \mathbf{I}_U & \cdot & \cdot & -\mathbf{I}_U \\ \cdot & \mathbf{I}_B & \cdot & \cdot & \cdot & \cdot & \cdot \\ \cdot & \cdot & \cdot & \mathbf{I}_U & \cdot & \cdot & \cdot \\ \cdot & \cdot & \cdot & \cdot & \mathbf{I}_B & \cdot & \cdot \\ \cdot & \cdot & \cdot & \cdot & \cdot & \mathbf{I}_L & \cdot \end{bmatrix}$$

$$= [\mathbf{b}^T, \mathbf{l}^T, -\mathbf{u}^T] \tag{11.6-11}$$

If we partition $\mathbf{x}^0$, $\mathbf{y}^0$, and $\mathbf{z}^0$ to conform to the above matrix partitions, namely, as

$$\mathbf{x}^0 = \begin{bmatrix} \mathbf{x}_B^0 \\ \mathbf{x}_L^0 \\ \mathbf{x}_U^0 \end{bmatrix} \qquad \mathbf{y}^0 = \begin{bmatrix} \mathbf{y}_B^0 \\ \mathbf{y}_U^0 \end{bmatrix} \qquad \mathbf{z}^0 = \begin{bmatrix} \mathbf{z}_B^0 \\ \mathbf{z}_L^0 \end{bmatrix} \qquad \mathbf{l} = \begin{bmatrix} \mathbf{l}_B \\ \mathbf{l}_L \\ \mathbf{l}_U \end{bmatrix} \qquad \mathbf{u} = \begin{bmatrix} \mathbf{u}_B \\ \mathbf{u}_L \\ \mathbf{u}_U \end{bmatrix} \tag{11.6-12}$$

then (11.6-11) leads immediately to the following equations:

$$(\mathbf{x}_B^0)^T(\mathbf{B}^0)^T + (\mathbf{x}_L^0)^T(\mathbf{N}_L^0)^T + (\mathbf{x}_U^0)^T(\mathbf{N}_U^0)^T$$
$$= \mathbf{b}^T \Rightarrow \mathbf{x}_B^0 = (\mathbf{B}^0)^{-1}(\mathbf{b} - \mathbf{N}_L^0\mathbf{x}_L^0 - \mathbf{N}_U^0\mathbf{x}_U^0)$$
$$(\mathbf{x}_B^0)^T + (\mathbf{y}_B^0)^T = \mathbf{l}_B^T \Rightarrow \mathbf{y}_B^0 = \mathbf{l}_B - \mathbf{x}_B^0$$
$$(\mathbf{x}_L^0)^T = \mathbf{l}_L^T \Rightarrow \mathbf{x}_L^0 = \mathbf{l}_L$$
$$(\mathbf{x}_U^0)^T + (\mathbf{y}_U^0)^T = \mathbf{l}_U^T \Rightarrow \mathbf{y}_U^0 = \mathbf{l}_U - \mathbf{x}_U^0$$
$$-(\mathbf{x}_B^0)^T + (\mathbf{z}_B^0)^T = -\mathbf{u}_B^T \Rightarrow \mathbf{z}_B^0 = \mathbf{x}_B^0 - \mathbf{u}_B$$
$$-(\mathbf{x}_L^0)^T + (\mathbf{z}_L^0)^T = -\mathbf{u}_L^T \Rightarrow \mathbf{z}_L^0 = \mathbf{x}_L^0 - \mathbf{u}_L$$
$$-(\mathbf{x}_U^0)^T = -\mathbf{u}_U^T \Rightarrow \mathbf{x}_U^0 = \mathbf{u}_U.$$

Using $\mathbf{x}_L^0 = \mathbf{l}_L$ and $\mathbf{x}_U^0 = \mathbf{u}_U$, we can write these equations more simply as

$$\mathbf{x}_B^0 = (\mathbf{B}^0)^{-1}(\mathbf{b} - \mathbf{N}_L^0\mathbf{l}_L - \mathbf{N}_U^0\mathbf{u}_U) \tag{11.6-13a}$$

$$\mathbf{y}_B^0 = \mathbf{l}_B - \mathbf{x}_B^0 \qquad \mathbf{z}_B^0 = \mathbf{x}_B^0 - \mathbf{u}_B \tag{11.6-13b}$$

and

$$\mathbf{y}_U^0 = (\mathbf{l}_U - \mathbf{u}_U) \le \mathbf{0} \qquad \mathbf{z}_L^0 = (\mathbf{l}_L - \mathbf{u}_L) \le \mathbf{0}. \tag{11.6-13c}$$

The role played by the above variables in a primal or dual setting should now be clear. Vectors $\mathbf{x}_B^0$, $\mathbf{x}_L^0$, and $\mathbf{x}_U^0$ represent multipliers of the n general constraints of (D) and correspond to primal variables that are respectively basic, nonbasic at lower bound, and nonbasic at upper bound in (P). Vectors $\mathbf{y}_B^0$ and $\mathbf{z}_B^0$ are multipliers in (D) corresponding to the bound constraints $\boldsymbol{\lambda}_B \ge \mathbf{0}$, $\boldsymbol{\mu}_B \ge \mathbf{0}$ of (D) and are slack variables for the bound constraints $\mathbf{l}_B \le \mathbf{x}_B^0$ and $\mathbf{x}_B^0 \le \mathbf{u}_B$ of (P). Finally, $\mathbf{y}_U^0$ and $\mathbf{z}_L^0$ are multipliers corresponding to the active bounds $\boldsymbol{\lambda}_U = \mathbf{0}$ and $\boldsymbol{\mu}_L = \mathbf{0}$ of (D) and slack variables for the "opposite" bounds $\mathbf{l}_U \le \mathbf{x}_U^0$ and $\mathbf{x}_L^0 \le \mathbf{u}_L$. Recall that the $\mathbf{x}_U^0$ variables are at their upper primal bounds and the $\mathbf{x}_L^0$ variables are at their lower primal bounds. To complete the picture, we could define $\mathbf{y}_L^0 \equiv \mathbf{0}$ and $\mathbf{z}_U^0 \equiv \mathbf{0}$ as multipliers for the inactive constraints of (D), namely, $\boldsymbol{\lambda}_L \ge \mathbf{0}$ and $\boldsymbol{\mu}_U \ge \mathbf{0}$ and slack variables for the active primal bounds of (P), namely, $\mathbf{l}_L = \mathbf{x}_L^0$ and $\mathbf{x}_U^0 = \mathbf{u}_U$.

Because we are *maximizing* the objective of (D), an active constraint with *positive* multiplier must be relaxed to obtain further improvement. From (11.6-13), we see that any $(\mathbf{y}_B^0)_p > 0$ or $(\mathbf{z}_B^0)_p > 0$ provides a suitable candidate. In the primal setting, any basic constraint that violates its lower bound or upper bound may be dropped from the basis. *Suppose it corresponds to* $(\mathbf{y}_B^0)_p > 0$.

An improving direction $\bar{\mathbf{d}}$ is then computed in the usual way from

$$\begin{bmatrix} (\mathbf{B}^0)^T & \mathbf{I}_B & \cdot & \cdot & -\mathbf{I}_B & \cdot & \cdot \\ (\mathbf{N}_L^0)^T & \cdot & \mathbf{I}_L & \cdot & \cdot & -\mathbf{I}_L & \cdot \\ (\mathbf{N}_U^0)^T & \cdot & \cdot & \mathbf{I}_U & \cdot & \cdot & -\mathbf{I}_U \\ \cdot & \mathbf{I}_B & \cdot & \cdot & \cdot & \cdot & \cdot \\ \cdot & \cdot & \cdot & \mathbf{I}_U & \cdot & \cdot & \cdot \\ \cdot & \cdot & \cdot & \cdot & \mathbf{I}_B & \cdot & \cdot \\ \cdot & \cdot & \cdot & \cdot & \cdot & \mathbf{I}_L & \cdot \end{bmatrix} \begin{bmatrix} \bar{\mathbf{d}}_\pi \\ \bar{\mathbf{d}}_{\lambda_B} \\ \bar{\mathbf{d}}_{\lambda_L} \\ \bar{\mathbf{d}}_{\lambda_U} \\ \bar{\mathbf{d}}_{\mu_B} \\ \bar{\mathbf{d}}_{\mu_L} \\ \bar{\mathbf{d}}_{\mu_U} \end{bmatrix} = \begin{bmatrix} \cdot \\ \cdot \\ \cdot \\ \mathbf{e}_p \\ \cdot \\ \cdot \\ \cdot \end{bmatrix} \tag{11.6-14}$$

This gives

$$\bar{\mathbf{d}}_{\lambda_B} = \mathbf{e}_p \qquad (\mathbf{B}^0)^T \bar{\mathbf{d}}_\pi + \bar{\mathbf{d}}_{\lambda_B} = \mathbf{0}, \tag{11.6-15}$$
$$(\mathbf{N}_L^0)^T \bar{\mathbf{d}}_\pi + \bar{\mathbf{d}}_{\lambda_L} = \mathbf{0} \qquad (\mathbf{N}_U^0)^T \bar{\mathbf{d}}_\pi - \bar{\mathbf{d}}_{\mu_U} = \mathbf{0}.$$

Let us define

$$\bar{\mathbf{d}}_{\sigma_N} = \begin{bmatrix} \bar{\mathbf{d}}_{\lambda_L} \\ -\bar{\mathbf{d}}_{\mu_U} \end{bmatrix}. \tag{11.6-16}$$

Then (11.6-15) simplifies to

$$\bar{\mathbf{d}}_\pi = -(\mathbf{B}^0)^{-T}\mathbf{e}_p \tag{11.6-17a}$$

$$\bar{\mathbf{d}}_{\sigma_N} = (-\mathbf{N}^0)^T\bar{\mathbf{d}}_\pi. \tag{11.6-17b}$$

Finally, we must compute the largest step $t \geq 0$, so that the vector

$$\begin{bmatrix} \boldsymbol{\pi}^0 \\ \boldsymbol{\sigma}_N^0 \end{bmatrix} + t\begin{bmatrix} \bar{\mathbf{d}}_\pi \\ \bar{\mathbf{d}}_{\sigma_N} \end{bmatrix}$$

remains feasible. Then, from (11.6-17), this is equivalent to

$$\boldsymbol{\sigma}_L^0 + t(\mathbf{N}_L^0)^T(\mathbf{B}^0)^{-T}\mathbf{e}_p \geq \mathbf{0}, \tag{11.6-18a}$$

$$\boldsymbol{\sigma}_U^0 + t(\mathbf{N}_U^0)^T(\mathbf{B}^0)^{-T}\mathbf{e}_p \leq \mathbf{0}. \tag{11.6-18b}$$

From (11.6-17a),

$$(\mathbf{B}^0)^T\bar{\mathbf{d}}_\pi = -\mathbf{e}_p \tag{11.6-19a}$$

and let us define $\mathbf{w}$ as

$$\mathbf{w} = \begin{bmatrix} \mathbf{w}_L \\ \mathbf{w}_U \end{bmatrix} = \begin{bmatrix} (\mathbf{N}_L^0)^T \\ (\mathbf{N}_U^0)^T \end{bmatrix}\bar{\mathbf{d}}_\pi = (\mathbf{N}^0)^T\bar{\mathbf{d}}_\pi. \tag{11.6-19b}$$

Then we can write (11.6-18) as

$$\boldsymbol{\sigma}_L^0 - t\mathbf{w}_L \geq \mathbf{0}, \tag{11.6-20a}$$

$$\boldsymbol{\sigma}_U^0 - t\mathbf{w}_U \leq \mathbf{0}. \tag{11.6-20b}$$

Therefore

$$\bar{\theta}_p = \min_{j \in \bar{E}} \sigma_j^0/w_j \tag{11.6-21a}$$

and

$$\bar{S} = \left\{s \mid s = \operatorname*{argmin}_{j \in \bar{E}} \sigma_j^0/w_j\right\}, \tag{11.6-21b}$$

where

$$\bar{E} \equiv \{j \mid j \in L \quad \text{and} \quad w_j > 0 \quad \text{or} \quad j \in U \quad \text{and} \quad w_j < 0\}. \tag{11.6-22}$$

If $\bar{S} = \varphi$, then the dual is unbounded and the primal is infeasible. Otherwise, pick any $s \in \bar{S}$ and $\bar{\theta}_p = \sigma_s^0/w_s$. This defines the variable to enter the basis.

Exercise 11.6-1. Show when the exiting variable is determined by $(\mathbf{z}_B^0)_p > 0$ (see paragraph just before (11.6-14)) that

$$\bar{E} = \{j \mid j \in L \quad \text{and} \quad w_j < 0 \quad \text{or} \quad j \in U \quad \text{and} \quad w_j > 0\}.$$

Exercise 11.6-2. Extend the dual simplex algorithm of Sec. 11.4-2 to the problem with bounds, namely (11.6-1), considered in this section.

Finally, it is important to note that it is no coincidence that the *constraints of the dual* (*11.6-2*) *so closely resemble the optimality criterion* (*1.4-15*) *of the primal* (See also Sec. 11.3.) Indeed, an optimal solution of the primal problem always has an associated (basic feasible) solution that is optimal for the dual. Suppose the primal problem (e.g., the right-hand side vector) is modified so that the previous optimal solution is no longer even feasible for the (modified) primal but the associated dual solution remains feasible for the (modified) dual. Then we are in a position to initiate the dual simplex algorithm using the latter, are thereby recover optimality.

11.6-2. Further Considerations of Implementation

In order to implement the dual simplex algorithm in a practical manner, we can draw extensively upon the techniques discussed in Part II of this book. Problem setup and basis handling, using the LU factorization, can be carried out as discussed in Chapters 3 through 6. A primary use of the dual simplex algorithm is to perform sensitivity analysis, once an optimal solution to (P) has been found using the primal simplex algorithm. This produces a basic feasible solution for the dual, as we have noted above, so we shall not worry about the problem of finding an initial basic feasible solution for (D). Clearly analogous techniques to those of Chapter 8 could be developed specifically for the dual, if needed.

The choice of exiting variable (see, for example, Step D2 of the dual simplex algorithm of Sec. 11.4-2 and its extension discussed in Sec. 11.6-1) could be obtained directly from a step like *FORMC* in the implementation of Sec. 9.1, i.e., when checking primal feasibility. This corresponds to *pricing* in the dual problem. Strategies that restrict the feasibility check to only a subset of the basic variables would correspond to partial pricing on the dual and require suitable extension of *FORMC*. Choice of the entering variable is based on the reduced costs σ_j^0, $j \in N$ and the vector $\mathbf{w}$. The reduced costs are typically obtained initially (along with the starting solution) from the primal simplex algorithm, and are thereafter updated during consecutive iterations of the dual simplex algorithm; the vector $\mathbf{w}$ is obtained by an operation analogous to *BTRAN*, giving $\bar{\mathbf{d}}_\pi$ (see (11.6-19a)), followed by an operation analogous to *PRICE*, giving $\bar{\mathbf{d}}_\pi^T\mathbf{a}_j$, $j \in N$. However, the latter operation is applied to *every* variable so that partial pricing procedures used in *PRICE* are not applicable here, i.e., this component simplifies and is the counterpart of the *FORMC* operation becoming potentially more elaborate. Finally, the entering variable is determined from an operation analogous to *CHUZR* for the feasible case, as discussed in Chapter 7. We may also observe that extensions to the dual simplex algorithm that permit nonbasic variables to be pegged between their bounds could be developed in an analogous manner to extensions of the primal simplex algorithm (see Sec. 2.2-1).

Exercise 11.6-3. Formulate an implementation analogous to that of Sec. 9.1 for the dual simplex algorithm of Sec. 11.4-2.

Exercise 11.6-4. Formulate an implementation analogous to that of Sec. 9.1 for the dual simplex algorithm of Exercise 11.6-2.

Sensitivity analysis can be performed in this setting in much the same manner as was discussed in Sec. 11.5. When a linear program is optimized, the optimal solution, say $\mathbf{x}^*$, is primal feasible and has an associated dual feasible solution. Many perturbations of the problem preserve either primal feasibility or dual feasibility of these solutions. Thus the appropriate algorithm (primal or dual simplex) or a combination of the two can be applied in order to recover optimality. We are primarily concerned with developing the main algorithms of linear programming, and we therefore refer the reader to one of the numerous texts (for example, Chvatal [1983]) that discuss, in detail, their use for sensitivity and parametric analysis, in a practical setting.

Notes

Sec. 11.2. The theorems in this section follow Shapiro [1979].

Sec. 11.4. The dual simplex algorithm is due to Lemke [1954]. Here we give the *revised* form of the algorithm.

12
The Decomposition Principle and the Simplex Method

12.1. Introduction

Large-scale linear programs are usually very sparse, a feature that must be taken into consideration when designing effective solution techniques. In addition to this "microstructure," practical linear programs often exhibit a great deal of "macrostructure," i.e., their formulation often results in a natural partition of the constraints and variables into subsets in such a way that the matrices corresponding to different subsets follow an overall regularity pattern. Typical examples are shown in Figure 12.1.

The idea behind the *decomposition principle* (Dantzig and Wolfe [1960]) *is to reformulate a given linear program* (*or possibly even to define it from the outset*) *in terms of a set of interrelated linear programs, each usually of smaller dimension than the original linear program and determined explicitly* (*or implicitly*) *by a partition of its constraints and variables. These linear programs are solved in a suitably coordinated sequence, thereby yielding a solution to the original linear program.*

Throughout the main body of this chapter we consider the following linear program.

$$\begin{aligned} \text{minimize } & \mathbf{c}^T\mathbf{x} \\ \text{s.t. } & \mathbf{A}^1\mathbf{x} = \mathbf{b}^1 \\ & \mathbf{A}^2\mathbf{x} = \mathbf{b}^2 \\ & \mathbf{x} \geq \mathbf{0}, \end{aligned} \qquad (12.1\text{-}1)$$

where $\mathbf{A}^1 \in R^{m_1 \times n}$, $\mathbf{A}^2 \in R^{m_2 \times n}$ and the other vectors in (12.1-1) are of matching dimensions. Also let

$$\mathbf{A} \equiv \begin{bmatrix} \mathbf{A}^1 \\ \mathbf{A}^2 \end{bmatrix}, \mathbf{b} \equiv \begin{bmatrix} \mathbf{b}^1 \\ \mathbf{b}^2 \end{bmatrix}.$$

Concentrating on (12.1-1) and its dual will enable us to present the main ideas, without having to introduce cumbersome notation associated with the more general case as depicted in Figure 12.1. (We shall comment

Figure 12.1 LP matrices with block structures

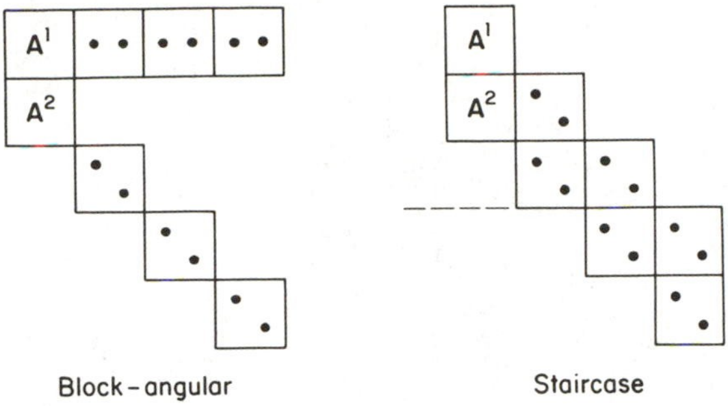

briefly on the more general case in Sec. 12.6.) It is also important to emphasize that decomposition may, in principle, be applied to *any* linear program, because constraints and variables can always be partitioned, albeit artifically, into subsets.

12.2. Decomposition Method with Row Partitions

Consider just the second set of constraints of (12.1-1), whose feasible solutions define a polyhedral set

$$P = [\mathbf{x} \mid \mathbf{A}^2\mathbf{x} = \mathbf{b}^2, \mathbf{x} \geq \mathbf{0}]. \tag{12.2-1a}$$

Recall from the representation theorem 1.3-2 that any member $\mathbf{x}$ of P can be expressed as a linear convex combination of its extreme points (vertices) $\mathbf{x}^j$, $j = 1, \ldots, n_p$ and a nonnegative linear combination of its extreme directions $\mathbf{r}^j$, $j = 1, \ldots, n_r$.

$$\mathbf{x} = \sum_{j=1}^{n_p} s_j \mathbf{x}^j + \sum_{j=1}^{n_r} t_j \mathbf{r}^j \tag{12.2-1b}$$

$$\sum_{j=1}^{n_p} s_j = 1 \qquad s_j \geq 0 \qquad t_j \geq 0. \tag{12.2-1c}$$

The reader will recall that we quoted but did not prove Theorem 1.3-2. Now, however, we have the simplex method and the convergence result of Theorem 2.5-1 at our disposal. This provides us with a powerful proof technique that we can use to establish the representation theorem 1.3-2 as follows.

Proof of Representation Theorem 1.3-2. Suppose that there do *not* exist numbers satisfying (12.2-1c) such that $\mathbf{x} \in P$ can be expressed in the form (12.2-1b), namely,

$$\begin{bmatrix} \mathbf{x} \\ 1 \end{bmatrix} = \sum_{j=1}^{n_p} s_j \begin{bmatrix} \mathbf{x}^j \\ 1 \end{bmatrix} + \sum_{j=1}^{n_r} t_j \begin{bmatrix} \mathbf{r}^j \\ 0 \end{bmatrix} \tag{12.2-2}$$

Then, by Lemmas 1.4-4 and 1.4-5 (Minkowski–Farkas), there exists a vector $\begin{bmatrix}\mathbf{d}\\ \theta\end{bmatrix}$, $\mathbf{d} \in R^n$, $\theta \in R^1$ such that

$$[(\mathbf{x}^j)^T \mid 1]\begin{bmatrix}\mathbf{d}\\ \theta\end{bmatrix} \geq 0 \qquad \text{for all } j$$
$$(\mathbf{r}^j)^T\mathbf{d} \geq 0 \qquad \text{for all } j \tag{12.2-3}$$
$$[\mathbf{x}^T \mid 1]\begin{bmatrix}\mathbf{d}\\ \theta\end{bmatrix} < 0.$$

Consider now the linear program defined by P and the cost vector $\mathbf{d}$, namely,

$$\begin{aligned} \text{minimize } & \mathbf{d}^T\mathbf{y} \\ \text{s.t. } & \mathbf{A}^2\mathbf{y} = \mathbf{b}^2 \\ & \mathbf{y} \geq \mathbf{0}. \end{aligned} \tag{12.2-4}$$

(We use the variable $\mathbf{y}$ in place of $\mathbf{x}$, to avoid conflict of notation.)

Variable $\mathbf{x}$ is feasible for (12.2-4) by assumption. Also from (12.2-3), $\mathbf{d}^T\mathbf{r}^j \geq 0$ for all extreme directions of P, so the optimal value of (12.2-4) cannot be unbounded from below. It follows from Theorem 2.5-1 that (12.2-4) has an optimal solution at a basic feasible solution (vertex, extreme point) of (12.2-4). Because this is some $\mathbf{x}^{j^*}$, it follows from the first relation of (12.2-3) that $(\mathbf{x}^{j^*})^T\mathbf{d} \geq -\theta$. But from the third relation of (12.2-3), $\mathbf{x}^T\mathbf{d} < -\theta$. This contradicts optimality in (12.2-4) of $\mathbf{x}^{j^*}$, since $\mathbf{x}$ is feasible and has a lower objective value.

Therefore $\mathbf{x}$ must be expressible in the form (12.2-1b) and (12.2-1c), and this completes the proof. ■

If we were to explicitly enumerate (by brute force) all extreme points and rays of P, we could substitute the expression for $\mathbf{x}$ given by (12.2-1b) in the first set of constraints of (12.1-1). This gives the *full master problem*, namely,

$$\begin{aligned} \text{minimize } & \sum_{j=1}^{n_p} (\mathbf{c}^T\mathbf{x}^j)s_j + \sum_{j=1}^{n_r} (\mathbf{c}^T\mathbf{r}^j)t_j \\ \text{s.t. } & \sum_{j=1}^{n_p} (\mathbf{A}^1\mathbf{x}^j)s_j + \sum_{j=1}^{n_r} (\mathbf{A}^1\mathbf{r}^j)t_j = \mathbf{b}^1 \\ & \sum_{j=1}^{n_p} s_j = 1 \\ & s_j \geq 0 \qquad t_j \geq 0. \end{aligned} \tag{12.2-5}$$

The last equality constraint of (12.2-5) is often called the *convexity* constraint and the remaining constraints are called the *nonconvexity* constraints. Clearly the full master problem (12.2-5) is *equivalent* to the original linear program (12.1-1) in that they have the same set of feasible and optimal solutions.

Example 12.2-1. Show, formally, that every solution of (12.2-5) has an associated solution of (12.1-1) with the same objective value, and vice versa.

Exercise 12.2-2. Show by means of an example that the correspondence between solutions of (12.1-1) and (12.2-5) is not one to one.

The foregoing exercises will make it clear that the polyhedral sets defined by the constraints of (12.1-1) and (12.2-5) can be quite different, i.e., the two linear programs are not *combinatorially* equivalent and can have different numbers of extreme points, different numbers of extreme directions, different "diameters," and so on.

We have reduced the original linear program (12.1-1) to a linear program (12.2-5) of smaller row dimension, whose columns are defined by an enumeration of all the extreme points (basic feasible solutions) and extreme directions of the polyhedral set (12.2-1a). If we had all the computer resources in the universe (and a little bit more besides), the enumeration could be done by a straightforward procedure—start with any basic feasible solution, generate all its neighbors and insert them in a list, take each one in turn and generate all neighbors and insert them in the list if they are not already in it, and so on. This is obviously totally impractical. We now turn to a technique for generating columns of (12.2-5) as and when they are needed.

12.2-1. Decomposition Algorithm Using Delayed Column Generation (Dantzig–Wolfe Algorithm)

Suppose that we have some *subset* of the extreme points and directions, say $\mathbf{x}^j$, $j = 1, \ldots, k$ and $\mathbf{r}^j$, $j = 1, \ldots, l$, initially available. Let K denote an iteration count.

Write the full master problem in the following form.

$$
\begin{aligned}
&\text{minimize } \sum_{j=1}^{k} (\mathbf{c}^T\mathbf{x}^j)s_j + \sum_{j=1}^{l} (\mathbf{c}^T\mathbf{r}^j)t_j + \sum_{j=k+1}^{n_p} (\mathbf{c}^T\mathbf{x}^j)s_j + \sum_{j=l+1}^{n_r} (\mathbf{c}^T\mathbf{r}^j)t_j \\
&\quad \text{s.t.} \\
&\boldsymbol{\pi}_1^K\!: \sum_{j=1}^{k} (\mathbf{A}^1\mathbf{x}^j)s_j + \sum_{j=1}^{l} (\mathbf{A}^1\mathbf{r}^j)t_j + \sum_{j=k+1}^{n_p} (\mathbf{A}^1\mathbf{x}^j)s_j + \sum_{j=l+1}^{n_r} (\mathbf{A}^1\mathbf{r}^j)t_j = \mathbf{b}^1 \\
&\rho^K\!: \sum_{j=1}^{k} s_j \qquad\qquad\qquad\qquad\quad + \sum_{j=k+1}^{n_p} s_j \qquad\qquad\qquad\qquad\quad = 1 \\
&\qquad s_j \geq 0 \qquad t_j \geq 0.
\end{aligned}
\qquad (12.2\text{-}6)
$$

Let us drop the variables s_j, $j > k$, and t_j, $j > l$ from (12.2-6) and solve the resulting *restricted master problem* RM_K by the simplex method. We shall assume that it has a feasible solution. Suppose $\boldsymbol{\pi}_1^K$ and ρ^K are the components of the *optimal* price vector associated with the nonconvexity and the convexity constraints, respectively, of (12.2-6). Also, if the optimal solution of the restricted master RM_K is given by s_j^*, $j = 1, \ldots, k$ and t_j^*, $j = 1, \ldots, l$, the solution of the original linear program (12.1-1) is

$$\mathbf{x}^* = \sum_{j=1}^{k} s_j^* \mathbf{x}^j + \sum_{j=1}^{l} t_j^* \mathbf{r}^j. \tag{12.2-7a}$$

The optimality conditions for (12.2-6) imply that the associated reduced costs are nonnegative, namely,

$$\mathbf{c}^T\mathbf{x}^j - (\boldsymbol{\pi}_1^K)^T\mathbf{A}^1\mathbf{x}^j - \rho^K \geq 0 \qquad j = 1, \ldots, k \tag{12.2-7b}$$

$$\mathbf{c}^T\mathbf{r}^j - (\boldsymbol{\pi}_1^K)^T\mathbf{A}^1\mathbf{r}^j \geq 0 \qquad j = 1, \ldots, l. \tag{12.2-7c}$$

Note that both expressions in (12.2-7b and 12.2-7c) contain a term of the form $(\mathbf{c}^T - (\boldsymbol{\pi}_1^K)^T\mathbf{A}^1)\mathbf{z}$ where $\mathbf{z}$ denotes an extreme point or direction of P that belongs to the subset used to define the restricted master. In order now to verify that the solution of the restricted master problem is optimal for the full master problem, we must verify that expressions of the form (12.2-7b and 12.2-7c) hold for *every* extreme point and direction of P. This can be done *implicitly*, by solving the following linear program, known as the *subproblem*.

$$\begin{aligned} (SP_K)\text{: minimize } & (\mathbf{c}^T - (\boldsymbol{\pi}_1^K)^T\mathbf{A}^1)\mathbf{x} \\ \boldsymbol{\pi}_2^K\text{: } & \mathbf{A}^2\mathbf{x} = \mathbf{b}^2 \\ & \mathbf{x} \geq \mathbf{0}. \end{aligned} \tag{12.2-8}$$

Since the polyhedral set P defined by the constraints of (12.2-8) is assumed to be nonempty, we know that this linear program either achieves its optimum at an extreme point, say $\mathbf{x}^{k+1}$, or it has an unbounded optimal solution, see Theorem 2.5-1 and its corollary. Let us consider each possibility in turn.

1. The optimum is achieved at, say, $\mathbf{x}^{k+1}$. Then

$$(\mathbf{c}^T - (\boldsymbol{\pi}_1^K)^T\mathbf{A}^1)\mathbf{r}^j \geq 0 \qquad \text{for all } j. \tag{12.2-9}$$

Otherwise the optimum would be unbounded from below. Suppose, in addition,

$$(\mathbf{c}^T - (\boldsymbol{\pi}_1^K)^T\mathbf{A}^1)\mathbf{x}^{k+1} - \rho^K \geq 0. \tag{12.2-10}$$

Then clearly this must also be true for all extreme points of P, so

$$(\mathbf{c}^T - (\boldsymbol{\pi}_1^K)^T\mathbf{A}^1)\mathbf{x}^j - \rho^K \geq 0 \qquad j = 1, \ldots, n_p. \tag{12.2-11}$$

Now the expressions in (12.2-9) and (12.2-11) are, of course, the reduced costs of the columns of the full master problem, which includes the ones in the dotted box of (12.2-6), and these inequalities therefore imply that none of the latter columns are improving columns relative to the current solution of the restricted master problem. The solution $\mathbf{x}^*$ is therefore optimal for the full master problem.

If, on the other hand, (12.2-10) is not true, then $\mathbf{x}^{k+1}$ must be one of the columns in the dotted box of (12.2-6) and it has a negative reduced cost. Note that columns $\mathbf{x}^1$ through $\mathbf{x}^k$ are already in the restricted master and have nonnegative reduced costs relative to the optimal price vector $\boldsymbol{\pi}_1^K$ and ρ^K. Therefore $\mathbf{x}^{k+1}$ can now be profitably introduced into the restricted master by adding a column of the form

$$\begin{bmatrix} \mathbf{c}^T\mathbf{x}^{k+1} \\ \mathbf{A}^1\mathbf{x}^{k+1} \\ 1 \end{bmatrix} \tag{12.2-12}$$

The new restricted master can then be reoptimized and the foregoing cycle repeated.

2. The other possibility to consider is that (12.2-8) has an unbounded optimum. (At an initial reading, this possibility may be skipped by assuming that P is bounded.) In this case, the simplex method will discover an extreme direction or ray of (12.2-8), say $\mathbf{r}^{l+1}$, which will be of the form,

$$\mathbf{r}^{l+1} = \begin{bmatrix} -(\mathbf{B}_s^*)^{-1}(\mathbf{A}^2)_{m_2+k} \\ \mathbf{e}_k \end{bmatrix} \tag{12.2-13}$$

Here $\mathbf{B}_s^*$ is a basis matrix of the subproblem (12.2-8) which, without loss of generality, we assume to be the first m_2 columns of $\mathbf{A}^2$, and $(\mathbf{A}^2)_{m_2+k}$ is the kth *nonbasic* column, corresponding to this basis. (12.2-13) is an expression of the form (2.2-20). Because we have an unbounded optimum associated with the ray $\mathbf{r}^{l+1}$, we must have

$$(\mathbf{c}^T - (\boldsymbol{\pi}_1^K)^T\mathbf{A}^1)\mathbf{r}^{l+1} < 0. \tag{12.2-14}$$

Therefore, again, we have found a column of the dotted box of (12.2-6) that is improving relative to the current basis of the restricted master. We introduce this column into the restricted master, namely,

$$\begin{bmatrix} \mathbf{c}^T\mathbf{r}^{l+1} \\ \mathbf{A}^1\mathbf{r}^{l+1} \\ 0 \end{bmatrix} \tag{12.2-15}$$

Again the new restricted master can be reoptimized and the cycle continued.

Note that the column (12.2-15) does *not* necessarily define an extreme direction of the new restricted master, i.e., if $\mathbf{B}_M^*$ was the optimal basis of the restricted master associated with the price vector $\begin{bmatrix} \boldsymbol{\pi}_1^K \\ \rho^K \end{bmatrix}$, then

$$(\mathbf{B}_M^*)^{-1} \begin{bmatrix} \mathbf{A}^1 \mathbf{r}^{l+1} \\ 0 \end{bmatrix}$$

does not necessarily correspond to an extreme direction of the new restricted master problem.

We see each time the subproblem is solved that we generate a new column or terminate the process. At each cycle we solve a *restricted master* RM_K that has $m_1 + 1$ rows (not counting the objective) and we *coordinate* its solution with that of a linear program (the *subproblem*) that has m_2 rows and that provides new columns for the restricted master. Thus, in essence, decomposition using row partitions replaces the original linear program (12.1-1), which has $m_1 + m_2$ rows, by a master program with smaller row dimension, namely $m_1 + 1$, and an unspecified number of columns. The latter are generated from the *subproblem* (also of smaller dimension) as and when needed, through a *column generation procedure,* and used to define the restricted master problem.

The cycle of the algorithm (Dantzig–Wolfe [1960]) based on the above method is summarized in the flowchart of Figure 12.2. At each cycle, K is increased by 1. Also at the end of each cycle, either k or l increases by 1.

Lemma 12.2-1. Suppose $\boldsymbol{\pi}_2^K$ is the optimal price vector of SP_K, (12.2-8). Then $\boldsymbol{\pi}_1^K$, $\boldsymbol{\pi}_2^K$ give a feasible solution of the dual of the original linear program (12.1-1).

Figure 12.2 Cycle of Dantzig–Wolfe algorithm

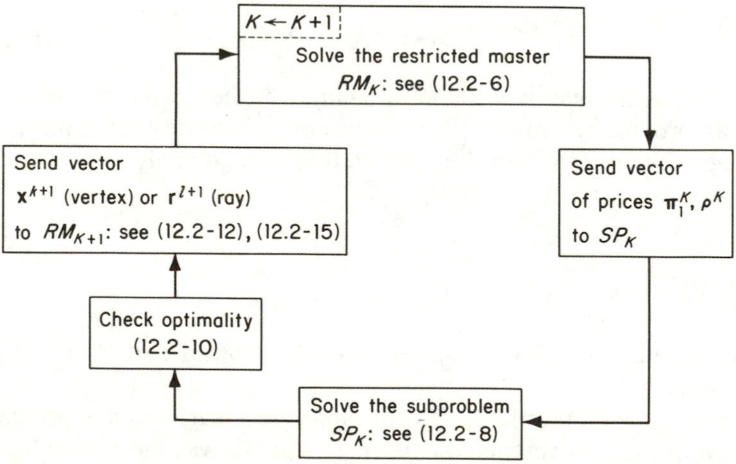

Proof. The constraints of the dual of (12.1-1) are defined by

$$(\mathbf{A}^1)^T\boldsymbol{\pi}_1 + (\mathbf{A}^2)^T\boldsymbol{\pi}_2 \leq \mathbf{c}. \tag{12.2-16}$$

Since $\boldsymbol{\pi}_2^K$ is associated with the optimal solutionof SP_K,

$$\mathbf{c} - (\mathbf{A}^1)^T\boldsymbol{\pi}_1^K - (\mathbf{A}^2)^T\boldsymbol{\pi}_2^K \geq \mathbf{0}.$$

Clearly, therefore, $\boldsymbol{\pi}_1 = \boldsymbol{\pi}_1^K$, $\boldsymbol{\pi}_2 = \boldsymbol{\pi}_2^K$ give a feasible solution of (12.2-16). ■

We see, therefore, that each cycle of the above algorithm provides a feasible solution to the original primal problem (see (12.2-7a)) and, as we have just demonstrated, a feasible solution to its dual (when the subproblem is optimized at each cycle). However, these two solutions do not satisfy the complementary slackness conditions (11.2-14). We see that the Dantzig–Wolfe decomposition algorithm is summarized by the third row of Table 11.2, i.e., it maintains primal and dual feasibility of the original linear program and looks for complementary slackness.

Exercise 12.2-3. Apply the decomposition principle to the following linear programs.

$$\begin{aligned} \text{(a) maximize } & \mathbf{b}^T\mathbf{y} \\ \text{s.t. } & \mathbf{A}_1^T\mathbf{y} \leq \mathbf{c}_1 \\ & \mathbf{A}_2^T\mathbf{y} \leq \mathbf{c}_2 \\ \text{(b) minimize } & \mathbf{c}^T\mathbf{x} \\ \text{s.t. } & \mathbf{A}^1\mathbf{x} = \mathbf{b}^1 \\ & \mathbf{A}^2\mathbf{x} = \mathbf{b}^2 \\ & \mathbf{l} \leq \mathbf{x} \leq \mathbf{u} \end{aligned}$$

where, in each case, the first set of rows defines the master problem.

Exercise 12.2-4. Show that the points generated by the Dantzig–Wolfe decomposition algorithm can lie in the *interior* of the feasible polytope defined by the constraints of (12.1-1).

12.2-2. Convergence of the Decomposition Algorithm

Under the usual nondegeneracy assumptions, convergence follows directly from the convergence of the simplex algorithm. Practical experience confirms that decomposition algorithms often converge quite rapidly to the neighborhood of a solution and then exhibit a long "tail" during which little progress is made. In exact arithmetic this must be a consequence of either the changed combinatorial structure (see Exercise 12.2-2), or the fact that the solution is approached through the interior (see Exercise 12.2-4), so that considerable "cleaning up" is required to obtain a vertex, i.e., to exactly satisfy the complementary slackness relations. In finite precision arithmetic, slowed convergence could be the result of numerical difficulties, as we discuss further in Sec. 12.5.

12.3. Decomposition Method with Column Partitions

We now turn to an alternative view of the above decomposition, but this time on the dual of (12.1-1), namely,

$$\begin{aligned} \text{maximize } & (\mathbf{b}^1)^T\boldsymbol{\pi}_1 + (\mathbf{b}^2)^T\boldsymbol{\pi}_2 \\ \text{s.t. } & (\mathbf{A}^1)^T\boldsymbol{\pi}_1 + (\mathbf{A}^2)^T\boldsymbol{\pi}_2 \le \mathbf{c}. \end{aligned} \tag{12.3-1}$$

Suppose we *fix* the prices $\boldsymbol{\pi}_1$ at some values, say, $\boldsymbol{\pi}_1^K$. The resulting linear program is

$$\begin{aligned} \text{maximize } & (\mathbf{b}^2)^T\boldsymbol{\pi}_2 \\ \text{s.t. } & (\mathbf{A}^2)^T\boldsymbol{\pi}_2 \le \mathbf{c} - (\mathbf{A}^1)^T\boldsymbol{\pi}_1^K. \end{aligned} \tag{12.3-2}$$

We must, of course, ensure feasibility of (12.3-2), and we therefore restrict $\boldsymbol{\pi}_1^K$ to lie in the set

$$S = [\boldsymbol{\pi}_1 \mid \text{there exists } \boldsymbol{\pi}_2 \text{ s.t. } (\mathbf{A}^2)^T\boldsymbol{\pi}_2 \le \mathbf{c} - (\mathbf{A}^1)^T\boldsymbol{\pi}_1]. \tag{12.3-3}$$

The optimal value of (12.3-2) is a function of $\boldsymbol{\pi}_1^K$, say $\psi(\boldsymbol{\pi}_1^K)$. S and $\psi(\boldsymbol{\pi}_1^K)$ are more conveniently characterized in terms of the dual of (12.3-2), namely,

$$\begin{aligned} \psi(\boldsymbol{\pi}_1^K) = \min\ & (\mathbf{c}^T - (\boldsymbol{\pi}_1^K)^T\mathbf{A}^1)\mathbf{x} \\ \text{s.t. } & \mathbf{A}^2\mathbf{x} = \mathbf{b}^2 \\ & \mathbf{x} \ge \mathbf{0}. \end{aligned} \tag{12.3-4}$$

From the representation theorem, in particular, (12.2-1b), we can define a program equivalent to (12.3-4), namely,

$$\begin{aligned} \psi(\boldsymbol{\pi}_1^K) = \min \sum_{j=1}^{n_p} [(\mathbf{c}^T - (\boldsymbol{\pi}_1^K)^T\mathbf{A}^1)\mathbf{x}^j]s_j + \sum_{j=1}^{n_r} [(\mathbf{c}^T - (\boldsymbol{\pi}_1^K)^T\mathbf{A}^1)\mathbf{r}^j]t_j \\ \text{s.t. } \sum_{j=1}^{n_p} s_j = 1 \\ s_j \ge 0 \qquad t_j \ge 0 \qquad \text{for all } j. \end{aligned} \tag{12.3-5}$$

Two facts immediately emerge.

1. $(\mathbf{c}^T - (\boldsymbol{\pi}_1^K)^T\mathbf{A}^1)\mathbf{r}^j \ge 0$. Otherwise (12.3-5) would have an unbounded optimum and (12.3-2) would be infeasible. We therefore must have the feasibility requirement

$$S = [\boldsymbol{\pi}_1 \mid (\mathbf{c}^T - (\boldsymbol{\pi}_1^K)^T\mathbf{A}^1)\mathbf{r}^j \ge 0 \qquad j = 1, \ldots, n_r]. \tag{12.3-6}$$

2. With $\boldsymbol{\pi}_1^K \in S$, we then have the optimality requirement

$$\psi(\boldsymbol{\pi}_1^K) = \min_{1 \le j \le n_p} (\mathbf{c}^T - (\boldsymbol{\pi}_1^K)^T \mathbf{A}^1)\mathbf{x}^j. \tag{12.3-7}$$

The cost associated with $\boldsymbol{\pi}_1^K$ is $(\mathbf{b}^1)^T \boldsymbol{\pi}_1^K$ in the objective of (12.3-1). The best we can achieve after making this decision is given by the optimum of (12.3-2), namely $\psi(\boldsymbol{\pi}_1^K)$ in (12.3-7). The total cost is thus $(\mathbf{b}^1)^T \boldsymbol{\pi}_1^K + \psi(\boldsymbol{\pi}_1^K)$ and we seek to optimize this subject to $\boldsymbol{\pi}_1^K \in S$. Therefore our linear program (12.3-1) is equivalent to

$$\underset{\boldsymbol{\pi}_1^K \in S}{\text{maximize}}\ (\mathbf{b}^1)^T \boldsymbol{\pi}_1^K + \min_{1 \le j \le n_p} (\mathbf{c}^T - (\boldsymbol{\pi}_1^K)^T \mathbf{A}^1)\mathbf{x}^j. \tag{12.3-8}$$

Using (12.3-6), we can equivalently write this as follows:

$$\begin{aligned} \text{maximize}\ & (\mathbf{b}^1)^T \boldsymbol{\pi}_1 + \rho \\ \text{s.t.}\ & \rho \le (\mathbf{c}^T - (\boldsymbol{\pi}_1)^T \mathbf{A}^1)\mathbf{x}^j \qquad j = 1, \ldots, n_p \\ & 0 \le (\mathbf{c}^T - (\boldsymbol{\pi}_1)^T \mathbf{A}^1)\mathbf{r}^j \qquad j = 1, \ldots, n_r. \end{aligned} \tag{12.3-9}$$

The first set of constraints in (12.3-9) are often called *optimality cuts* and the second set of constrants are often called *feasibility cuts,* for reasons that are apparent from the above discussion, in particular, (12.3-6) and (12.3-7).

The linear program (12.3-9) is the full master program that is equivalent to (12.3-1) and we immediately see that it is the dual of (12.2-5). Also, subproblems (12.2-8) and (12.3-4) are identical. *Conceptually,* the two decomposition approaches discussed in this and the previous section are the same; note, however, that the former is *column oriented* in that vertices or extreme directions of the subproblem define columns of the master program (12.2-5), while the latter is *row oriented,* in that vertices or extreme directions of the subproblem define rows of the master (12.3-9). Again, an explicit enumeration in (12.3-9) of vertices and extreme directions of the subproblem would be hopelessly impractical.

12.3-1. Decomposition Algorithm Using Delayed Row Generation (Benders' Algorithm)

A practical procedure based on defining a restricted master from (12.3-9) and generating successive rows of (12.3-9) from the subproblem (12.3-4) as and when they are needed, could be organized along very similar lines to that of Sec. 12.2-1 (see, in particular, the flowchart of Figure 12.2). Thus we might solve a restricted master defined from (12.3-9) with $\mathbf{x}^j$, $j = 1, \ldots, k$ defining the optimality cuts and $\mathbf{r}^j$, $j = 1, \ldots, l$ defining the feasibility cuts. This gives the dual of RM_K defined from (12.2-6). Suppose that $\boldsymbol{\pi}_1^K$ and ρ^K are its solution. Then the subproblem is solved to yield either a new optimality cut $\mathbf{x}^{k+1}$ or feasibility cut $\mathbf{r}^{l+1}$, and the process repeated. It terminates when (12.2-10) is satisfied, indicating that the current optimal solution of the restricted master satisfies *all* constraints of the full master program (12.3-9). (Since the restricted master defined from (12.3-9) is row oriented, we might solve it by the version of the simplex algorithm discussed in Sec. 2.3.) Such a procedure is usually known as Benders' decomposition algorithm, see Benders [1962]. The decompositions, as developed in Sec. 12.2 and as developed here, are the same in principle, but the organization of the computations in the Dantzig–Wolfe algorithm and Benders' algorithm differ from one another.

The Dantzig–Wolfe algorithm applied to the primal (12.1-1) and Benders' algorithm applied to the dual (12.3-1), with the same initialization, develop an identical sequence of iterates. Their relationship is much the same as that between the primal simplex algorithm and the dual simplex algorithm.

Exercise 12.3-1. Develop the analogue of the flowchart of Figure 12.2, for Benders' algorithm.

12.4. Variants

There are a number of variants on the foregoing ideas. In this section we develop a brief categorization that will give the reader some perspective on the extensive literature on decomposition.

Decomposition of a linear program can proceed along the following main lines.

12.4-1. Price-Directed Decomposition

This is precisely the form of decomposition discussed in Sec. 12.2 and 12.3 and summarized in Figure 12.2. The decomposition is driven by the *price* vector $\boldsymbol{\pi}_1^K$ and, indeed, the development in Sec. 12.3 is directly in these terms. Each cycle generates an *activity* or setting of the $\mathbf{x}$ variables of the primal problem corresponding to a vertex or an extreme direction of the subproblem, and this is used to define a restricted master column (Dantzig–Wolfe) or a restricted master row (Benders').

12.4-2. Activity-Directed Decomposition

Here we reverse the partitioning of primal and dual in the developments of Secs. 12.2 and 12.3. Thus we apply the procedure of Sec. 12.3 to the primal problem now *partitioned by columns* instead of by rows, as follows.

$$\begin{aligned} \text{minimize}\ & \mathbf{c}_1^T\mathbf{x}_1 + \mathbf{c}_2^T\mathbf{x}_2 \\ \text{s.t.}\ & \mathbf{A}_1\mathbf{x}_1 + \mathbf{A}_2\mathbf{x}_2 = \mathbf{b} \\ & \mathbf{x}_1 \geq \mathbf{0} \qquad \mathbf{x}_2 \geq \mathbf{0} \end{aligned} \tag{12.4-1}$$

where $\mathbf{A} = [\mathbf{A}_1, \mathbf{A}_2]$.

We have thus reversed the roles played by price and activity, but in other respects the decomposition procedure closely follows the discussion of Secs. 12.2 and 12.3, in particular, Figure 12.2. It should be clear that Benders' algorithm now applied to (12.4-1) is conceptually equivalent to the Dantzig–Wolfe algorithm applied to the dual of (12.4-1). (See also Exercise 12.2-3a.)

Benders' algorithm is most commonly applied within the setting of *column* partitions of the *primal*, namely (12.4-1), while Dantzig–Wolfe is most commonly applied within the setting of *row* partitions of the *primal*, namely (12.1-1).

12.4-3. Cost-Directed Decomposition

Note in price-directed and activity-directed decomposition that there is an implicit hierarchy, with the partition corresponding to the subproblem playing a subordinate role to the partition corresponding to the master. We can seek to give the partitions of the original linear program a more equal role in (12.3-1) by fixing the amount of the *right-hand side,* namely, the budget (cost) available to the first partition of (12.3-1) rather than by fixing the levels of the corresponding variables, namely, the prices. Thus, if $\boldsymbol{\chi}_1$ is allocated to the first partition of (12.3-1) and $\boldsymbol{\chi}_2 \leq \mathbf{c} - \boldsymbol{\chi}_1$ to the second partition, then we obtain the equivalent master program

$$\begin{aligned} \text{minimize} \quad & \psi_1(\boldsymbol{\chi}_1) + \psi_2(\boldsymbol{\chi}_2) \\ \text{s.t.} \quad & \boldsymbol{\chi}_1 + \boldsymbol{\chi}_2 \leq \mathbf{c} \\ & \boldsymbol{\chi}_1 \in S_1 \qquad \boldsymbol{\chi}_2 \in S_2 \end{aligned} \tag{12.4-2}$$

where, in complete analogy to (12.3-3), we have

$$S_i \equiv [\boldsymbol{\chi}_i \mid \text{there exists } \boldsymbol{\pi}_i \text{ s.t. } (\mathbf{A}^i)^T \boldsymbol{\pi}_i \leq \boldsymbol{\chi}_i] \qquad i = 1, 2$$

and for $\boldsymbol{\chi}_i \in S_i$,

$$\psi_i(\boldsymbol{\chi}_i) \equiv \min_{\boldsymbol{\pi}_i} [(\mathbf{b}^i)^T \boldsymbol{\pi}_i \mid (\mathbf{A}^i)^T \boldsymbol{\pi}_i \leq \boldsymbol{\chi}_i] \qquad i = 1, 2.$$

S_i and $\psi_i(\boldsymbol{\chi}_i)$ can be characterized in terms of the extreme points and directions of the sets

$$P_i \equiv [\mathbf{x} \mid \mathbf{A}^i \mathbf{x} = \mathbf{b}^i \qquad \mathbf{x} \geq \mathbf{0}] \qquad i = 1, 2,$$

and a row-oriented procedure developed (i.e., one that generates successive rows of a restricted master problem) along lines analogous to Sec. 12.3. Note that there are now *two* subproblems, defined by the two partitions of the original program (12.3-1) with a restricted master, say *RM,* playing a coordinating role.

A conceptually equivalent development could equally well be carried out within the setting of (12.1-1), to obtain a column-oriented procedure (i.e., one that generates successive columns of a master program that is dual to the above master *RM*) along lines analogous to Sec. 12.2.

12.4-4. Resource-Directed Decomposition

This time we work with (12.4-1) instead of (12.3-1). We noted earlier that activity-directed decomposition is essentially the same as price-directed decomposition, when we partition by columns instead of by rows and reverse the roles played by price and activity. In a similar vein, resource-directed decomposition is essentially the same as cost-directed decomposition when we partition by columns instead of by rows, and reverse the roles of cost and resource (the vector $\mathbf{b}$ in (12.4-1)). Applied to (12.4-1), it would lead to a row-oriented procedure analogous to the one just outlined for cost-directed decomposition in which successive rows are generated for a restricted master program that coordinates two subproblems.

Again the development could be carried out on the dual of (12.4-1), leading to a column-oriented procedure in which successive columns of a coordinating restricted master program are generated.

12.4-5. Summary

There is, in fact, a unifying theme to all these variants. Each can be described in terms of the basic decomposition procedure (column oriented as in Sec. 12.2 or row oriented as in Sec. 12.3) on either the primal program or the dual program. These are partitioned either by rows or by columns, with the partitioning being followed, in some cases, by a further *problem transformation,* which introduces new "linking variables" or "linking constraints."

Suppose, for example, in resource-directed decomposition that we make the following transformation of (12.4-1):

$$
\begin{array}{rlllr}
\text{minimize} & \mathbf{c}_1^T\mathbf{x}_1 + \mathbf{c}_2^T\mathbf{x}_2 & & & \\
\text{s.t.} & \mathbf{A}_1\mathbf{x}_1 & -\chi_1 & = \mathbf{0} & \\
& \qquad\qquad \mathbf{A}_2\mathbf{x}_2 & \qquad -\chi_2 & = \mathbf{0} & \qquad (12.4\text{-}3) \\
& & \chi_1 + \chi_2 & = \mathbf{b} & \\
& \mathbf{x}_1 \geq \mathbf{0} \qquad \mathbf{x}_2 \geq \mathbf{0} & & &
\end{array}
$$

where χ_1 and χ_2 are new linking variables and $\chi_1 + \chi_2 = \mathbf{b}$ are new linking constraints. It can then be readily demonstrated that resource-directed decomposition is obtained by applying the procedure of Sec. 12.3 to (12.4-3), with the columns corresponding to the linking variables defining the master and the remaining columns defining the subproblem. The latter separates into two independent subproblems, corresponding to the $\mathbf{x}_1$ and $\mathbf{x}_2$ variables.

Exercise 12.4-1. Show that cost-directed decomposition can be obtained by introducing new linking variables and constraints into (12.1-1) and applying the Dantzig–Wolfe procedure of Sec. 12.2.

Exercise 12.4-2. Decomposition by Tenders. Consider the following transformation of (12.4-1).

$$\begin{aligned} \text{minimize}\quad & \mathbf{c}_1^T\mathbf{x}_1 && + \mathbf{c}_2^T\mathbf{x}_2 \\ \text{s.t.}\quad & \mathbf{A}_1\mathbf{x}_1 - \boldsymbol{\chi} && = \mathbf{0} \\ & && \boldsymbol{\chi} + \mathbf{A}_2\mathbf{x}_2 = \mathbf{b} \\ & \mathbf{x}_1 \geq \mathbf{0} && \mathbf{x}_2 \geq \mathbf{0}. \end{aligned}$$

Apply the Dantzig–Wolfe algorithm of Sec. 12.2 to this transformed problem, with the first set of rows defining the master. Show that the procedure is identical to the primal simplex algorithm, *with a nonstandard partial and multiple pricing strategy*. Show that this equivalence to the simplex algorithm is no longer true when there are bounds on $\mathbf{x}_2$ of the form, $\mathbf{l}_2 \leq \mathbf{x}_2 \leq \mathbf{u}_2$, or bounds on $\boldsymbol{\chi}$ of the form $\mathbf{l}_\chi \leq \boldsymbol{\chi} \leq \mathbf{u}_\chi$.

12.5. Numerical Behavior of Decomposition Algorithms

We have observed that the full master problem is not combinatorially equivalent to the original linear program from which it is derived. Now we show, by means of examples, that it may also have very different *numerical* characteristics. Indeed, from a numerical standpoint, decomposition is a rather fascinating process, whose properties are far from well understood at present.

In each of the following examples, we shall be concerned with the full master problem as developed in Sec. 12.2.

Example 12.5-1. Scaling. A linear program can be well scaled in its original form and become badly scaled after the decomposition principle is applied. Consider the following case.

$$\begin{array}{ll} \text{minimize} & x + y \\ \textit{subproblem rows} & \left\{\begin{array}{l} x + e'y \geq 0 \\ x - \ ey \geq 0 \\ x + \ ey \leq 1 \\ x - e'y \leq 1 \end{array}\right. \\ \textit{master rows} & \left\{\begin{array}{r} y \leq 1 \\ -y \leq 1 \end{array}\right. \end{array} \qquad (12.5\text{-}1)$$

where e is small and $\frac{1}{2} \leq e' \leq 1$. If the simplex algorithm is applied to this problem, then every basis matrix $\mathbf{B}$ is well conditioned, i.e., has a reasonable condition number $\|\mathbf{B}\|_2 \, \|\mathbf{B}^{-1}\|_2$. Small variations in the coefficients will produce small variants in the extreme points of the problem, and the problem is well scaled.

Suppose, however, we decompose this linear program as in Sec. 12.2 (Dantzig–Wolfe) into a master and subproblem. The extreme points of the subproblem are

$$\begin{bmatrix}0\\0\end{bmatrix} \quad \begin{bmatrix}1\\0\end{bmatrix} \quad \begin{bmatrix}1/2\\1/(2e)\end{bmatrix} \quad \begin{bmatrix}1/2\\-1/(2e')\end{bmatrix} \tag{12.5-2}$$

and the full master problem becomes

$$\begin{aligned}
\text{minimize} \quad & s_2 + \frac{e+1}{2e}s_3 + \frac{e'-1}{2e'}s_4 \\
& \frac{1}{2e}s_3 - \frac{1}{2e'}s_4 + y_1 = 1 \\
& -\frac{1}{2e}s_3 + \frac{1}{2e'}s_4 + y_2 = 1 \\
& s_1 + s_2 + s_3 + s_4 = 1 \\
& s_i \geq 0 \quad y_1 \geq 0 \quad y_2 \geq 0.
\end{aligned} \tag{12.5-3}$$

When e is small, this program has basis matrices with large condition numbers. The structural columns differ widely in magnitude, and their reduced costs, in a pricing operation of the simplex algorithm applied to (12.5-3), can give misleading information about the value of introducing a particular column into the basis. The master program is now badly scaled.

Example 12.5-2. Correlation of Error. Ill-conditioned basis matrices need not, of course, give solutions $\mathbf{x}^j$ with large elements, but when slightly perturbed, they usually give solutions, say $\mathbf{x}_c^j$, that differ substantially from $\mathbf{x}^j$. This next example has some ill-conditioned subproblem basis matrices, but the (unique) optimal solution of the original linear program is well conditioned. Let $\mathbf{x}^*$ denote this optimal solution and let s_j^c denote the components of the (exact) optimal solution of the master problem with columns defined by $\mathbf{x}_c^j$. We demonstrate that the substantial errors in s_j^c (when compared to the optimal solution s_j^* of the master program with columns defined by $\mathbf{x}^j$) are so *correlated* with the substantial errors in $\mathbf{x}_c^j$ that the optimal solution $\mathbf{x}^c = \sum_j s_j^c \mathbf{x}_c^j$, obtained by decomposition, remains *close* to the true solution $\mathbf{x}^* = \sum_j s_j^* \mathbf{x}^j$.

Consider the linear program

$$
\begin{array}{lll}
\text{minimize} & & x_4 \\
\textit{subproblem rows} & \begin{cases} x_1 \quad + (1+e_1)x_3 & = a_1 \\ x_1 + x_2 + (1+e_2)x_3 & = a_2 \end{cases} & \\
\textit{master rows} & \begin{cases} x_3 + x_4 = a_3 \\ x_4 = a_4 \end{cases} & \\
& x_i \geq 0. &
\end{array}
\qquad (12.5\text{-}4)
$$

Let e_1 and e_2 be small, $e_2 > e_1$, $a_2 > a_1 > a_2(1+e_1)/(1+e_2)$. Then there are only two feasible extreme points of the subproblem, and these are

$$
\mathbf{x}^1 = \begin{bmatrix} a_1 \\ a_2 - a_1 \\ 0 \end{bmatrix} \quad \text{and} \quad \mathbf{x}^2 = \begin{bmatrix} a_1 - (1+e_1)k \\ 0 \\ k \end{bmatrix}
$$

where $k = (a_2 - a_1)/(e_2 - e_1)$.

The full master problem is

$$
\begin{array}{ll}
\text{minimize} & x_4 \\
\text{s.t.} & \dfrac{a_2 - a_1}{e_2 - e_1} s_2 + x_4 = a_3 \\
& x_4 = a_4 \\
& s_1 + s_2 = 1 \\
& s_1 \geq 0 \quad s_2 \geq 0 \quad x_4 \geq 0.
\end{array}
\qquad (12.5\text{-}5)
$$

This has the solution

$$
s_1^* = 1 - \frac{a_3 - a_4}{k} \qquad s_2^* = \frac{a_3 - a_4}{k} \qquad x_4^* = a_4
$$

where we assume that $a_3 > a_4 > 0$ and that a_3 and a_4 are chosen so that s_1^*, $s_2^* > 0$. The optimal solution $\mathbf{x}^* = s_1^*\mathbf{x}^1 + s_2^*\mathbf{x}^2$ is as follows.

$$
\mathbf{x}^* = \begin{bmatrix} a_1 - (1+e_1)(a_3 - a_4) \\ (a_2 - a_1) - (a_3 - a_4)(e_2 - e_1) \\ a_3 - a_4 \\ a_4 \end{bmatrix} \qquad (12.5\text{-}6)
$$

Now solving the linear program (12.5-4) directly yields the unique optimal solution and the associated 4×4 basis matrix is well conditioned. This implies that $\mathbf{x}^*$ is relatively insensitive to small changes in the matrix or right-hand side elements, as can be directly seen from (12.5-6). Suppose, however, that a_1, a_2, e_1, and e_2 are obtained by *truncating* $\bar{a}_1$, $\bar{a}_2$, $\bar{e}_1$, and $\bar{e}_2$, which are not machine-representable numbers. Suppose also that the foregoing derivation is carried out with $\bar{a}_1$, $\bar{a}_2$, $\bar{e}_1$, and $\bar{e}_2$ in place of a_1, a_2, e_1, and e_2, and denote the quantities corresponding to k, s_1^*, s_2^*, $\mathbf{x}^1$, and $\mathbf{x}^2$ by $\bar{k}$, $\bar{s}_1$, $\bar{s}_2$, $\bar{\mathbf{x}}^1$, and $\bar{\mathbf{x}}^2$. Then the quantity k, which determines $\mathbf{x}^2$, can be drastically different from $\bar{k} = (\bar{a}_2 - \bar{a}_1)/(\bar{e}_2 - \bar{e}_1)$, which determines $\bar{\mathbf{x}}^2$. This is because the basis matrix that determines $\mathbf{x}^2$ is ill conditioned. The corresponding quantities $\bar{s}_1$ and $\bar{s}_2$ will also be quite different from s_1^* and s_2^*. However, the errors in $\bar{s}_1$ and $\bar{s}_2$ will be so correlated with the errors in $\bar{\mathbf{x}}^1$ and $\bar{\mathbf{x}}^2$ that $\bar{\mathbf{x}} = \sum_{j=1}^{2} \bar{s}_j \bar{\mathbf{x}}^j$ does *not* change drastically, and remains close to $\mathbf{x}^*$.

Example 12.5-3. Reconstruction of Solution. Consider a linear program of the form

$$
\begin{array}{lll}
\text{minimize} & \mathbf{c}_1^T\mathbf{x}_1 + \mathbf{c}_2^T\mathbf{x}_2 & \\
\text{s.t.} & & \\
\textit{subproblem rows} & \{\mathbf{A}_1\mathbf{x}_1 \qquad\quad = \mathbf{b}^1 & \\
\textit{master rows} & \{\mathbf{B}_1\mathbf{x}_1 + \mathbf{A}_2\mathbf{x}_2 = \mathbf{b}^2 & \\
\mathbf{x}_1 \geq \mathbf{0} \qquad \mathbf{x}_2 \geq \mathbf{0}. & &
\end{array}
\tag{12.5-7}
$$

If this is decomposed using Dantzig–Wolfe decomposition, with the partition of rows indicated above, the solution recovered from the master is of the form

$$
\begin{bmatrix} \sum_j s_j^* \mathbf{x}_1^j \\ \mathbf{x}_2^* \end{bmatrix}
$$

where $\mathbf{x}_1^j$ denotes an extreme point of the subproblem which, for simplicity, we assume to be bounded, and $\mathbf{x}_2^*$ comes directly from the master.

To facilitate data handling, many implementations of the Dantzig–Wolfe algorithm form master columns $\mathbf{B}_1\mathbf{x}_1^j$ directly during the solution of the subproblem and do not preserve the corresponding extreme points $\mathbf{x}_1^j$. In this case, the $\mathbf{x}_1$ component of the optimal solution, say $\mathbf{x}_1^*$, is recovered or *reconstructed*, by solving a linear program of the following form.

$$
\begin{aligned}
\text{minimize } & \mathbf{c}_1^T\mathbf{x}_1 \\
\text{s.t. } & \mathbf{A}_1\mathbf{x}_1 = \mathbf{b}^1 \\
& \mathbf{B}_1\mathbf{x}_1 = \mathbf{b}^2 - \mathbf{A}_2\mathbf{x}_2^* \\
& \mathbf{x}_1 \geq \mathbf{0}.
\end{aligned}
\tag{12.5-8}
$$

Our example illustrates numerical difficulties that arise with this method of reconstruction of the solution.

Consider the following linear program,

$$\begin{array}{ll}
\text{maximize} & x_4 \\
\text{s.t.} & \\
\textit{subproblem rows} & \begin{cases} x_1 \quad\quad + (1+e_1)x_3 = a_1 \\ \quad\quad x_2 + (1+e_2)x_3 = a_2 \end{cases} \\
\textit{master rows} & \{x_1 + x_2 + 2(1+e_3)x_3 + x_4 = a_3 \\
 & x_i \geq 0.
\end{array}$$

There can only be two feasible extreme points of the subproblem, and these are given by

$$\mathbf{x}_1^1 = \begin{bmatrix} a_1 \\ a_2 \\ 0 \end{bmatrix} \qquad \mathbf{x}_1^2 = \begin{bmatrix} a_1 - a_2 \dfrac{1+e_1}{1+e_2} \\ 0 \\ \dfrac{a_2}{1+e_2} \end{bmatrix}$$

where we assume that a_1 and a_2 are chosen so that $\mathbf{x}_1^1 \geq \mathbf{0}$, $\mathbf{x}_1^2 \geq \mathbf{0}$. The basis matrices corresponding to these solutions are well conditioned. The master problem is

$$\begin{array}{ll}
\text{maximize} & x_4 \\
\text{s.t.} &
\end{array}$$

$$(a_1 + a_2)s_1 + \left[a_1 - a_2\frac{1+e_1}{1+e_2} + \frac{2(1+e_3)a_2}{1+e_2}\right]s_2 + x_4 = a_3 \tag{12.5-9}$$

$$s_1 + \qquad\qquad s_2 \qquad = 1$$

$$s_1, s_2, x_4 \geq 0.$$

Assume that $2e_3 < e_1 + e_2$. Then the coefficient of s_2 is less than the coefficient of s_1, and the optimal solution of (12.5-9) is $s_1^* = 0$, $s_2^* = 1$, with x_4^* determined by (12.5-9). The optimal solution of the linear program, obtained by taking the linear convex combination of extreme points is

$$\mathbf{x}^* = \begin{bmatrix} a_1 - a_2\dfrac{1+e_1}{1+e_2} \\ 0 \\ \dfrac{a_2}{1+e_2} \\ a_3 - \dfrac{a_1 + a_2(1+2e_3-e_1)}{1+e_2} \end{bmatrix} \tag{12.5-10}$$

The basis matrix corresponding to the variables x_1, x_3, and x_4 in the original linear program is well conditioned and the solution, $\mathbf{x}^*$ can alternatively be obtained from this basis. If, however, we use the method of reconstruction (12.5-8), we must solve the system

$$\begin{bmatrix} 1 & 0 & 1+e_1 \\ 0 & 1 & 1+e_2 \\ 1 & 1 & 2(1+e_3) \end{bmatrix} \begin{bmatrix} x_1 \\ x_2 \\ x_3 \end{bmatrix} = \begin{bmatrix} a_1 \\ a_2 \\ a_1 + (1+2(e_3-e_1))a_2/(1+e_2) \end{bmatrix}$$

This is an ill-conditioned problem. Numerical error will usually result in this reconstructed solution being substantially different from the true well-conditioned solution (12.5-10). The former may even be infeasible.

In implementations of the simplex algorithm for solving large-scale linear programs, it is still quite common to trade off stability in the basis update for efficiency in data handling, in particular, a reduction in the number of I/O operations involved in moving the basis representation between main and secondary storage, as discussed in Chapter 10. By monitoring the basis update, it is possible to detect instability and then initiate a stable refactorization of the corresponding basis matrix. In decomposition algorithms, however, it imperative that columns of the master program be developed in a numerically stable manner. We must be able to assert that each *computed* column of the master, using the subproblem basis matrix corresponding to it, say $\mathbf{B}^j$, is the one obtained by *exact* computation using a perturbed basis matrix $\mathbf{B}^j + \boldsymbol{\delta}\mathbf{B}^j$, where $\|\boldsymbol{\delta}\mathbf{B}^j\|/\|\mathbf{B}^j\|$ is small. Otherwise, the optimal solution of the computed master may bear little resemblance to the optimal solution of the original program. Note when stable techniques are employed that different columns of the master would require different perturbations $\boldsymbol{\delta}\mathbf{B}^j$ of the corresponding basis matrices. Thus one cannot, in general, claim that the *computed* full master problem can be obtained by applying the decomposition principle to a slight perturbation of the original linear program, for example, to the problem (12.5-7), with each matrix $\mathbf{A}_i$ and $\mathbf{B}_i$ replaced by $\mathbf{A}_i + \boldsymbol{\delta}\mathbf{A}_i$ and $\mathbf{B}_i + \boldsymbol{\delta}\mathbf{B}_i$ respectively, where $\|\boldsymbol{\delta}\mathbf{A}_i\|/\|\mathbf{A}_i\|$ and $\|\boldsymbol{\delta}\mathbf{B}_i\|/\|\mathbf{B}_i\|$ are small.

As we have noted in the previous examples, even when stable techniques are employed, the columns of the computed master problem can differ substantially from the true ones. The implication of Example 12.5-2 is that errors in the extreme points (or directions) defining the columns of the computed master can be *correlated* with the errors in the solutions obtained from the computed master, thus *preserving* the optimal solution. The optimal solution itself must, of course, be relatively insensitive to perturbations of the original data. However, if one is to take advantage of correlation of error, Example 12.5-3 shows that careful attention must be paid to the method of reconstruction of the solution.

Finally, as noted in Example 12.5-1, the master problem can become badly scaled. Some of the techniques of Chapter 7 that seek scale invariance may be of particular value, when used in implementations of decomposition algorithms.

The numerical behavior of decomposition algorithms has only relatively recently come under scrutiny, and rigorous analysis is not yet available. The numerical behavior of decomposition algorithms is, of course, closely related to their convergence properties. Some of the long "tail" convergence characteristics noted earlier may be the result of numerical difficulties discussed in this section.

12.6. Practical Details of Implementation

The decomposition principle and the algorithms resulting from it represent *an overall approach to LP problem solving* that requires careful attention to formulation (see Sec. 12.4), to numerical behavior (see Sec. 12.5), and to questions of strategy (as we shall elaborate on below). These techniques are therefore more effective in the hands of an experienced practitioner and are less amenable to "black box" use, in contrast to techniques discussed in earlier chapters.

The structures most amenable to decomposition are the two shown in Figure 12.1. In the case of the block angular LP matrix, the linking rows would define the master and each of the remaining diagonal blocks would define a different subproblem. In principle, the decomposition would be performed precisely along the lines of Secs. 12.2 and 12.3. However, the details and, in particular, the notation are a good deal more cumbersome. For details, see Lasdon [1970]. For the staircase structure, the most common approach is to use decomposition *recursively*. For example, the last set of constraints in Figure 12.1 is used to define the subproblem and the remaining constraints define the master. This master is, in turn, further decomposed into an (inner) master and subproblem, using the partition indicated by the dashed line in Figure 12.1, and so on. The procedure, known as *nested decomposition,* was first suggested by Dantzig [1963] and further developed by several authors, see, in particular, Glassey [1971] and Ho and Manne [1974].

All the techniques discussed in Part II for problem setup, basis handling, and selection strategies (in particular, initialization by using a suitable cost vector and associated price vector) are directly applicable to the implementation of a decomposition algorithm. In addition, there are a number of specialized implementation techniques for decomposition, in particular, strategies that are the analogue of partial and multiple pricing (namely, *not* requiring a subproblem to be pushed to optimality at every cycle, and extracting several subproblem solutions at each cycle), purging strategies for deciding which nonbasic columns to keep in the restricted master, techniques for subproblem setup that avoid having to split rows of the original problem matrix explicitly, and techniques for solution reconstruction (see Example 12.5-3). It would take us too far afield to discuss them here and the interested reader should consult Ho [1974].

Notes

Secs. 12.2–12.3. For further details, see Lasdon [1970].

Sec. 12.5. The examples are taken from Nazareth [1984]. For a discussion of convergence in the presence of numerical error, see Ho [1984].

13
The Homotopy Principle and the Simplex Method

13.1. Introduction

In this final chapter, we consider an approach to solving linear programs based on the *homotopy principle,* which can be stated as follows. *The linear program to be solved is first deformed to one that is trivial and has a unique optimal solution. Beginning with the solution to the trivial problem, a route of (optimal) solutions is followed as the system is deformed back to the original linear program.*

We shall deal with a linear program of the form,

$$\begin{aligned} \text{minimize } & \mathbf{c}^T\mathbf{x} \\ \text{s.t. } & \mathbf{A}\mathbf{x} = \mathbf{b} \\ & \mathbf{x} \geq \mathbf{0}, \end{aligned} \tag{13.1-1}$$

where $\mathbf{A}$, $\mathbf{b}$, and $\mathbf{c}$ are defined in the usual way. Let us assume that it has an optimal solution $\mathbf{x}^*$.

Let us suppose that $\mathbf{B}^0$ is some basis matrix defined by m linearly independent columns $B_0 = \{\beta_1^0, \ldots, \beta_m^0\}$. The corresponding basic solution need *not* be primal or dual feasible. Let $\mathbf{N}^0$ denote the nonbasic columns, defined by the index set $N_0 = \{\eta_1^0, \ldots, \eta_{n-m}^0\}$. Thus

$$\begin{aligned} \mathbf{x}_{B_0}^0 &= (\mathbf{B}^0)^{-1}\mathbf{b} \\ (\boldsymbol{\pi}_{B_0}^0)^T &= \mathbf{c}_{B_0}^T(\mathbf{B}^0)^{-1} \\ \boldsymbol{\sigma}_{N_0}^0 &= \mathbf{c}_{N_0} - (\mathbf{N}^0)^T\boldsymbol{\pi}_{B_0}^0, \end{aligned} \tag{13.1-2}$$

and $\mathbf{x}_{B_0}^0$ and $\boldsymbol{\sigma}_{N_0}^0$ may have some negative components. We can, however, ensure that $\mathbf{x}_{B_0}^0 \geq \mathbf{0}$ and $\boldsymbol{\sigma}_{N_0}^0 \geq \mathbf{0}$ by adding a sufficiently large number $\mu \geq 0$ to each negative component of these vectors. Stated in terms of the cost vector and right-hand side vector of the original linear program, this would mean replacing $\mathbf{b}$ and $\mathbf{c}$ by $\mathbf{b}_\mu$ and $\mathbf{c}_\mu$ defined as follows:

$$\mathbf{b}_\mu = \mathbf{b} + \mu\mathbf{B}^0\boldsymbol{u} \tag{13.1-3a}$$

$$\mathbf{c}_\mu = \mathbf{c} + \mu\mathbf{v} \tag{13.1-3b}$$

where $\boldsymbol{u}$ is an m-vector with a unit element in each position i where $(\mathbf{x}^0_{B_0})_i < 0$ and zero elsewhere. (To avoid any possible confusion with the vector of upper bounds, $\mathbf{u}$, we use $\boldsymbol{u}$ in (3.1-13a).) $\mathbf{v}$ is an n-vector with a unit element in each position where $\sigma^0_j < 0$, $j \in N_0$, and zero in all other positions, including the ones corresponding to the basic variables. $\mu \geq 0$ is a scalar parameter. Also, for convenience, let us make the definition

$$\mathbf{w} = B^0\boldsymbol{u}. \tag{13.1-4}$$

When we make the replacements defined by (13.1-3) in the linear program (13.1-1), we obtain the deformed system

$$\begin{aligned} (LP_\mu)\text{: minimize } & \mathbf{c}^T_\mu \mathbf{x} \\ \text{s.t. } & \mathbf{A}\mathbf{x} = \mathbf{b}_\mu \\ & \mathbf{x} \geq \mathbf{0}. \end{aligned} \tag{13.1-5}$$

For this linear program, the solution, price vector, and reduced costs corresponding to the basis matrix $\mathbf{B}^0$ are

$$\begin{aligned} \mathbf{x}^\mu_{B_0} &= (\mathbf{B}^0)^{-1}\mathbf{b}_\mu = \mathbf{x}^0_{B_0} + \mu\boldsymbol{u} \\ (\boldsymbol{\pi}^\mu_{B_0})^T &= \mathbf{c}^T_{B_0}(\mathbf{B}^0)^{-1} \\ \boldsymbol{\sigma}^\mu_{N_0} &= \boldsymbol{\sigma}^0_{N_0} + \mu\mathbf{v}_{N_0}, \end{aligned} \tag{13.1-6}$$

where $(\mathbf{v}_{N_0})_k = v_{\eta_k}$, $\eta_k \in N_0$. We can ensure $\mathbf{x}^\mu_{B_0} > \mathbf{0}$ and $\boldsymbol{\sigma}^\mu_{N_0} > \mathbf{0}$ by choosing $\mu = \mu_0$ to be sufficiently large. We then have a nondegenerate unique optimal solution to LP_μ. Indeed, this was the purpose behind the foregoing construction.

Associated with any other basis, say $\mathbf{B}^1$, of (13.1-5) or (13.1-1) with defining index sets $B_1 = \{\beta^1_1, \ldots, \beta^1_m\}$ and $N_1 = \{\eta^1_1, \ldots, \eta^1_{n-m}\}$ and nonbasic columns $\mathbf{N}^1$, we similarly have the following expressions (see also (13.1-8) and (13.1-9)).

$$\begin{aligned} \mathbf{x}^\mu_{B_1} &= (\mathbf{B}^1)^{-1}\mathbf{b}_\mu = (\mathbf{B}^1)^{-1}\mathbf{b} + \mu(\mathbf{B}^1)^{-1}\mathbf{w} = \mathbf{x}^1_{B_1} + \mu\mathbf{w}^1_{B_1} \\ (\mathbf{c}^\mu_{B_1})^T &= \mathbf{c}^T_{B_1} + \mu\mathbf{v}^T_{B_1} \\ (\boldsymbol{\pi}^\mu_{B_1})^T &= (\mathbf{c}^\mu_{B_1})^T(\mathbf{B}^1)^{-1} = \mathbf{c}^T_{B_1}(\mathbf{B}^1)^{-1} + \mu\mathbf{v}^T_{B_1}(\mathbf{B}^1)^{-1} = \boldsymbol{\pi}^1_{B_1} + \mu\boldsymbol{\pi}^v_{B_1} \\ \boldsymbol{\sigma}^\mu_{N_1} &= (\mathbf{c}_{N_1} + \mu\mathbf{v}_{N_1}) - (\mathbf{N}^1)^T\boldsymbol{\pi}^\mu_{B_1} = (\mathbf{c}_{N_1} - (\mathbf{N}^1)^T\boldsymbol{\pi}^1_{B_1}) + \mu(\mathbf{v}_{N_1} - (\mathbf{N}^1)^T\boldsymbol{\pi}^v_{B_1}). \end{aligned} \tag{13.1-7}$$

Thus

$$\boldsymbol{\sigma}^\mu_{N_1} = \boldsymbol{\sigma}^1_{N_1} + \mu\boldsymbol{\sigma}^v_{N_1}.$$

In the foregoing expressions, $\mathbf{v}_{B_1}$ and $\mathbf{v}_{N_1}$ denote the components of $\mathbf{v}$ corresponding to the basic and nonbasic variables respectively, and we make the usual definitions,

$$\mathbf{x}^1_{B_1} \equiv (\mathbf{B}^1)^{-1}\mathbf{b} \qquad (\boldsymbol{\pi}^1_{B_1})^T \equiv \mathbf{c}^T_{B_1}(\mathbf{B}^1)^{-1} \qquad \boldsymbol{\sigma}^1_{N_1} \equiv \mathbf{c}_{N_1} - (\mathbf{N}^1)^T\boldsymbol{\pi}^1_{B_1}. \tag{13.1-8}$$

Analogously, we also make the definitions

$$\mathbf{w}^1_{B_1} \equiv (\mathbf{B}^1)^{-1}\mathbf{w} \qquad \boldsymbol{\pi}^v_{B_1} \equiv \mathbf{v}^T_{B_1}(\mathbf{B}^1)^{-1} \qquad \boldsymbol{\sigma}^v_{N_1} \equiv \mathbf{v}_{N_1} - (\mathbf{N}^1)^T\boldsymbol{\pi}^v_{B_1}. \tag{13.1-9}$$

Therefore the basic solution, price vector, and reduced costs of LP_{μ}, associated with the basis $\mathbf{B}^1$ and an arbitrary value of $\mu \geq 0$, may be formed from the usual quantities (13.1-8) for the original linear program, by adding to them μ times the corresponding quantities formed from the right-hand side $\mathbf{w}$ and cost $\mathbf{v}$. The latter are also defined in the standard way (13.1-9). (In an actual computation, these could be obtained more efficiently by updating over successive iterations.) For the basis $\mathbf{B}^0$ and the starting value μ_0, we also have an optimal solution for LP_{μ_0}, as noted above.

13.2. The Self-Dual Simplex Algorithm

To solve (13.1-1), we now develop an algorithm based on the simplex method and the homotopy principle, which traces a path of optimal solutions of LP_{μ}, as μ is progressively reduced from μ_0. The aim, of course, is to find the solution on the path for $\mu = 0$, which provides an optimal solution for the linear program (13.1-1).

We proceed as follows. As μ is progressively reduced from the value μ_0 with starting basis matrix $\mathbf{B}^0$, we see from (13.1-6) that $\mathbf{B}^0$ continues to define an optimal solution until some threshold value is reached, say $\mu = \mu_1$, where an element of either $\mathbf{x}_{B_0}^{\mu_1}$ or $\boldsymbol{\sigma}_{N_0}^{\mu_1}$ goes from nonnegative to negative. Let us consider these two possibilities.

1. If an element of $\boldsymbol{\sigma}_{N_0}^{\mu_1}$, say $\boldsymbol{\sigma}_s^{\mu_1}$, is the first to cross its threshold, then we define x_s to be the entering variable in an iteration of the *primal simplex algorithm* applied to LP_{μ_1}, with the various quantities needed being defined by (13.1-7) through (13.1-9). The exiting variable $x_{\beta_p^0}$ corresponding to the pth column of $\mathbf{B}^0$ is found by the usual *CHUZR* rule; see Sec 2.4, in particular, Exercise 2.4-1. This gives a new basis matrix, say $\mathbf{B}^1$, and index sets B_1 and N_1. If there is no blocking variable at this step, then LP_{μ_1} has an unbounded optimal solution. Thus its dual is infeasible, so that

$$\mathbf{A}^T\boldsymbol{\pi} \leq \mathbf{c}_{\mu_1} = \mathbf{c} + \mu_1\mathbf{v} \tag{13.2-1}$$

is infeasible. Since $\mu_1\mathbf{v} \geq \mathbf{0}$, it follows that the constraints $\mathbf{A}^T\boldsymbol{\pi} \leq \mathbf{c}$ are also infeasible. Therefore (13.1-1) must either be infeasible or unbounded; see Corollary 11.2-1 and also the exercise that follows it. When we assume that (13.1-1) has an optimal solution $\mathbf{x}^*$, this situation will not occur.

2. If an element of $\mathbf{x}_{B_0}^{\mu_1}$, say $(\mathbf{x}_{B_0}^{\mu_1})_p$, is the first to cross its threshold, then we define $x_{\beta_p^0}$ to be the exiting variable in an iteration of the *dual simplex algorithm* applied to LP_{μ_1}, with the various quantities needed being defined by (13.1-7) through (13.1-9). The entering variable x_s is found by the usual rule, see Step D5 of the algorithm of Sec. 11.4-2. This

again gives a new basis matrix, say $\mathbf{B}^1$, and index sets B_1 and N_1. If there is no blocking variable at this step, then the dual of LP_{μ_1} has an unbounded optimal solution. Thus the primal problem LP_{μ_1} is infeasible, so there is no solution to the constraints

$$\mathbf{Ax} = \mathbf{b}_{\mu_1} = \mathbf{b} + \mu_1 \mathbf{w} \qquad \mathbf{x} \geq \mathbf{0}. \tag{13.2-2}$$

Thus $\mathbf{Ax} = \mathbf{b}$, $\mathbf{x} \geq \mathbf{0}$ must also be infeasible. To see this, suppose that these constraints have a feasible solution, say $\bar{\mathbf{x}}$. Then using $\mathbf{w} = (\mathbf{B}^0)\boldsymbol{u}$, we can construct a feasible solution to (13.2-2) of the form $\bar{\mathbf{x}} + \mu_1 \bar{\boldsymbol{u}}$, where $\bar{\boldsymbol{u}}$ has a unit element in columns corresponding to indices in B_0 and zeros elsewhere. This gives the necessary contradiction. Therefore, when there is no blocking variable, the original problem (13.1-1) is infeasible and its dual is either infeasible or has an unbounded optimum.

Let us make the assumption that *no ties occur* when making the previous choice, i.e., only one of (1) or (2) occurs and that the corresponding element is uniquely defined. Let us also assume that aside from this degeneracy in primal or dual, there are no other degeneracies associated with the two successive basis matrices, before and after the pivot operation, respectively. We shall show in the theorem below that after the basis change, the quantities $\mathbf{x}_{B_1}^{\mu_1}$ and $\boldsymbol{\sigma}_{N_1}^{\mu_1}$ satisfy

$$\mathbf{x}_{B_1}^{\mu_1} > \mathbf{0} \tag{13.2-3a}$$

$$\sigma_j^{\mu_1} > 0, \qquad j \in N_1, \qquad j \neq \beta_p^0, \tag{13.2-3b}$$

and $\sigma_{\beta_p^0}^{\mu_1}$ is increasing from zero, as μ decreases from μ_1. We can therefore repeat the foregoing procedure, again reducing progressively from μ_1 until some new threshold, say $\mu = \mu_2$, when an element of $\mathbf{x}_{B_1}^{\mu_2}$ or $\boldsymbol{\sigma}_{N_1}^{\mu_2}$ defined by (13.1-7) goes from nonnegative to negative. This again requires a basis change. The process is continued until μ attains the value zero, at which point we have found an optimal solution $\mathbf{x}^*$ to (13.1-1).

The above algorithm is called the *self-dual* (*parametric*) *simplex algorithm,* because the same algorithm would be obtained if we had begun with the dual problem. This should be clear from the symmetric way in which the primal and dual are treated. Since we deal with basic solutions throughout, the complementary slackness conditions (11.2-14) are satisfied. However, basic solutions on the path of solutions are not necessarily primal or dual feasible for the original linear program. We see therefore that we are in a situation summarized by the last row of Table 11.2.

Exercise 13.2-1. If $\mathbf{x}^0$ in (13.1-2) is primal feasible, so that we can take $\mathbf{w} = \mathbf{0}$, show that the self-dual simplex algorithm is equivalent to the primal simplex algorithm with the largest-coefficient rule for choosing the entering variable.

Exercise 13.2-2. Discuss the necessary modifications to the foregoing development when there are bounds present on the $\mathbf{x}$ variables of the form $\mathbf{l} \leq \mathbf{x} \leq \mathbf{u}$.

We conclude with two results that characterize the path of solutions and establish convergence of the self-dual simplex algorithm.

Lemma 13.2-1. Given two successive threshold values, say μ_i and μ_{i+1} on the path of solutions, the solution associated with the corresponding basis $\mathbf{B}^i$ (with index sets B_i and N_i) is optimal for LP_μ, for all values of μ in the range $\mu_{i+1} \leq \mu \leq \mu_i$, and the path is piecewise linear.

Proof. The first part follows directly from (13.1-7), because $\mathbf{x}_{B_i}^{\mu_{i+1}} \geq \mathbf{0}$ and $\boldsymbol{\sigma}_{N_i}^{\mu_{i+1}} \geq \mathbf{0}$ imply that these relations also hold for all values of μ greater than μ_{i+1}. Piecewise-linearity follows from the first relation of (13.1-7). ■

Theorem 13.2-1. With each basis pair $\mathbf{B}^i$ and $\mathbf{B}^{i+1}$ associated with the threshold level μ_{i+1}, let us assume that the only degeneracy in $LP_{\mu_{i+1}}$ is the one that arises in making the choice between (1) and (2) in the formulation of the self-dual simplex algorithm of this section. Then the algorithm converges in a finite number of iterations.

Proof. Suppose that case (1) occurs and x_s is the entering variable with the pth column of $\mathbf{B}^i$ being replaced. Let us use a simpler notation in which we write σ_j^μ for reduced costs associated with $\mathbf{B}^i$ and $\bar{\sigma}_j^\mu$ for reduced costs associated with $\mathbf{B}^{i+1}$.

From (11.4-8) with the appropriate changes of notation,

$$\bar{\sigma}_j^\mu = \sigma_j^\mu - \frac{\sigma_s^\mu}{(-\mathbf{a}_s^T(\mathbf{B}^i)^{-T}\mathbf{e}_p)}(-\mathbf{a}_j^T(\mathbf{B}^i)^{-T}\mathbf{e}_p) \qquad \text{for all } j.$$

Thus

$$\bar{\sigma}_j^\mu = \sigma_j^\mu - \frac{(\tilde{\mathbf{a}}_j)_p}{(\tilde{\mathbf{a}}_s)_p}\sigma_s^\mu \qquad \text{for all } j.$$

Also, $\sigma_s^{\mu_{i+1}} = 0$ and for some value of μ slightly smaller than μ_{i+1}, $\sigma_s^\mu = \delta < 0$. $\sigma_j^{\mu_{i+1}} > 0$, $j \neq s$ and hence $\sigma_j^\mu > 0$ for $j \neq s$ and μ sufficiently close to μ_{i+1}. Finally, note that $\tilde{\mathbf{a}}_{\beta_p^i} = \mathbf{e}_p$. Thus

$$\bar{\sigma}_j^\mu = \sigma_j^\mu - [(\tilde{\mathbf{a}}_j)_p/(\tilde{\mathbf{a}}_s)_p]\delta > 0 \qquad j \in N_{i+1} \qquad j \neq \beta_p^i \quad \text{and} \quad \delta \text{ sufficiently small,}$$

$$\bar{\sigma}_{\beta_p^i}^\mu = -[1/(\tilde{\mathbf{a}}_s)_p]\delta > 0 \qquad \text{for } \delta < 0.$$

By the nondegeneracy assumption on the basic variables before and after the basis change,

$$\mathbf{x}_{B_{i+1}}^\mu > \mathbf{0}.$$

Therefore we can reduce μ from μ_{i+1} to some new level μ_{i+2}. Also, no basis can repeat itself. This is a consequence of Lemma 13.2-1 and completes the proof of the theorem. ■

Exercise 13.2-1. Investigate the convergence of the self-dual algorithm, when the uniqueness assumption of Theorem 13.2-1 is relaxed.

13.3. Practical Details of Implementation

Most of the implementation techniques of Part II may be utilized here. However, there is no analogue of *partial pricing* for the self-dual simplex algorithm as formulated above, although variants may be possible. The merit of the self-dual algorithm is that it does not require a starting primal or dual basic *feasible* solution. Since it subsumes both the primal and dual simplex algorithms, it can also perform its own sensitivity analysis.

Notes

Sec. 13.1. A discussion of the homotopy principle may be found, for example, in Eaves [1979]. See also Nazareth [1986b].

Sec. 13.2. The algorithm and proof of termination is a reformulation, in matrix terms, of the tableau version given in Dantzig [1963]. The self-dual simplex algorithm is related to Newton's method and is of great importance as a means for investigating the average-case complexity of the simplex method, see Smale [1983]. Worst-case behavior of the self-dual simplex algorithm is discussed in Murty [1980].

Bibliography

Adler, I., and Megiddo, N. 1985. A simplex algorithm whose average number of steps is bounded between two quadratic functions of the smaller dimension. *Journal of the Association for Computing Machinery 32,* 871–895.

Bartels, R. H. 1971. A stabilization of the simplex method. *Numerische Mathematik 16,* 414–434.

Bartels, R. H., and Golub, G. H. 1969. The simplex method of linear programming using LU decomposition. *Communications of the Association for Computing Machinery 12,* 266–268.

Bazaraa, M. S., and Jarvis, J. J. 1977. *Linear Programming and Network Flows.* Wiley, New York.

Beale, E. M. L. 1955. Cycling in the dual simplex algorithm. *Naval Research Logistics Quarterly 2,* 269–275.

Benichou, M., Gauthier, J. M., Hentges, G., and Ribiere, G. 1977. The efficient solution of large-scale linear programming problems—Some algorithmic techniques and computational results. *Mathematical Programming 13,* 280–322.

Benders, J. F. 1962. Partitioning procedures for solving mixed-variables programming problems. *Numerische Mathematik 4,* 238–252.

Bland, R. G. 1977. New finite pivoting rules for the simplex method. *Mathematics of Operations Research 2,* 103–107.

Borgwardt, K. H. 1982. The average number of pivot steps required by the simplex method is polynomial. *Zeitschrift fur Operations Research 26,* 157–177.

Chvatal, V. 1983. *Linear Programming.* W. H. Freeman and Co., New York.

Dantzig, G. B. 1963. *Linear Programming and Extensions.* Princeton University Press, Princeton, N.J.

Dantzig, G. B. 1982. Reminiscences about the origins of linear programming. In: A. Bachem, M. Grotschel, and B. Korte (Eds.), *Mathematical Programming: The State of the Art, Bonn 1982.* Springer-Verlag, Berlin, 1983, pp. 78–86.

Dantzig, G. B. 1985. Impact of linear programming on computer development. Technical Report SOL 85–7, Systems Optimization Laboratory, Department of Operations Research, Stanford University, Stanford, Calif.

Dantzig, G. B., and Wolfe, P. 1960. Decomposition principle for linear programs. *Operations Research 8,* 101–111.

Duff, I. S. 1981. On algorithms for obtaining a maximum transversal. *ACM Transactions on Mathematical Software 7,* 315–330.

Duff, I. S., and Reid, J. K. 1978. An implementation of Tarjan's algorithm for the block triangularization of a matrix. *ACM Transactions on Mathematical Software 4,* 137–147.

Eaves, B. C. 1979. A view of complementary pivot theory (or solving equations with homotopies). In: *Constructive Approaches to Mathematical Models.* Academic Press, New York, 1979, pp. 153–170.

Fletcher, R., and Matthews, S. P. J. 1984. Stable modification of explicit LU factors for simplex updates. *Mathematical Programming 30,* 267–284.

Forrest, J. J. H., and Tomlin, J. A. 1972. Updating triangular factors of the basis to maintain sparsity in the product form simplex method. *Mathematical Programming 2,* 263–278.

Forsythe, G. E. 1970. Pitfalls in computation, or why a math book isn't enough. *The American Mathematical Monthly 9,* 931–956.

Forsythe, G. E., and Moler, C. B. 1967. *Computer Solution of Linear Algebraic Systems.* Prentice-Hall, New Jersey.

Gay, D. M., 1978, On combining the schemes of Reid and Saunders for sparse LP bases. In I. S. Duff and G. W. Stewart (Eds.). *Sparse Matrix Proceedings, 1978.* SIAM, Philadelphia, 1978, pp. 313–334.

George, A., and Ng, E. 1984. Symbolic factorization for sparse Gaussian elimination with partial pivoting. Technical Report CS-84-43, Department of Computer Science, University of Waterloo, Ontario, Canada.

Gill, P. E., Murray, W., and Wright, M. H. 1981. *Practical Optimization.* Academic Press, New York.

Gill, P. E., Murray, W., Saunders, M. A., and Wright, M. H. 1986. Maintaining LU factors of a general sparse matrix. Technical Report SOL 86–8, Systems Optimization Laboratory, Department of Operations Research, Stanford University, Stanford, Calif.

Givens, W. 1954. Numerical computation of the characteristic values of a real symmetric matrix. Technical Report ORNL-1574, Oak Ridge National Laboratory, Oak Ridge, Tennessee.

Glassey, C. R. 1971. Dynamic LP's for production scheduling. *Operations Research 19,* 45–56.

Goldfarb, D., and Reid, J. K. 1977. A practical steepest-edge simplex algorithm. *Mathematical Programming 12,* 361–371.

Greenberg, H. 1978a. A tutorial on matricial packing. In: H. Greenberg (Ed.), *Design and Implementation of Optimization Software.* Sijthoff and Noordhoff, 1978, pp. 109–142.

Greenberg, H. 1978b. Pivot selection tactics. In H. Greenberg (Ed.), *Design and Implementation of Optimization Software.* Sijthoff and Noordhoff, 1978, pp. 143–174.

Greenberg, H., and Kalan, J. 1975. An exact update for Harris' TREAD. *Mathematical Programming Study 4,* 26–29.

Grunbaum, B. 1967. *Convex Polytopes.* Wiley, New York.

Harris, P. M. J. 1975. Pivot selection methods in the Devex LP code. *Mathematical Programming Study 4,* 30–57.

Hellerman, E., and Rarick, D. 1971. Reinversion in the preassigned pivot procedure. *Mathematical Programming 1,* 195–216.

Hellerman, E., and Rarick, D. 1972. The partitioned preassigned pivot procedure. In D. J. Rose and R. A. Willoughby (Eds.). *Sparse Matrices and Their Applications.* Plenum Press, New York, pp. 67–76.

Ho, J. 1974. Nested decomposition for large-scale programs with the staircase structure. Technical Report SOL 74-4, Systems Optimization Laboratory, Department of Operations Research, Stanford University, Stanford, Calif.

Ho, J. 1984. Convergence behaviour of decomposition algorithms for linear programs. *Operations Research Letters 3,* 91–94.

Ho, J., and Manne, A. 1974. Nested decomposition for dynamic models. *Mathematical Programming 6,* 121–140.

Isaacson, E., and Keller, H. B. 1966. *Analysis of Numerical Methods.* Wiley, New York.

Karmarkar, N. 1984. A new polynomial-time algorithm for linear programming. *Combinatorica 4,* 373–395.

Khachiyan, L. G. 1979. A polynomial algorithm for linear programming. *Doklady*

Akademiia Nauk SSSR 244, 1093–1096. [English translation: *Soviet Mathematics Doklady 20,* 191–194.]

Klee, V., and Minty, G. J. 1972. How good is the simplex algorithm? In O. Shisha (Ed.). *Inequalities—III.* Academic Press, New York, pp. 159–175.

Knuth, D. E. 1968. *The Art of Computer Programming. Vol. 1: Fundamental Algorithms.* Addison-Wesley, Reading, Mass.

Lemke, C. E. 1954. The dual method for solving the linear programming problem. *Naval Research Logistics Quarterly 1,* 36–47.

Lasdon, L. S. 1970. *Optimization Theory for Large Systems.* Macmillan, London.

Marsten, R. E. 1981. The design of the XMP linear programming library. *ACM Transactions on Mathematical Software 7,* 481–497.

Murtagh, B. A., and Saunders, M. A. 1978. Large-scale linearly constrained optimization. *Mathematical Programming 14,* 41–72.

Murtagh, B. A., and Saunders, M. A. 1983. Minos 5.0 User's Guide. Technical Report SOL-83-20, Systems Optimization Laboratory, Department of Operations Research, Stanford University, Stanford, Calif.

Murty, K. G. 1980. Computational complexity of parametric linear programming. *Mathematical Programming 19,* 213–219.

Nazareth, J. L. 1984. Numerical behaviour of LP algorithms based upon the decomposition principle. *Linear Algebra and its Applications 57,* 181–189.

Nazareth, J. L. 1986a. Implementation aids for optimization algorithms that solve sequences of linear programs. *ACM Transactions on Mathematical Software 12,* 307–323.

Nazareth, J. L. 1986b. Homotopy Techniques in linear programming. *Algorithmica, 1,* 529–535.

Nazareth, J. L. 1987. Pricing criteria in linear programming. Technical Report PAM–382, Center for Pure and Applied Mathematics, University of California, Berkeley, Calif.

Orchard-Hays, W. 1968. *Advanced Linear Programming Computing Techniques.* McGraw-Hill, New York.

Orchard-Hays, W. 1978. History of mathematical programming systems. In H. Greenberg (Ed.). *Design and Implementation of Optimization Software.* Sijthoff and Noordhoff, 1978, pp. 1–26.

Osterby, O., and Zlatev, Z. 1983. *Direct Methods for Sparse Matrices.* Lecture Notes in Computer Science, 157, Springer-Verlag, Berlin.

Reid, J. K. 1976. Fortran subroutines for handling sparse linear programming bases. Technical Report R8269, Atomic Energy Research Establishment, Harwell, England.

Reid, J. K. 1982. A sparsity-exploiting variant of the Bartels–Golub decomposition for linear programming bases. *Mathematical Programming 24,* 55–69.

Sargent, R. W. H., and Westerberg, A. W. 1964. "Speed-up" in chemical engineering design. *Transactions of the Institute of Chemical Engineers 42,* 190–197.

Saunders, M. A. 1976. A fast, stable implementation of the simplex method using Bartels–Golub updating. In J. Bunch and D. J. Rose (Eds.). *Sparse Matrix Computations,* Academic Press, New York, pp. 213–226.

Saunders, M. A. 1980. Large-scale linear programming. Notes for the tutorial conference *Practical Optimization.* Systems Optimization Laboratory, Department of Operations Research, Stanford University, Stanford, Calif.

Shapiro, J. F. 1979. *Mathematical Programming: Structures and Algorithms.* Wiley, New York.

Simonnard, M. 1966. *Linear Programming.* Prentice-Hall, Englewood Cliffs, N.J. [English translation by W. S. Jewell.]

Solow, D. 1984. *Linear Programming: An Introduction to Finite Improvement Algorithms.* North-Holland, New York and Amsterdam.

Smale, S. 1976. A convergent process of price adjustment and global Newton methods. *Journal of Mathematical Economics 3,* 107–120.

Smale, S. 1983. The problem of the average speed of the simplex method. In A. Bachem, M. Grotschel, and B. Korte (Eds.). *Mathematical Programming: The State of the Art, Bonn, 1982.* Springer-Verlag, Berlin, 1983, pp. 530–539.

Tarjan, R. E. 1972. Depth-first search and linear graph algorithms. *SIAM Journal on Computing 1,* 146–160.

Tomlin, J. A. 1972a. Pivoting for size and sparsity in linear programming inversion routines. *Journal of the Institute of Mathematics and its Applications 10,* 289–295.

Tomlin, J. A. 1972b. Modifying triangular factors of the basis in the simplex method. In D. J. Rose and R. A. Willoughby (Eds.). *Sparse Matrices and Their Applications.* Plenum Press, New York, pp. 77–85.

Tomlin, J. A. 1975a. LPM1 User's Guide. Systems Optimization Laboratory, Department of Operations Research, Stanford University, Stanford, Calif.

Tomlin, J. A. 1975b. On scaling linear programming problems. *Mathematical Programming Study 4,* 146–166.

Tomlin, J. A. 1975c. An accuracy test for updating triangular factors. *Mathematical Programming Study 4,* 142–145.

Wegner, P. 1968. *Programming Languages, Information Structures and Machine Organization.* McGraw-Hill, New York.

Wilkinson, J. H. 1963. *Rounding Errors in Algebraic Processes.* Prentice-Hall, Englewood Cliffs, N.J.

Wilkinson, J. H. 1965. *The Algebraic Eigenvalue Problem.* Clarendon Press, Oxford.

Wolfe, P. 1962. The reduced-gradient method. Unpublished manuscript, Rand Corporation, Santa Monica, Calif.

Wolfe, P. 1963. A technique for resolving degeneracy in linear programming. *Journal of the Society for Industrial and Applied Mathematics 11,* 205–211.

Von Neumann, J., and Goldstine, H. H. 1947. Numerical inverting of matrices of high order. *Bulletin of the American Mathematical Society 53,* 1021–1099.

Zangwill, W. I. 1969. *Nonlinear Programming: A Unified Approach.* Prentice-Hall, Englewood Cliffs, N.J.

Index